Waste Management and Treatment

This book covers environmental and economic waste treatment with resource recovery strategies covering mining, urban, agricultural, industrial, and sewage wastes. It includes waste management, life-cycle assessment, recycling, and recovering valuable materials from blast furnace slags, iron ore, coal, and bauxite mining. Wastewater recycling, reuse, treatment methods, and economic gain through nanotechnology are covered, along with biochemical study cycles in mining waste and tailing, landfill stabilization, energy, and nutrient recovery from household, urban, and hazardous waste.

Features:

- Includes details on the generation and composition of hazardous wastes.
- Focuses on practices in waste management followed in developing countries.
- Illustrates the concept of energy and resource recovery.
- Explains ethical, innovative management strategies dealing with solid, biomedical, and chemical wastes.
- Discusses sustainable nanotechnology and engineered materials for treatment of biological and chemical pollutants in wastewater.

This book is aimed at researchers and graduate students in environmental engineering and management.

Waste Management and Treatment

Advances and Innovations

Edited by
Ankur Rajpal, Moharana Choudhury, Srijan Goswami,
Arghya Chakravorty, and Vimala Raghavan

CRC Press
Taylor & Francis Group
Boca Raton London New York

CRC Press is an imprint of the
Taylor & Francis Group, an **informa** business

Designed cover image: Shutterstock

First edition published 2025
by CRC Press
2385 NW Executive Center Drive, Suite 320, Boca Raton FL 33431

and by CRC Press
4 Park Square, Milton Park, Abingdon, Oxon, OX14 4RN

CRC Press is an imprint of Taylor & Francis Group, LLC

ISBN: 978-1-032-19256-7 (hbk)
ISBN: 978-1-032-19257-4 (pbk)
ISBN: 978-1-003-25837-7 (ebk)

DOI: 10.1201/9781003258377

Typeset in Times LT Std
by Apex CoVantage, LLC

Dedication

The editors of the book dedicate the book to their family members and colleagues.

Contents

Preface

Since the genesis of the Earth, nature has had its mechanisms to deal with naturally occurring pollutants or contaminants. Early humans used natural resources like air, water, shelter, and food to satisfy their basic needs. The residues generated using these resources were generally compatible with the environment or readily assimilated by it. Primitive humans usually ate raw plants and animals without disturbing the atmosphere with the smoke from their fires. When fire became common, the small amount of smoke generated by these activities was rapidly dispersed and easily assimilated by the atmosphere. Early civilizations drank from the same rivers and water bodies where they bathed and deposited their wastes. However, the impact of using natural water resources was relatively low since natural cleansing mechanisms or self-purification processes efficiently restored the water's natural quality. Early humans used natural shelters like caves or built their homes from wood, mud, or animal skins. They left behind items readily decomposed by microorganisms and absorbed by nature.

With the rise of the Industrial Revolution, humans were better able to satisfy their natural needs for air, water, food, and shelter. Automobiles, appliances, and processed food and beverages became popular as necessities by the early twentieth century. Meeting these acquired needs has become a significant thirst in modern industrial society. Unlike natural conditions, acquired needs are usually met by the products that must be processed, manufactured, or refined. The production, distribution, and use of such products typically result in more complex residues, many of which are incompatible with the environment or readily assimilated by it. This residue is less consistent with the environment and less likely to be incorporated into the biosphere readily. As modern societies ascend the socioeconomic ladder, the list of acquired needs or luxuries increases, and so do the complexities of the production chain and the amounts and varieties of pollutants they generate. People who work in environmental engineering and science are very concerned about how the modern human population affects the environment.

In the present state of the environment, the exponential rise in industrialization and urbanization has resulted in the contamination of the air one breathes, the water one drinks, and even the food and the soil and water in which the food grows. Environmental contamination has also resulted from various industrial operations like fugitive emissions, accidental spills, leaks, dumping of hazardous substances, etc. These are the consequences of using synthetic organic chemicals such as solvents, pesticides, refrigerants, chemical intermediates, etc. Over the years, the excessive use of these synthetic chemicals, which are not degradable by natural means, has become hazardous or toxic.

The present book is edited to provide comprehensive, multidimensional concepts and approaches related to recent innovations and advances in waste management and treatment in one place for easy access and understanding. The editorial team has made their best effort and concisely emphasized the basic and advanced concepts. This edited collection represents the combined efforts of the editorial team, all the contributing authors, and the peer reviewer panel from reputed institutions and organizations worldwide.

Ankur Rajpal
Moharana Choudhury
Srijan Goswami
Arghya Chakravorty
Vimala Raghavan

Acknowledgements

The editors of the book are grateful to the CRC Press—Taylor and Francis Group for giving them the opportunity to work on Waste Management and Treatment: Advances and Innovations.

The book's editors thank all the contributing authors for their collaboration and selflessly sharing their valuable knowledge, the key ingredient in completing of the present book.

The book's editors wish to thank all the reviewers (who chose to stay anonymous) for their expert opinion, constructive criticism, and their selfless dedication towards the development of knowledge, the scientific community, and the development of society as a whole.

About the Editors

Dr. Ankur Rajpal has a Ph.D. in environmental science, and currently, he is working as Project Scientist-C and is responsible for international and national wastewater, sludge, and solid waste management-related projects at the Indian Institute of Technology Roorkee (IITR), India. Earlier, he worked at Hong Kong Baptist University (HKBU), Hong Kong, as a postdoc fellow. He also worked as Senior Research Fellow on the Central Pollution Control Board (CPCB), MoEF, Delhi, India.

Dr. Rajpal is a fellow and life member of several professional bodies and organizations such as IWA, ANSF, etc. During past years, he has published more than 30 peer-reviewed articles in international journals. He is an active reviewer of several international journals and has participated in more than 25 international and national conferences. His research achievement has included many academic honours, such as Young Scientist Award (2019) from national level agencies. International travel grants from DST and CSIR, India, and several best paper awards. His research interests focus on problems related to wastewater and bio-waste treatment, enhancement of nutrient recovery, and micro-pollutant removal from the waste and wastewater.

Moharana Choudhury is a researcher and environmentalist from Guwahati, Assam, India. He has completed his education as M.Sc., PGDGIS, PGDEM, and PGCGARD from Gurukul Kangri Deemed to be University, Haridwar, Uttarakhand, North Eastern Hill University, Shillong Meghalaya, Tezpur University, Assam, and NIRD Hyderabad, respectively. Choudhury has worked on several scientific projects in various government departments and has more than 11 years of research and working experience in the field of the environment. He began his career as a research intern at the National Institute of Disaster Management in New Delhi, followed by the North Eastern Space Application Center/ISRO Unit (Department of Space, Government of India) as Junior Research Fellow. Choudhury has authored or edited five books with internationally reputed publishing houses, contributed more than 31 articles in national and internationally reputed journals and proceedings, and more than 26 book chapters. He has also written many articles and scientific reports in popular magazines. Choudhury is working on various environmental conservation activities in Assam and the country. He has worked on several initiatives to integrate environmentally sustainable projects for iconic heritage places, wetlands, and ecotourism spots in Assam, India, especially for cleanliness and beating single-use plastic pollution missions. He contributes to various environmental research, management, social media mass awareness, and research studies. He is also well known in India for his efforts to implement environmental education and safeguard the privileges of environmental degree holders in India.

Srijan Goswami, Author and Editor of several Medical Science Books (PG-Research category), is pursuing a Ph.D. in biotechnology from the Department of Biotechnology, University of North Bengal, India, under supervision of Dr. Manab Deb Adhikari. He is a Physician of Biochemic System of Medicine and also holds a specialization in medical biotechnology and clinical nutrition. He received his professional training on community nutrition and child care, medical microbiology, medical biochemistry, and diabetes counselling from reputed medical colleges in Kolkata, India. He is experienced in teaching at diverse range of Paramedical and Lifesciences Departments of reputed colleges and institutions. He is presently associated as Research Coordinator, Biomedical Research and Allied Sciences Division, Voice of Environment, Assam, India.

Dr. Goswami, has around 11 years of experience in the field of teaching and research and has published four books (and six in pipeline), and approximately 40 research papers in national and international journals. He has independently and successfully guided around 10 undergraduate and postgraduate students to complete their research work and publications in the past three years. Dr. Goswami has received four Letters of Commendation for Global Community Services and the Certificate of Honor for his contribution in the field of science and knowledge from the *International Journal of Agricultural Research, Sustainability and Food Sufficiency* (IJARSFS), *International Journal of Advances in Medical Sciences and Biotechnology* (IJAMSB), and *International Journal of Health, Safety and Environment* (IJHSE) in 2019, 2020, 2021, and 2023. The *Current Research in Nutrition and Food Science* (Enviro Research Group) and *Academia Scholarly Journals* have awarded him the Certificate of Excellence as a Reviewer in 2020, 2021, and 2023. He is actively associated as Editorial Review Board of 15 international journals.

Arghya Chakravorty is an early career researcher, doctoral fellow at the Centre for Nanotechnology Research, Vellore Institute of Technology – Vellore, India; and European Union's Horizon 2020-funded visiting researcher at the Institute of Physical Chemistry, Warsaw, Poland. He holds six Indian patents, while his research interest lies in interdisciplinary material science and nanobiotechnology. Recently he has edited two books, *Handbook of Porous Carbon Materials* for Springer Nature; and *Waste Management and Treatment:Advances and Innovations* for Taylor & Francis. As an early researcher, he has also published a dozen research and review papers and book chapters with international repute. Before starting his research journey, he earned his B.Sc. Honors (Microbiology) from Midnapore College, India, and a first-class M.Sc. (Microbiology) from Hemvati Nandan Bahuguna Garhwal University, A Central University, India. He is an active member of the American Chemical Society, Royal Society of Chemistry, Solid Waste Association of North America, and British Society for Antimicrobial Chemotherapy.

Dr. Vimala Raghavan is an enthusiastic and innovative Associate Professor at the Centre for Nanotechnology Research, Vellore Institute of Technology, Vellore, India, and has an h-index (Google Scholar) of 22. She received her Ph.D. degree in environmental biotechnology from VIT, Vellore, India. She has guest-edited several special issues of SCI- and SCOPUS-indexed journals, while her research interest is in various niche areas of nanotechnology, like hybrid nanobiomaterials, synthesis, and applications in therapeutics, diagnostics, environmental remediation, and energy sector. She has guided six Ph.D. scholars and more than 30 master's project students. She is currently guiding three Ph.D. scholars as well as serving as an investigator of several internationally and nationally funded research projects. She has more than 80 international and national journal publications and has contributed in many book chapters and many conference papers.

Contributors

Jose K. Antony
University of the People
USA

Julekha Bagum
Centre for Life Sciences
Vidyasagar University
Midnapore, West Bengal, India

Preetam Bhardwaj
Centre for Nanotechnology Research
Vellore Institute of Technology
Vellore, India

Jaya Bharti
Department of Psychology
A.N.D.N.N.M. Mahavidyalaya College
CSJM University
Harshnagar, Kanpur, India

Bhuvaneswari P
PG and Research Department of Microbiology
Cauvery College for Women (Autonomous)
Trichirappalli, India

Mcdonald Choudhury
Institute of Environmental Studies
Kurukshetra University
Kurukshetra, Haryana, India

Himadri Tanaya Das
Centre of Excellence for Advanced Materials and Applications
Utkal University
Odisha, India

Nigamananda Das
Department of Chemistry
Utkal University
Bhubaneswar, Odisha, India

Payaswini Das
CSIR-Institute of Minerals and Materials Technology
Bhubaneswar, Odisha, India

Einstein Mariya David
VIT School of Agricultural Innovations and Advanced Learning (VAIAL)
Vellore Institute of Technology
Vellore, India
and
School of Biosciences and Technology
Vellore Institute of Technology
Vellore, India

Anil Kumar Dikshit
Environmental Science and Engineering (ESE) Department
Indian A. Varalakshmi–VIT School of Agricultural Innovations and Advanced Learning (VAIAL)
Vellore Institute of Technology
Vellore, India

Swapnamoy Dutta
Department of Green Energy Technology
Pondicherry University
Puducherry, India

Saiyyeda Ferdous
VIT School of Agricultural Innovations and Advanced Learning (VAIAL)
Vellore Institute of Technology
Vellore, India

Abraham Francis
Social Work and Human Services College of Arts, Society, and Education (CASE)

Division of Tropical Environments and Societies (DTES)
James Cook University
Douglas, Townsville, Australia

Amar P. Garg
School of Biological Engineering and Life Sciences
Shobhit Institute of Engineering and Technology (Deemed-to-be-University)
Modipuram, Meerut, India

Shyamapada Ghosh
Department of Botany and Forestry
Vidyasagar University
West Bengal, India

Kumar Gourav
UIC, Chandigarh University
Chandigarh, Punjab, India

Annika Durve Gupta
Department of Biotechnology
B.K. Birla College (Autonomous)
Kalyan, Thane, India

Vandana Gupta
Department of Environmental Sciences
B.K. Birla College (Autonomous)
Kalyan, Thane, India

Tanmoy Jana
Department of Microbiology
Vidyasagar University
Midnapore, West Bengal, India

Pankaj Kanti Jodder
Urban and Rural Planning Discipline
Khulna University
Khulna, Bangladesh

Emilda K. Joseph
Department of Commerce
Christ College (Affiliated to MG University)
Puliyanmala, Kattappana, Idukki, Kerala, India

Tomy K. Kallarakal
School of Commerce, Finance and Accountancy
CHRIST (Deemed to be University)
Bengaluru, Karnataka, India

Sheeja Karalam
Department of Sociology and Social Work
Christ University
Bangalore, India

M. Karan
VIT School of Agricultural Innovations and Advanced Learning (VAIAL)
Vellore Institute of Technology
Vellore, India

Absar Ahmad Kazmi
Department of Civil Engineering
Indian Institute of Technology Roorkee
Roorkee, India

Anwar Khursheed
Department of Civil Engineering
College of Engineering
King Saud University
Riyadh, KSA

Martin Klein
Catholic University of Applied Sciences of North Rhine
Westphalia, Cologne, Germany

Garima Kohli
The Business School
University of Jammu
Jammu & Kashmir, India

Joseph Varghese Kureethara
Christ University
Bangalore, India

Rabeya Sultana Leya
Urban and Rural Planning Discipline
Khulna University
Khulna, Bangladesh

Aniruddha Mondal
Division of Materials Science
Department of Engineering Sciences and Mathematics
Luleå University of Technology
Luleå, Sweden

Abhisek Nalluri
Department of Materials Science and Engineering
Center for Fuel Cell Innovation
Huazhong University of Science and Technology
Wuhan, China

Goldie Oza
National Laboratory for Microfluidics and Nanofluidics (LABMyN)
Centro de Investigación y Desarrollo Tecnológico en Electroquímica (CIDETEQ)
Queretaro, Mexico

Rajchandar Padmanaban
Centre of Geographic Studies
Institute of Geography and Spatial Planning
University of Lisbon
Lisbon, Portugal

Theivasigamani Parthasarathi
VIT School of Agricultural Innovations and Advanced Learning (VAIAL)
Vellore Institute of Technology
Vellore, India

Sonali Zankar Patil
Department of Botanical Sciences
B.K. Birla College (Autonomous)
Kalyan, Thane, India

Karthikeyan Ramesh
School of Biosciences and Technology
Vellore Institute of Technology
Vellore, India
And
Centre for Nanotechnology Research
Vellore Institute of Technology
Vellore, India

Rehab A. Rayan
Department of Epidemiology
High Institute of Public Health
Alexandria University
Alexandria, Egypt

Aditi Roy
Department of Botany (Environmental Sciences)
University of Lucknow
U.P., India

Subhrajeet Sahoo
Centre for Life Sciences
Vidyasagar University
Midnapore, West Bengal, India

Muhammad Sajid
Faculty of Materials and Chemical Engineering
Yibin University
Yibin, China

G. Sajith Kumar
VIT School of Agricultural Innovations and Advanced Learning (VAIAL)
Vellore Institute of Technology
Vellore, India

Arijit Samanta
Department of Microbiology
Vidyasagar University
Midnapore, West Bengal, India

Siva Sankar Sana
School of Chemical Engineering and Technology
North University of China
Taiyuan, China

Sangeetha K
PG and Research Department of Microbiology
Cauvery College for Women (Autonomous)
Trichirappalli, India

V. Santhosh Kumar
VIT School of Agricultural Innovations and Advanced Learning (VAIAL)
Vellore Institute of Technology
Vellore, India

Bristi Sarkar
Sociology Discipline
Khulna University
Khulna, Bangladesh

Shivani Singh
Central Institute of Petrochemicals Engineering and Technology
Chennai, India

Swati Singh
CSIR-National Botanical Research Institute
Lucknow, India

Md. Sohel Rana
Urban and Rural Planning Discipline
Khulna University
Khulna, Bangladesh

Ghazal Srivastava
Department of Civil Engineering
Indian Institute of Technology Roorkee
Roorkee, Uttarakhand, India

Steffi PF
Microbiology PG and Research Department of Microbiology
Cauvery college for women (Autonomous)
Trichy, Tamil Nadu, India

Karuppasamy Sudalaimuthu
Department of Civil Engineering
SRM Institute of Science and Technology
Kattankulathur, India

G. Sunil Kumar
VIT School of Agricultural Innovations and Advanced Learning (VAIAL)
Vellore Institute of Technology
Vellore, India

K. V. Suresh Babu
University of Cape Town
Cape Town, South Africa

Thamilmaraiselvi B
PG and Research Department of Microbiology
Cauvery College for Women (Autonomous)
Trichirappalli, India

Lijo Thomas
Christ University
Bangalore, India

Dhayalan Vaithiyanathan
Department of Agricultural Engineering
Hindusthan College of Engineering and Technology
Coimbatore, Tamil Nadu, India

Shivkumar Vanjari
Environmental Science and Engineering (ESE) Department
Indian Institute of Technology Bombay
Powai Mumbai, Maharashtra, India

A. Varalakshmi
VIT school of Agricultural Innovations and Advanced Learning (VAIAL)
Vellore Institute of Technology
Vellore, India

Bindi Varghese
Christ College (Affiliated to MG University)
Puliyanmala, Kattappana, Idukki, Kerala, India

M. Vigneshwaran
VIT School of Agricultural Innovations and Advanced Learning (VAIAL)
Vellore Institute of Technology
Vellore, India

Theresa Nithila Vincent
Department of Commerce
Christ University
Bangalore, India

1 Introduction to Environmental Waste
A Serious Issue

Julekha Bagum, Subhrajeet Sahoo, Arijit Samanta, Abhisek Nalluri, Siva Sankar Sana, Preetam Bhardwaj, Karthikeyan Ramesh, Vimala Raghavan, Amar P. Garg, and Mcdonald Choudhury

1.1 INTRODUCTION

Environmental waste is a significant cause of pollution. Wastes are inconsumable ingredients. Waste is any material discarded after using necessary substances or that is defective, useless, and of no value. Waste products may become a joint product, by-product, or resource through a development that increases their waste value above zero. Solid and liquid, poisonous and non-poisonous wastes are generated on a daily basis in our homes, classrooms, workplaces, emergency clinics, and projects. The aggregate amount of waste produced (town, new, hazardous) exceeds 4 billion tons worldwide each year (Hoornweg & Bhada-Tata, 2012; Ahmed et al., 2018). Likewise, massive waste's global consequences are growing extraordinarily rapidly as free waste management costs would increase from the current annual $205.4 billion to around $375.5 billion per year by 2025 (Mavropoulos & Newman, 2015), which is very high compared to other types of waste. Approximately 10 million tons of oil returns to streams and seas per year, alongside 500 billion tons of new waste (Skenderovic et al., 2015). Industrial waste not only pollutes soil and water but also pollutes the air. A billion tons of aerosols and ashes are spreading into the atmosphere from industrial facilities and transport systems. With the creation of urban communities, the waste issue is serious. Dumping grounds such as Leogang in Shanghai, China; Sudokwon in Seoul; the now-full Jardim Gramacho in Rio de Janeiro, Brazil; and Bordo Poniente in Mexico City are seeking the title of the biggest in the world. Growing sites regularly receive more than 10,000 tons of waste every day. The development of urban communities such as Shenzhen in China adds up to 2,000 or more waste incinerator stocks worldwide. With the most significant ready for processing capacity of more than 5,000 tons per day, pressures on air pollution, debris removal, and costs are rising (Hoornweg et al., 2013; Simatele et al., 2017; Fernández-Aracil et al., 2018). As urban dwellers become increasingly affluent, the calculation of the waste they create is approaching its limit. The biochemical waste decomposition process unfavorably affects the environment. Many rats, mice, and insects are reproducing on them, which contributes to the spread of infection and

DOI: 10.1201/9781003258377-1

contaminates the whole situation. This thing carries out the problem of protecting the environment through waste management (Skenderovic et al., 2015; Ye et al., 2020; Choudhury & Choudhury, 2014: Bai & Ogbourne, 2016; Sana et al., 2020; Knust et al., 2020; Murray et al., 2020; Lopes et al., 2016; Fernándes-Fernández et al., 2019; Evans, 2016). Almost all countries in the world are producing more waste per person than any other. Nobody is immune to daily waste disposal-related problems. Like any nearby state and government guidelines that administer its administration, how waste is handled often depends on its source and quality. Practices differ greatly for urban and rural living arrangements and ventures as well as for developing and developing nations (Willis et al., 2018; Zhang & Zeng, 2019; Chen & Liu, 2019).

1.2 TYPES OF WASTES

Owing to the advancement of invention and the technological transition, using examples of individuals has all-inclusively changed in the twentieth century. Scatter utilization has increased in natural assets and products. For this, massive quantities of various forms of waste are delivered to the planet every day, generating a worrying problem (Hoornweg et al., 2013; Sana et al., 2020; Pepper et al., 2017). Commonly, the wastes may be categorized as follows:

1.2.1 Solid Wastes

Solid waste is not limited to very solid wastes, but various sturdy wastes are fluid, semi-strong, or vaporous. Solid wastes are the materials that have been discarded after their use (i.e., disposed of, destroyed, burned, bogusly recycled) (Abdel-Shafy & Mansour, 2018; Kaza et al., 2018; Sharma et al., 2020; Kumar & Samadder, 2017).

1.2.2 Liquid Wastes

Liquid waste can be characterized as sewage, fats, oils or oil (FOG), the fuel used, liquids or mud, and hazardous residential fluids such as wastewater from food preparation and assembly industries as well as from other sectors. For example, wastewater may contain natural substances and supplements that provide agribusiness incentives or may be unsafe due to the pathogens and synthetic compounds contained therein (Momayez et al., 2018; Sabourin-Provost & Hallenbeck, 2009; Lee et al., 2018; Wang et al., 2020).

1.2.3 Gaseous Wastes

Gaseous waste is a waste item created by various human exercises in the gas structure – e.g., preparation, manufacture, material utilization, or natural processes. These wastes, discharged from industrial plants, automobiles, copying non-renewable energy sources, and mixed in the climate, contain CO, CO_2, SO_2, NO_2, O3, and CH_4, respectively (Sydney et al., 2021; Yang et al., 2019; Alves et al., 2019).

1.3 CLASSIFICATION OF THE WASTE DEPENDING ON THEIR CHEMICAL, BIOLOGICAL, AND PHYSICAL PROPERTIES

1.3.1 Biodegradable Wastes

Biodegradable waste is any product that breaks down naturally by water, the sun's rays, radiation, oxygen, microorganisms, and other environmental factors. In the process, organic matter is down into simpler units. The issue will eventually return to the soil by decomposition. In this way, the land is getting nourishment (Kumar et al., 2018; Beneroso et al., 2016).

1.3.2 Non-Biodegradable Wastes

A non-biodegradable substance is a sort of material that can't be separated by familiar creatures and is an essential source of contamination (Karthikeyan et al., 2020; Velvizhi et al., 2020).

1.3.2.1 Hazardous Wastes

Hazardous wastes have properties like ignitability, destructiveness, reactivity, and harmfulness. That makes them, like the earth, risky and potentially unsafe for human well-being. Dangerous wastes can be solids, gases, fluids, or slimes. They can be side effects of assembling forms or dismissing business items, such as liquids and pesticides for cleaning (Hossain et al., 2020; Yang et al., 2018).

1.3.2.2 Non-Hazardous Wastes

Waste materials not explicitly supposed to be hazardous underneath federal law have been considered as non-hazardous wastes. They include woods, papers, plastics, glasses, metals, chemicals, and other materials generated by commercial, industrial, agricultural, and residential sources. Inappropriate management of them possesses essential risks to the atmosphere and human well-being (Leonelli et al., 2017; Zhang et al., 2016; Lecomte et al., 2019).

1.4 CLASSIFICATION OF WASTE BASED ON SOURCE

1.4.1 Urban Waste

Urban Waste has described as the waste ready from the urban areas and towns' mechanical and local locations. If properly recycled, compensated, or supervised, this waste can cause significant health and ecological problems. As per a report by Down To Earth, "More than 377 million urban people live in 7,935 cities and urban areas and generate approximately 62 million tons of heavy civil waste per year. Just 43 million tons (MT) of waste has produced, of which 11.9 MT has been recycling, and 31 MT has been dumping in landfill locales". Urban waste contains municipal solid waste (MSW), construction and demolition waste, and municipal wastewater (sewage sludge) (Lavanya, 2018; Choudhury & Dutta, 2017; Noor et al., 2020; Khandelwal et al., 2019).

1.4.2 Municipal Solid Waste

Municipal solid wastes (MSW) are called the trash accumulated from private houses, markets, lanes, and various places, mostly in urban regions and discarded by municipal bodies. There is an extraordinary assortment of vast sums, an incredible assortment, complex segments, and wellsprings of them; removing those turn out to be extremely troubling (Figure 1.1-B).

1.4.3 Waste from Construction and Debris

Development and demolition (C&D) wastes indicate as the stable waste created by constructing, restoring, redesigning, renovating, modifying, or dismantling private, industrial, governmental, or institutional systems, business offices, and networks,

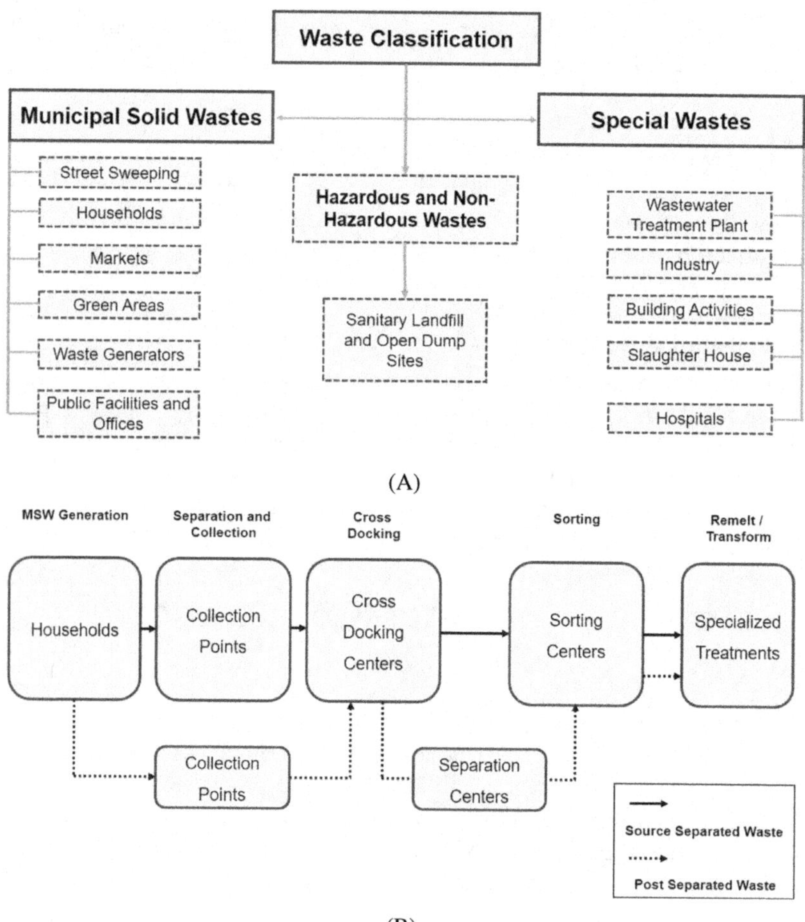

FIGURE 1.1 (A) Categories of solid waste sources in developing country [Adapted with Permission from (Ferronato et al. 2019) (2019, Elsevier)] License Number: 4840940873084; License date: Jun 02, 2020 (1.13a.m). (B) A Flowchart for Recycling of Municipal Solid Waste [Adapted with permission from (Bing et al. 2016, Elsevier)].

such as stretches, streets, caves, dams, air terminals, and rail lines. Flotsam and jetsam construction and destruction (C&D) indicate a vital segment of urban waste that accounts for around 15–30% of all massive waste by weight.

1.4.4 Municipal Wastewater (Sewage Sludge)

Sewage sludge is generated from both industrial and municipal wastewater treatment. Crude metropolitan wastewater experiences primary, necessary, auxiliary, and now and again, further action to generate awarded wastewater and a concentrated stream of solids in a city ooze fluid (Carrère et al., 2010; Qi et al., 2020). Since the physical-concoction forms associated with the procedure, the mud contained significant metals. Natural mixes adopt inadequately biodegradable, just as potentially pathogenic, forms of life (viruses, bacteria, microorganisms, and so on).

1.4.5 Industrial Waste

Unwanted or residual materials that result from industrial operations are called industrial waste. Waste discharge from power plants, concrete production lines, paint projects, metallurgical plants, material enterprises, industrial plants, mining activities, food preparation firms, and oil companies are warm-force facilities. It can be substantial, semi-solid, fluid, or vaporous; it may be toxic waste or dangerous waste. Risky waste can be harmful, damaging, ignitable, reactive, or radioactive. Mechanical trash may end up contaminating the soil, ground, or near water-bodies in the ocean. Modern waste has also mixed with municipal waste, which makes detailed evaluations disturbing (Ahmad et al., 2019; Banerjee et al., 2021).

1.5 DIFFERENT TYPES OF INDUSTRIAL WASTE

1.5.1 Agricultural Wastes

Agricultural waste is classified as:

- Crop-waste paddy husks, sugar cane biogases, rice straw, cotton stalk, new straw.
- Animal waste: animal excreta, animal carcasses.
- Processing-waste materials: as packaging materials, bottles, etc.
- Hazardous waste: pesticides, insecticides.

1.5.2 Commercial Wastes

Waste generated from any business or trading activity is termed industrial wastes. This activity could be related to entertainment, recreation, education, sport, or any other activity with a commerce element. As a result of planning events, it does exclude agricultural, family, or urban waste (Coskuner et al., 2020).

1.5.3 Mining

The debris is created from the mine, and there are three kinds to be found in enormous volume: waste rock, tailings, and mine-water. The synthetic compounds in the handling stage – for instance, cyanide and others – offer impressive danger due to their high poisonousness (Alekseenko et al., 2020).

1.5.4 Radioactive Wastes

Radioactive waste is a hazardous waste containing radioactive material, which is a side effect of various forms of atomic innovation. Dangerous waste production involves nuclear research, nuclear medication, atomic force, development, manufacturing, and nuclear weapons reprocessing (Stefanovsky et al., 2016).

1.5.5 Biomedical Wastes

Biomedical waste (BMW) includes waste created from emergency clinics, neurotic concentrate, social insurance offices, and research laboratories for wellness. Such wastes consist of solids, liquids, sharp instruments, and laboratory wastes, causing infections if discarded without proper maintenance. There are various categories of biomedical wastes. Under the subcategory of solid wastes, there are tubes, catheters, masks, surgical staples, scrubs, disposable gowns, sutures, wound dressings, etc. In the liquid forms of biomedical wastes, tissues, body fluids, organs, cells, and tissue cultures have been observed. Sharp objects such as blades, scalpels, razors, lancets, slides, cuvettes, needles, and syringes are also a part of biomedical wastes. Different types of pharmacological concoctions and medicines can also have characterized as biomedical wastes (Obi et al., 2016; Rahman et al., 2020; Rajan & Robin, 2019).

1.6 EFFECT OF WASTES ON ENVIRONMENT

Waste-related outcomes fluctuate widely and have affected the substances or synthetic substances found in the garbage and how they monitor it. Even though there is no information to legitimately interface waste drifts with effects on human well-being and the earth, waste management can result in waste and synthetic substances entering the environment (Sana et al., 2020; Lijó et al., 2017).

By definition, dangerous waste can adversely influence human well-being and the earth, which is why it has been so carefully managed. Hazardous wastes are either explicitly recorded as perilous by the EPA (US Environmental Protection Agency) or a state or display at least one of the accompanying attributes: ignitability, destructiveness, reactivity, or poisonousness. Age and the board of risky wastes can defile the land, air, and water and influence human well-being and natural conditions.

Municipal solid waste landfills are the third-biggest wellspring of human-related methane emanations in the United States, representing around 16% of these outflows in 2016 (USEPA, 1999).

The EPA is intrigued because gas outflows can be influenced by reusing and changing item use. For instance, reusing office paper or aluminum can decrease natural impacts (e.g., by diminishing the need to gather trees or mine bauxite to deliver aluminum).

Waste management and disposal are overall issues. Poor, obsolete, and illicit acts of urban and hazardous waste-removal influences have neighborhood networks basically in all nations; this incorporates unlawful trans-limit exchange, for the most part from industrialized nations (Marsili et al., 2017).

Apart from particular solid wastes, different kinds of human-made wastes are progressively unsafe to the earth. Wireless, for example, is made of lead, mercury, and plastic; thus, a large number of them get tossed as trash. This sort of electronic trash creates ecological issues. Electronic waste is quickly turning into a major danger to our condition. Another problem is whether the garbage is dangerous to remove.

Solid waste in urban zones has become a huge issue in numerous nations where urbanization is quick. As populaces and mechanical-assembling exercises increment in those territories, increasingly more waste will produce likewise. A lot of solid waste in urban territories has occupied too many land environments of people (Su et al., 2017).

As municipal solid waste can't diffuse quickly, the contamination threat they pose to the earth has long-term potential. The unsafe parts in solid wastes can make potentially dangerous nature, individuals, and creatures. If we eat contaminative plants or creatures as food, the contaminants will focus on the human body and cause mischief to human well-being. These ecological risk impacts can spread from people to the entire species; lastly, break the entire eco-equalization and harm our common assets (Obi et al., 2016; Fernandes et al., 2019).

Unpredictable natural results include food contamination, risky food-cleanliness, and debased farmland because of conceivably enduring and harmful synthetic compounds in bundling materials or jugs (Obi et al., 2016; Singh et al., 2020; Liu et al., 2019).

Due to excessive utilization of inorganic compost, the abundance divide has been holding in the dirt. A bit enters lakes and waterways because of either surface overflow or the received water-system framework. This situation brings about the contamination of surface water; a segment enters the groundwater, and a part dissipates or becomes de-nitrated, causing air contamination (Obi et al., 2016).

Air contamination incorporates scents radiating from confines coming about because of domestic animal wastes – the rotting procedure of fundamental issues in compost, creature pee, and excess nourishments (Obi et al., 2016).

An untreated and non-reusable waste source can create ozone-depleting substances while likewise effectively affecting the dirt's richness and causing water contamination (Obi et al., 2016).

In animal waste, water volume represents 75–95% of total volume, while the rest incorporates fundamental issues, inorganic issues, and numerous types of microscopic organisms and parasitic eggs. Those microbes and substances can extend ailments to people and are a source of many pessimistic consequences for the earth (Obi et al., 2016).

Demolition of wastes expanding airborne poisons has been created from the emptying exercises of waste. Incrementally, the chance of substantial metals and risky substances inside the waste flow leads to debasing both topsoil and underground water.

The effects of landfill and cremation are critical because of their potential for ozone-depleting-substance discharges (methane, carbon dioxide) and trans-limit movement of natural, miniaturized scale-contaminations (dioxins and furans) and unstable substantial metals. Waste issues are as old as humankind (Ermolaev et al., 2019; Pham et al., 2018).

Water released from industrial facilities has been tainted with unsafe synthetic compounds, radioactive resources, significant metals, or the usual slop, waste, and colors. This water contamination can slaughter untamed life and create mischief for the overall biological system, and the impacts of dirtied water have left a few plants and creature species jeopardized.

Construction waste raises the weight of landfill sites that are slowly becoming scarce. Furthermore, if the waste has not been properly controlled, materials such as solvents and synthetically compensated woods will contaminate the soil and water (Zhefeng et al., 2020; Polyakova et al., 2018).

The outflows from various industries include vaporous pollutants, such as arsenic, carbon dioxide (CO_2), nitrogen oxides, methane, etc. When there is a lot of rain, such gases cause a few ailments and natural hazards most of the time. A portion of the consequences of air pollution is the production of acid rainfall, exhaust cloud, and increased rates of respiratory problems among humans (Liu et al., 2017; Chakravorty et al., 2022).

When the dirt loses its fruitfulness and structure, attributable to the differing natural and counterfeit wonder, removing building wastes into landfills is among the counterfeit perspectives adding to soil contamination. Mechanical waste has various quantities of toxic materials and industrial concoctions with the ultimate aim that, when deposited in landfills, it collects in the topsoil, along these lines devaluing the fertility and organic activity of dirt due to soil destruction (Hale et al., 2021; Yuan et al., 2019).

Such suggestions ultimately add to their awkward nature, creating crop-efficiency problems along those lines. The synthetic concoctions and harmful materials harm soil accumulation in plants developed in such zones, messing with the well-being of the individuals who spend such yields (Hinojosa et al., 2021; Qian et al., 2018).

Industrial, poisonous, and synthetic wastes containing discarded water bodies or landfills are likewise liable for malignancies and harms to human cells. For example, an introduction to inorganic arsenic makes a tumor structure. Most importantly, mechanical toxins are liable for a great many diseases and unexpected losses around the globe (Cong, 2018; Ayiania et al., 2021).

The inclination of modern and assembling forms that continually requests creation assets and rehashed misuse of crude materials has prompted the pulverization of woodlands and the usual natural surroundings that help untamed life (Cong, 2018).

Thus, some untamed life species have confronted termination, while a few others remain profoundly imperiled. Mechanical wastes, mining, synthetic concoctions, discharges, or lucky breaks, fires, and oil slicks, before long, have likewise been prime supporters of natural life annihilation.

Waste burning, while keeping a strategic distance from landfill issues, can cause various dangerous ecological problems. For the most part, plastic produces poisons, which are similar to the dioxins it absorbs. The gases produced by cremation will contaminate the air and add to the acid downpour, causing impacts. Also, the incinerators create debris that, now and again, contains substantial metals, just as different poisons (Alves et al., 2019).

1.7 WASTE MANAGEMENT

Whenever done securely and capably, waste consumption can create clean vitality, which can be utilized rather than petroleum derivatives. This procedure gives a practical option, in contrast to putting garbage in landfills. As the issue of waste amassing arrives at overwhelming extents, an overall exertion is to control waste and to oversee it. Numerous nations and individual towns are putting forth an attempt to manage waste. Different types of garbage should be handled differently (Kumar et al., 2018; Choudhury & Dutta, 2017; Zhang et al., 2016; Lavanya, 2018; Noor et al., 2020; Khandelwal et al., 2019).

1.7.1 Agricultural-Waste Management

An agricultural-waste administration framework comprises six fundamental capacities – i.e., creation, assortment, stockpiling, treatment, move, and usage. For a particular strategy, these capacities might be consolidated, rehashed, wiped out, or masterminded (He et al., 2019; Fareed et al., 2020).

1.7.1.1 Production

Production is the capacity of the sum and nature of horticultural waste produced by a cultivating endeavor. The waste requires the executives, if the amounts created are sufficiently adequate, to turn into an asset concern. The waste-administration framework may need to oblige occasional varieties in the pace of creation. A record ought to have been kept of the information, suppositions, and counts used to decide the sort, consistency, volume, area, and timing of the waste created. The creation assessments ought to incorporate future extension (He et al., 2019; Fareed et al., 2020; Wei et al., 2020).

1.7.1.2 Collection

This procedure alludes to gathering waste from its root source. The AWMS plan ought to recognize the assortment strategy, area of the assortment focuses, booking of the gathering, work prerequisites, necessary gear or principal offices, the executives and establishment expenses of the parts, and the effect assortment has on the consistency of the waste (Liu et al., 2017).

1.7.1.3 Storage

Storage is the brief regulation of the waste. The storeroom of a waste administration framework is the device that gives the director authority over the planning and timing of the framework capacities. The waste-administration framework ought to recognize the capacity time frame; the necessary stockpiling volume; the sort, evaluate the size, area, and establishment cost of the storeroom; the administration cost of the capacity procedure; and the effect of the capacity on the consistency of the waste (Wei et al., 2020; Xu et al., 2018).

1.7.1.4 Treatment

Treatment is any capacity intended to decrease the waste's contamination potential, including physical, organic, and substance treatment. It incorporates exercises that are, here and there, considered pre-treatment; for example, the partition of solids. The arrangement should be comprised of an investigation of the attributes of the loss before treatment; an assurance of the ideal qualities of the accompanying trash techniques; the determination of the sort, evaluate the size, area, and the establishment cost of the treatment office; and the administration cost of the treatment procedure (Wei et al., 2020; Thines et al., 2017).

1.7.1.5 Transfer

This procedure has been alluding to the development and transportation of waste all through the framework. It incorporates the rubbish's exchange from the assortment point to the storeroom to the treatment office and the use site. The waste may require expulsion as a strong fluid or slurry, contingent upon the absolute, stable fixation (He et al., 2019; Wei et al., 2020).

The framework plan ought to incorporate an examination of a technique for transportation, separation between focuses, recurrence and planning, major hardware, and the establishment and the board expenses of the exchange framework.

1.7.1.6 Utilization

Utilizations incorporate reusing reusable waste items and reintroducing non-reusable waste items keen on the earth. Agrarian wastes possibly use as a wellspring of vitality, bedding, creature feed, mulch, the fundamental issue, or plant supplements. Appropriately rewarded, they can be attractive. A typical practice is to reuse the supplements in the loss through land application. A complete investigation of usage through land application incorporates choosing the fields, booking applications, structuring the circulation framework, choosing necessary hardware, deciding application rates and volumes, estimation of the reused items, and establishing and the board costs related to the use procedure (Xu et al., 2018; Nadar et al., 2018; Thines et al., 2017).

1.8 SOLID WASTE MANAGEMENT

1.8.1 Solid Waste Collection

The assortment incorporates all exercises related to the get-together of stable wastes and the pulling of the waste gathered to the area from where the assortment vehicle will arrive at the last car to the site of removal. There are three essential techniques for assortment.

1.8.1.1 Community-Stockpiling Point

The city rejects, taken to fixed storage receptacles and put away until the waste assortment organization gathers it day by day for this removal in a vehicle.

1.8.1.2 Curbside Collection

Ahead of assortment time, the reject has gotten holders and set the footway from where it has been gathering by the waste-assortment organization (Nemerow & Dasgupta, 1991).

1.8.1.3 Block Collection

Individuals acquire the waste compartments and hand it over to the assortment staff that exhausts it into the holding-up vehicle and returns the container to the people (Nemerow & Dasgupta, 1991).

1.8.1.4 Disposal of Solid Wastes

Before stable waste has eventually been discarded, it has improved the steady waste removal framework's proficiency and recoups useful resources out of the stable wastes. The handling procedures – for example, compaction – i.e., mechanical volume reduction or burning – i.e., warm volume decrease – and manual segment partition – i.e., manual arranging of the waste – are utilized to expand the effectiveness of stable waste administration (Nemerow & Dasgupta, 1991). The regularly utilized techniques for removal are as follows:

1.8.1.5 Salvage or Manual Component Separation

Before extreme departure, the manual detachment of dependable waste segments has been achieved to recover and reuse materials. Cardboard, newsprint, top-notch paper, glass, metals, wood and aluminum jars, and so forth are physically sifted through or rescued either for reusing or resale (Nemerow & Dasgupta, 1991).

1.8.1.6 Compaction or Mechanical Volume Reduction

After the partition of reusable or expendable articles, compactors have been used to compress the waste materials straightforwardly into enormous holders or to shape parcels that could then be positioned in huge compartments. Compaction expands the helpful existence of landfills (Nemerow & Dasgupta, 1991).

1.8.1.7 Incineration or Thermal Volume Reduction

Exceptionally flammable waste like plastics, cardboard, paper, elastic and combustible waste like containers, wood scrap, floor sweepings, food wastes, and so forth are subjected to burial – i.e., consuming at high temperatures. Burning results in air contamination; thus, legitimate control gear needs to be introduced to maintain a strategic distance from the earth's pollution.

1.8.1.8 Open Dumping

Open dumping of stable wastes is done in low-lying zones and edges of the towns and urban areas. Being less expensive, this technique for removal is utilized broadly in India.

1.8.1.9 Sanitary Landfilling or Controlled Tipping

The sterile land-filling includes removing city wastes on or in the upper layers of the world's mantle, particularly in corrupted zones needing rebuilding. In landfills, the stable wastes are compacted and extend in far segments, and each section has been consistently covered by a layer of soil (Nemerow & Dasgupta, 1991).

1.8.1.10 Pyrolysis or Destructive Distillation

In this removal strategy, stable wastes have warmed under anaerobic conditions (i.e., consuming without oxygen). The stable wastes' natural segments split up into unpredictable fluid and vaporous divisions (CO, CO2, CH4, tar, scorched carbon). In contrast to the profoundly exothermic procedure of ignition, pyrolysis is an exceptionally endothermic procedure, and that is the reason it is additionally called dangerous refining (Nemerow & Dasgupta, 1991).

1.8.1.11 Land Cultivating

In this waste-removal technique, the biodegradable construction-waste product is rewarded by the natural and physical concoction it forms outside of the dirt. The natural wastes are either applied on the land or infused beneath the dirt's surface with reflective gear, where they experience bacterial and chemical decay.

1.8.1.12 Recycling

The reuse of materials and its division from waste has been called reusing. Some procedures that recycle those are Alternate Fuel as Refuse-Derived Fuel (RDF), Plasma

Pyrolysis Technology (PPT), screening, dissolution, sterilization, de-inking, blanching process, and vitrification technology.

1.9 BIODEGRADABLE-WASTE MANAGEMENT

1.9.1 Aerobic Degradation Process

In the oxygen-consuming degradation process, natural materials such as trash, leaves, excrement, and so forth have eroded to give a humus item, generally called manure, that may be used as compost. This procedure takes significant time (Patwardhan, 2013).

1.9.2 Anaerobic Digestion

Anaerobic processing prompts a profoundly attractive item called methane. The initial step is to separate the mind-boggling natural resources nearby in the reject into natural corrosive and CO2.

1.9.3 Gasification

Gasification has included the warm disintegration of the fundamental gases at high temperatures, restricting the measuring of air. It has created a blend of flammable and non-burnable gases like carbon monoxide (CO) and carbon dioxide (CO2), respectively (Patwardhan, 2013).

1.9.4 Fertilizing the Soil

Feeding the land includes the breakdown of natural waste within sight of microorganisms, warmth, and dampness. Three sorts of living beings have engaged with treating the soil: microscopic organisms, parasites, and actinomycetes, which follow up on the waste to transform it into sugars, starch, and natural acids (Patwardhan, 2013).

1.9.5 Vermicomposting

Vermicomposting is a process where food material and waste from the kitchen, counting vegetables and peelings from organic items, documents, etc., may be converted into fertilizer by can-worm behavior. The anaerobic state is made visible all over by the introduction of natural waste (Rajendran et al., 2019). It has also been found in an ongoing report that landfills' regular organization comprises more than 70 sorts of natural contaminations in the landfill creek, counting phenol, fragrant hydrocarbons, and chlorinated aliphatic forth., which were solids harmful to microalgae cells (Paul et al., 2019).

1.9.6 Biodiesel Production

Waste cooking oil is isolated from kitchen waste, using a physical treatment. This cooking oil has applied to receive biodiesel (Soni et al., 2016).

1.9.7 Ethanol Production

In the wake of isolating oil, the rest has matured utilizing yeast, and ethanol has been delivered. From matured crush, hydrous ethanol has been refined out. Hydrous liquor is ethanol having around a 95% v/v ethanol focus. It has been dried out by expelling dampness (utilizing a sub-atomic strainer) and parchedness and is used as fuel ethanol (for mixing in gas) (Soni et al., 2016).

1.9.8 Bio-Methanation

Spent wash, produced in the wake of refining out hydrous ethanol, has blended in with non-kitchen biodegradable waste, and bio-methanation has been completed. Biogas created by bio-methanation has around 55% methane. After bio-methanation, a flood from a biodigester has been utilized as fluid fertilizer (Soni et al., 2016).

1.10 SEWAGE AND SLUDGE MANAGEMENT

"Waste adjustment Pond System (WSPS)" – i.e., "Slow Rate Trickling Filters" – is utilized where land accessibility isn't an issue and temperature in winter isn't shallow. "Actuated Sludge Process (ASP)", "BIOFOR", "Up-stream Aerobic Sludge Blanket (UASB)" give a high level of unwavering quality and have been utilized in zones with land requirements. High-rate Trickling Filters, "Facultative Aerated Lagoons (FAL)" – these advances have been received any place "ASP, BIOFOR, UASB" innovation can't be utilized. Likewise, hardly any strategies are classified as dependent on minor exhibitions, which have not generally been applied, similar to the UASB framework with FPU for downstream treatment (Rajpal et al., 2022; Patwardhan, 2013).

1.10.1 Construction and Demolition Waste the Board

Development and Demolition Waste Management may be characterized as the order related to capacity and isolation, assortment and transportation, reusing, and reuse removal.

1.10.2 Storage and Segregation

Development and demolition wastes are best put away at the source – i.e., at the purpose of age. It might be accomplished by handling the blended material to evacuate unknown substances. Net insulation of waste from building and demolition into street work materials, essential structural elements, saved parts of the structure, and site freedom waste is critical. New segregation is necessary to facilitate the reuse/reuse of materials such as wood, glass, cabling, plastic, plasterboard, etc. before destruction, to produce a reused total that meets the particular requirement.

1.10.3 Collection and Transportation

The garbage of building and demolition has been laid off in skips. The frame used lifts equipped with a pressure-driven crane, at that point, for successful and brief expulsion. On the off-chance trailers have been used, tractors could evacuate these at that point. Front-end loaders, combined with powerful tipper trucks, may be used to take care of immense quantities, and the time required for stacking and emptying has held to the surface.

1.10.3.1 Reusing and Reuse

Reusing or reuse is a primary method for handling such waste. Apart from mounting waste management concerns, numerous factors that underpin the appropriation of reuse/reuse techniques has decreased extraction of crude materials, reduced transport costs, improved benefits, and depleted natural impact.

The three kinds of accessible reusing plants are mobile, semi-mobile, and stationary plant.

1.10.3.2 Mobile Plant

The substance is squashed and screened at the mobile factory, and by attractive detachment, ferrous pollutions have been isolated. The material is transported to the site of destruction itself and is ideal for handling only non-polluted cement or stonework waste.

1.10.3.3 Semi-Mobile Plant

Evacuation of pollutants is done by hand in semi-portable facilities, and the element has also been screened at long last. Attractive detachment has been done for expelling the ferrous material. The quality of finished results is superior to that of a portable factory. The previously mentioned plants are not appropriate for preparing a well of mixed annihilating waste with slight problems such as metal, plastic, wood, etc.

1.10.3.4 Stationary Plant

Stationary plants have planned, for pulverizing completion, screening much like sanitizing to remove the pollutants. Issues that are important to consider when erecting fixed plants are plant location, street base, accessibility of land space, weigh-bridge arrangement, the need for a capacity zone, and so forth. Various kinds of smashers have been utilized in reusing plants; in particular, jaw-smasher, sway smasher, impeller-sway smasher.

1.10.3.5 Disposal

Being prevalently latent C&D waste doesn't make for compound or bio-synthetic contamination. After that, the most drastic effort should have been made to reuse trash, as previously explained. The material may be used for low-lying territories filling/leveling.

1.11 INDUSTRIAL WASTE-WATER TREATMENT

1.11.1 Oxidation Processes

The procedure will annihilate some mixtures and constituents by oxidation and by reducing the response. Propelled oxidation is synthetic oxidation with the receptive and transient hydroxyl radicals. The materials should be generated nearby, in a reactor where the wastewater radicals might touch the organics. Hydroxyl radicals can be incorporated into structures using bright radiation/hydrogen peroxide, ozone/hydrogen peroxide, bright radiation/ozone, Fenton reagent (ferrous iron and hydrogen peroxide), titanium dioxide/bright radiation, and by various methods.

1.11.2 Reduction/Oxidation

Redox responses synthetically transform the dangerous pollutants into nonhazardous or less harmful intensifications – that is, increasingly stable, less compact, and inactive. Substance redox is a full-scale, settled technique used for drinking water and wastewater sterilization and is a standard treatment for cyanide (oxidation) and chromium waste. For material redox, the target contaminant bunch is inorganic. The technique could also have been used alongside non-halogenated VOCs and SVOCs, hydrocarbons of fuel, and pesticides (Soni et al., 2016).

1.11.3 Chemical Precipitation

Metals don't debase in the earth. Compound precipitation is a widely used, tested method for removing wastewater of minerals and various inorganics, suspended solids, fats, oils, and certain other natural substances (counting organophosphates). The ionic metals have been transforming by the material's reaction between the solvent metal mixes and the advancing reagent into an insoluble structure (molecule). The particles produced by this response have been removing by settling and filtration from the arrangement (Soni et al., 2016).

1.11.4 Neutralization

The procedure may be used to treat corrosive wastewaters containing metals. The strategy involves expanding the corrosive waste's pH by including a soluble reagent, framing an accelerate, and gathering the hasten. The approaching arrangement is pH changed following the ideal variety for expediting metals as hydroxides (Soni et al., 2016).

1.11.5 Blending

The waste-inferred fuel has been consumed in connected concrete furnaces. This fuel mixing and warm decimation method securely frees the earth of risky waste. It moderates non-inexhaustible, regular assets and non-renewable energy sources by recouping the buildup (Paul et al., 2019).

1.11.6 Bioremediation

This involves the presentation of microorganisms or microscopic organisms into the normally devoured oils and different contaminants in wastewater, separating them into innocuous CO_2 and H_2O (Soni et al., 2016).

1.11.7 Filtration

Filtration strategies have been utilized for specific applications in the field of mechanical wastewater treatment. The three primary kinds of film-based filtration advances incorporate opposite assimilation, nano-filtration, and ultra-filtration. All these three layers have been applied onto a sponsorship material by covering a thin coating of a porous polymer or plastic. The yield is the most delightful product of the filtration process up to this stage. The inverse assimilation is the smallest; nano-filtration being a slightly more significant advance; ultra-filtration being a little bigger, again (Soni et al., 2016).

1.12 WASTE MANAGEMENT LAWS AND LEGISLATION

There are a few laws as rules and guidelines concerning the waste of the executives. Waste laws have usually tried to reduce or remove the unregulated distribution of waste materials into the surroundings in a way that can source natural or organic damage and implement regulations aimed at decreasing waste age and promoting or directing the reuse of waste. Different types of regulations are there in other countries.

The laws concerning Waste Management in India are:

- "The Water (Prevention and Control of Pollution) Act (1974)": Was a push to lessen and cease contamination in waterways (Rajaram & Das, 2008).
- "The Air (Prevention and Control of Pollution) Act (1981)": This has taken sufficient measures to preserve the characteristic properties of the earth, which include, in addition to other issues, protecting the quality of air and preventing air pollution (Kumar et al., 2009).
- "The Environment (Protection) Act (EPA), (1986)": It is the original Indian enactment to manage ecological assurance and its parts comprehensively (Rajaram & Das, 2008).
- "The Public Liability Insurance Act, (1991)": It went ahead of the impact points of the Bhopal Gas catastrophe. Its fundamental point was to give help to survivors of modern catastrophe casualties.
- "City Solid Wastes (Management and Handling) Rules in 2000": Making it obligatory for familiar specialists to set up waste preparing and removal offices, recognize sterile landfill locales, and improve existing dumpsites (Kumar et al., 2009).
- "Perilous Wastes (Management and Handling) Rules in 1989" (altered in 2000 and 2003) and "Hazardous Wastes (Management, Handling and Trans-limit Movement) Rules in 2008" (Sonak et al., 2008).
- "The Biomedical Waste (Management and Handling) Rules in 1998 (corrected in 2000 and 2003)": The Rules has dependent on the rule of isolation of general waste from BMW (Biomedical Waste) (Sharma, 1998).
- "The E-Waste (Management and Handling) Rules in May 2011": Got operational from May 2012 (Wath et al., 2010).
- The Ministry of Environment and Forests announced the draft "Plastics (Manufacture, Usage and Waste Management) Rules, 2009" to supplant the "Recycled Plastics Manufacture and Usage Rules, 1999" (changed in 2003) to direct the assembling and utilization of plastic convey packs. These Rules have finished as "Plastic Waste (Management and Handling) Rules 2011" and told on February 4, 2011 (Abazeri, 2014).
- "The Water (Prevention and Control of Pollution) Cess Act, 1977 Amended in 1991": The Water Cess Act accommodates the toll of a cess on water devoured by determined businesses (Shama, 1976).
- "The National Environment Tribunal Act, (1995)": An Act that accommodates the exacting danger of harm resulting from any mishap while, at the same time, coping with any hazardous material and creating a National

Environment Tribunal for the effective and rapid removal of cases arising from such a fiasco to mitigate and compensate for harm to individuals, property, and nature and matters associated notwithstanding that or accidental to that (Agarwal, 2005).

- "Batteries (Management and Handling) Rules 2001": Duties of producer, merchant, constructing agent and re-conditioner are referenced; for example, utilized batteries to be gathered back and sent uniquely to enrolled recyclers (Kumar & Singh, 2013).
- "The National Green Tribunal Act 2010" provides for a viable and rapid removal of cases relating to ecological insurance and the protection of forests and other standard assets, including the requirement of any lawful right identifying with nature and relieving and paying for damage to people and property (Gill, 2014).

1.12.1 Recent Environmental Legislation

- "Metropolitan Solid Waste (Management and Handling) Rules, 2016": Ministry of condition and woods (MoEF) has thought of draft rules for the executives and treatment of stable municipal waste (Joshi & Ahmed, 2016).
- "Bio-Medical Wastes (Management and Handling) Rules, 2016": Ministry of Environment and Forests has changed the Bio-Medical Waste (Management and Handling) Rules declared under the "Environment Protection Act of 1986".

(A)

i.	"Control of Pollution Act 1974 (COPA): the first UK statute regarding the disposal of waste to the land".
ii.	"Environmental Protection Act, 1990 (EPA)".
iii.	"Landfill (England and Wales) Regulations 2002".
iv.	"Waste and Emissions Trading Act 2003".

(B)

i.	"Waste Disposal Act, 1972".
ii	"Waste Management Act, 1986".
iii	"Closed Substance Cycle Waste Management Act 1994".

(C)

i.	"Nuisance Act, 1875".
ii.	"Waste Substances Act, 1977 (WSA)".
iii.	"Environmental Management Act 1993 (EMA)".
iv.	"Packaging and Packaging Waste Regulation 1997".

FIGURE 1.2 (A)—United Kingdom Waste Management Law; (B)—Germany Waste Management Law; (C)—Netherlands Waste Management Law; (D)—United States of America Waste Management; (E)—Australia Waste Management Law.

(D)

i.	"Solid Waste Disposal Act, 1965".
ii	"Resource Recovery Act, 1970".

(E)

i.	"Environment Protection Act 1970".
ii.	"Beverage Container Act, 1975".
iii.	"An amendment to the Act in 1986".

FIGURE 1.2 (Continued)

1.12.2 International Law on Waste management

International regulation incorporates understandings recognized with worldwide vehicle and removal of unsafe waste:

- "Convention on Civil Liability for Damage Caused during Carriage of Dangerous Goods by Road, Rail, and Inland Navigation Vessels (CRTD), Geneva, 1989" (Stokke & Thommessen, 2013).
- "Convention on the Control of Trans-boundary Movements of Hazardous Wastes and their Disposal, Basel, 1989" (Ongondo et al., 2011).
- "Conference on the Ban of the Import into Africa and the Control of Trans-boundary Movements and Management of Hazardous Wastes within Africa, Bamako, 1991" (Ongondo et al., 2011).
- "Convention on the Prior Informed Consent Procedure for Certain Hazardous Chemicals and Pesticides in International Trade, Rotterdam, 1998" (Kummer, 1999).
- "Meeting on the Trans-boundary Effects of Industrial Accidents, Helsinki, 1992" (Bosnjakovic, 1998).
- "European Agreement Concerning the International Carriage of Dangerous Goods by Inland Waterways (AND), Geneva, 2000" (UNECITC, 2014).
- "European Agreement Concerning the International Carriage of Dangerous Goods by Road (ADR), Geneva, 1957" (Unies, 1998).
- "FAO International Code of Conduct on the Distribution and Use of Pesticides, Rome, 1985" (Bergesen et al., 2018).
- "Minamata Convention on Mercury, Minamata 2013" (Mackey et al., 2014).
- "Stockholm Convention Stockholm Convention was on Persistent Organic Pollutants Stockholm, 2001" (Fiedler, 2007).
- "Convention to Ban the Importation into Forum Island Countries of Hazardous and Radioactive Wastes and to Control the Trans-boundary Movement and Management of Hazardous Wastes within the South Pacific Region, Waigani, 1995" (Bergesen et al., 2013).

1.12.3 Waste Management Laws of Different Countries

1.13 CONCLUSION AND FUTURE PROSPECTS

Environmental wastes result from varying anthropogenic activity. It increases day by day due to urbanization, the Industrial Revolution, and the enormous use of resources in a brainless way. Many waste management strategies have been developing to manage waste and their transformation from worthless, unusable substances to beneficial and usable elements. Despite that, having a particularly harmful impact leaves destructive influences on the environment and the ecosystem. People should spotlight this alarming problem, and we have to process and culture our practice more and more on waste management to evolve it. We have the critical power to change environmental health and retrieved it from a catabolic state to an anabolic state. Many countries are already showing their effort to manage waste sustainably. We, as a whole, need an across-the-board use of "modern environment" – planning the mechanical and urban framework to monitor materials and reuse of materials that associate steel, concrete, concoction, and paper firms into an automated biological system.

REFERENCES

Abazeri, M. (2014). *Rethinking waste in India: Innovative initiatives in waste management.* Sciences Po, PSLA.

Abdel-Shafy, H. I., & Mansour, M. S. (2018). Solid waste issue: Sources, composition, disposal, recycling, and valorization. *Egypt Journal of Petroleum*, *27*(4), 1275–1290.

Agarwal, V. K. (2005). Environmental laws in India: Challenges for enforcement. *Bulletin of the National Institute of Ecology*, *15*, 227–238.

Ahmad, T., Aadil, R. M., Ahmed, H., Ur Rahman, U., Soares, B. C., Souza, S. L., Pimentel, T. C., Scudino, H., Guimarães, J. T., Esmerino, E. A., & Freitas, M. Q. (2019). Treatment and utilization of dairy industrial waste: A review. *Trends in Food Science & Technology*, *88*, 361–372.

Ahmed, T., Shahid, M., Azeem, F., Rasul, I., Shah, A. A., Noman, M., Hameed, A., Manzoor, N., Manzoor, I., & Muhammad, S. (2018). Biodegradation of plastics: Current scenario and future prospects for environmental safety. *Environmental Science and Pollution Research*, *25*(8), 7287–7298.

Alekseenko, A. V., Drebenstedt, C., & Bech, J. (2022). Assessment and abatement of the eco-risk caused by mine spoils in the dry subtropical climate. *Environmental Geochemistry and Health*, *44*(5), 1581–1603.

Alves, C. A., Vicente, E. D., Evtyugina, M., Vicente, A., Pio, C., Amado, M. F., & Mahía, P. L. (2019). Gaseous and speciated particulate emissions from the open burning of wastes from tree pruning. *Atmospheric Research*, *226*, 110–121.

Ayiania, M., Terrell, E., Dunsmoor, A., Carbajal-Gamarra, F. M., & Garcia-Perez, M. (2019). Characterization of solid and vapor products from thermochemical conversion of municipal solid waste woody fractions. *Waste Management*, *84*, 277–285.

Bai, S. H., & Ogbourne, S. M. (2016). Glyphosate: Environmental contamination, toxicity and potential risks to human health via food contamination. *Environmental Science and Pollution Research*, *23*(19), 18988–19001.

Banerjee, A., Choudhury, M., Chakravorty, A., Raghavan, V., Biswas, B., Sana, S. S., . . . Ramakrishna, S. (2021). Microbiological degradation of organic pollutants from industrial wastewater. In *Nanobiotechnology for green environment* (pp. 83–114). CRC Press.

Beneroso Vallejo, D., Albero-Ortiz, A., Monzó-Cabrera, J., Díaz-Morcillo, A., Arenillas de la Puente, A., & Menéndez Díaz, J. Á. (2016). *Dielectric characterization of biodegradable wastes during pyrolysis.*

Bergesen, H. O., Parmann, G., & Thommessen, O. B. (2013). World Health Organization (WHO). In *Yearbook of international cooperation on environment and development 1999–2000* (pp. 257–258). Routledge.

Bergesen, H. O., Parmann, G., & Thommessen, Ø. B. (2018). European Agreement concerning the international carriage of dangerous goods by road (ADR). In *Yearbook of international cooperation on environment and development 1998–99* (pp. 94–95). Routledge.

Bing, X., Bloemhof, J. M., Ramos, T. R. P., Barbosa-Povoa, A. P., Wong, C. Y., & van der Vorst, J. G. (2016). Research challenges in municipal solid waste logistics management. *Waste Management, 48,* 584–592.

Bosnjakovic, B. (1998). UN/ECE strategies for protecting the environment with respect to international watercourses: The Helsinki and Espoo conventions. *World Bank Technical Paper*, 47–64.

Carrère, H., Dumas, C., Battimelli, A., Batstone, D. J., Delgenes, J. P., Steyer, J. P., & Ferrer, I. (2010). Pretreatment methods to improve sludge anaerobic degradability: A review. *Journal of Hazardous Materials, 183*(1–3), 1–15.

Chakravorty, A., Raghavan, V., Grace, A. N., Kim, S. Y., & Van Le, Q. (2022). Plants and microorganisms for phytoremediation of air. In *Current developments in biotechnology and bioengineering* (pp. 109–131). Elsevier.

Chen, J., Hua, C., & Liu, C. (2019). Considerations for better construction and demolition waste management: Identifying the decision behaviors of contractors and government departments through a game theory decision-making model. *Journal of Cleaner Production, 212,* 190–199.

Choudhury, M., & Choudhury, M. (2014). Trends of urban solid waste management in Agartala City, Tripura, India. *Universal Journal of Environmental Research & Technology, 4*(4).

Choudhury, M., & Dutta, J. (2017). A comparative study of municipal solid waste management status for three major towns of Upper Assam-India. *International Journal of Waste Resources, 7*(291), 2.

Cong, X. (2018). Air pollution from industrial waste gas emissions is associated with cancer incidences in Shanghai, China. *Environmental Science and Pollution Research, 25*(13), 13067–13078.

Coskuner, G., Jassim, M. S., Nazeer, N., & Damindra, G. H. (2020). Quantification of landfill gas generation and renewable energy potential in arid countries: Case study of Bahrain. *Waste Management & Research, 38*(10), 1110–1118.

Dove, M. R. (2021). *Bitter shade: The ecological challenge of human consciousness.* Yale Agrarian Studies.

Ermolaev, E., Sundberg, C., Pell, M., Smårs, S., & Jönsson, H. (2019). Effects of moisture on emissions of methane, nitrous oxide and carbon dioxide from food and garden waste composting. *Journal of Cleaner Production, 240,* 118165.

Evans, A. M. (2016). The speed of invasion: Rates of spread for thirteen exotic forest insects and diseases. *Forests, 7*(5), 99.

Fareed, A., Zaidi, S. B. A., Ahmad, N., Hafeez, I., Ali, A., & Ahmad, M. F. (2020). Use of agricultural waste ashes in asphalt binder and mixture: A sustainable solution to waste management. *Construction and Building Materials, 259,* 120575.

Fernandes, S. R., Silva, H. M., & Oliveira, J. R. (2019). Carbon dioxide emissions and heavy metal contamination analysis of stone mastic asphalt mixtures produced with high rates of different waste materials. *Journal of Cleaner Production, 226,* 463–470.

Fernández-Aracil, P., Ortuño-Padilla, A., & Melgarejo-Moreno, J. (2018). Factors related to municipal costs of waste collection service in Spain. *Journal of Cleaner Production, 175,* 553–560.

Fernández-Fernández, M., Naves, P., Witzell, J., Musolin, D. L., Selikhovkin, A. V., Paraschiv, M., Chira, D., Martínez-Álvarez, P., Martín-García, J., Muñoz-Adalia, E. J., & Altunisik, A. (2019). Pine pitch canker and insects: Relationships and implications for disease spread in Europe. *Forests*, *10*(8), 627.

Ferronato, N., Rada, E. C., Portillo, M. A. G., Cioca, L. I., Ragazzi, M., & Torretta, V. (2019). Introduction of the circular economy within developing regions: A comparative analysis of advantages and opportunities for waste valorization. *Journal of Environmental Management, 230,* 366–378.

Fiedler, H. (2007). National PCDD/PCDF release inventories under the Stockholm convention on persistent organic pollutants. *Chemosphere*, *67*(9), S96–S108.

Gill, G. N. (2014). The national green tribunal of India: A sustainable future through the principles of international environmental law. *Environmental Law Review*, *16*(3), 183–202.

Hale, L., Curtis, D., Leon, N., McGiffen Jr, M., & Wang, D. (2021). Organic amendments, deficit irrigation, and microbial communities impact extracellular polysaccharide content in agricultural soils. *Soil Biology and Biochemistry, 162*, 108428.

He, K., Zhang, J., & Zeng, Y. (2019). Knowledge domain and emerging trends of agricultural waste management in the field of social science: A scientometric review. *Science of the Total Environment, 670,* 236–244.

Hinojosa, M. B., Albert-Belda, E., Gómez-Muñoz, B., & Moreno, J. M. (2021). High fire frequency reduces soil fertility underneath woody plant canopies of Mediterranean ecosystems. *Science of The Total Environment, 752*, 141877.

Hoornweg, D., & Bhada-Tata, P. (2012). *What a waste: A global review of solid waste management.* Urban Development Series; Knowledge Papers No. 15, World Bank. © World Bank. https://openknowledge.worldbank.org/handle/10986/17388License: CC BY 3.0 IGO.

Hoornweg, D., Bhada-Tata, P., & Kennedy, C. (2013). Environment: Waste production must peak this century. *Nature News*, *502*(7473), 615.

Hossain, M. U., Wang, L., Chen, L., Tsang, D. C., Ng, S. T., Poon, C. S., & Mechtcherine, V. (2020). Evaluating the environmental impacts of stabilization and solidification technologies for managing hazardous wastes through life cycle assessment: A case study of Hong Kong. *Environment International, 145*, 106139.

Joshi, R., & Ahmed, S. (2016). Status and challenges of municipal solid waste management in India: A review. *Cogent Environmental Science*, *2*(1), 1139434.

Karthikeyan, S., Arun, P., & Thiyaneswaran, M. P. (2020, May). Summary of non-biodegradable wastes in concrete. In *AIP conference proceedings* (Vol. 2235, No. 1, p. 020020). AIP Publishing LLC.

Kaza, S., Yao, L., Bhada-Tata, P., & Van Woerden, F. (2018). *What a waste 2.0: A global snapshot of solid waste management to 2050.* World Bank Publications.

Khandelwal, H., Dhar, H., Thalla, A. K., & Kumar, S. (2019). Application of life cycle assessment in municipal solid waste management: A worldwide critical review. *Journal of Cleaner Production, 209,* 630–654.

Knust, B., Brown, S., de St. Maurice, A., Whitmer, S., Koske, S. E., Ervin, E., Patel, K., Graziano, J., Morales-Betoulle, M. E., House, J., & Cannon, D. (2020). Seoul virus Infection and spread in United States home-based ratteries: Rat and human testing results from a multistate outbreak investigation. *The Journal of Infectious Diseases*, *222*(8), 1311–1319.

Kumar, A., & Samadder, S. R. (2017). A review on technological options of waste to energy for effective management of municipal solid waste. *Waste Management, 69,* 407–422.

Kumar, S., Bhattacharyya, J. K., Vaidya, A. N., Chakrabarti, T., Devotta, S., & Akolkar, A. B. (2009). Assessment of the status of municipal solid waste management in metro cities, state capitals, class I cities, and class II towns in India: An insight. *Waste Management, 29*(2), 883–895.

Kumar, S., Negi, S., Mandpe, A., Singh, R. V., & Hussain, A. (2018). Rapid composting techniques in Indian context and utilization of black soldier fly for enhanced decomposition of biodegradable wastes-A comprehensive review. *Journal of Environmental Management, 227,* 189–199.

Kumar, U., Singh, D. N. (2013). Electronic waste management through EPR initiatives. *International Journal of Recent Trends in Electrical & Electronics Engineering, 2*(2), 34–43.

Kummer, K. (1999). Prior informed consent for chemicals in international trade: The 1998 Rotterdam convention. *Review of European Community and International Environmental Law, 8*, 323.

Lavanya, B. L. (2018). Technology adoption in futuristic solid waste management to empower village communities: A case study in southern district of Tamil Nadu. *International Journal of Research and Analytical Reviews, 6*(1), 164–170.

Lecomte, T., Mamindy-Pajany, Y., Lors, C., Lemay, M., Abriak, N. E., Bazin, C., & Vernus, E. (2019). A methodological approach for ecotoxicological characterization of non-hazardous sediments for their beneficial reuse. *Journal of Soils and Sediments*, 1–11.

Lee, S., Kim, Y., Park, J., Shon, H. K., & Hong, S. (2018). Treatment of medical radioactive liquid waste using Forward Osmosis (FO) membrane process. *Journal of Membrane Science, 556,* 238–247.

Leonelli, C., Kamseu, E., Lancellotti, I., & Barbieri, L. (2017). Geopolymerization as cold-consolidation techniques for hazardous and non-hazardous wastes. In *Key engineering materials* (Vol. 751, pp. 527–531). Trans Tech Publications Ltd.

Lijó, L., González-García, S., Bacenetti, J., & Moreira, M. T. (2017). The environmental effect of substituting energy crops for food waste as feedstock for biogas production. *Energy, 137,* 1130–1143.

Liu, B., Li, Y., Wang, Q., & Bai, S. (2019). Green fabrication of leather solid waste/thermoplastic polyurethanes composite: Physically de-bundling effect of solid-state shear milling on collagen bundles. *Composites Science and Technology, 181*, 107674.

Liu, Y., Sun, W., & Liu, J. (2017). Greenhouse gas emissions from different municipal solid waste management scenarios in China: Based on carbon and energy flow analysis. *Waste Management, 68,* 653–661.

Liu, Y., Yang, Y., Zhang, C., Huang, F., Wang, F., Yuan, J., . . . & Liu, L. (2020). Clinical and biochemical indexes from 2019-nCoV infected patients linked to viral loads and lung injury. *Science China Life Sciences, 63*, 364–374.

Lopes, P. C., Block, P., & König, B. (2016). Infection-induced behavioural changes reduce connectivity and the potential for disease spread in wild mice contact networks. *Scientific Reports, 6*(1), 1–10.

Mackey, T. K., Contreras, J. T., & Liang, B. A. (2014). The Minamata Convention on Mercury: Attempting to address the global controversy of dental amalgam use and mercury waste disposal. *Science of the Total Environment, 472,* 125–129.

Marsili, D., Fazzo, L., & Comba, P. (2009). Health risks from hazardous waste disposal: The need for international scientific cooperation [Rischi per la salute connessi allo smaltimento di rifiuti pericolosi: Necessità della cooperazione scientifica internazionale]. *European Journal of Oncology, 14*(3), 151–159.

Mavropoulos, A., & Newman, D. (2015). *Wasted health: The tragic case of dumpsites.* International Solid Waste Association.

Momayez, F., Karimi, K., & Horváth, I. S. (2018). Enhancing ethanol and methane production from rice straw by pretreatment with liquid waste from biogas plant. *Energy Conversion and Management, 178,* 290–298.

Murray, M. H., Fidino, M., Fyffe, R., Byers, K. A., Pettengill, J. B., Sondgeroth, K. S., Killion, H., Magle, S. B., Rios, M. J., Ortinau, N., & Santymire, R. M. (2020). City sanitation and socioeconomics predict rat zoonotic infection across diverse neighbour hoods. *Zoonoses and Public Health, 67*(6), 673–683.

Nadar, S. S., Rao, P., & Rathod, V. K. (2018). Enzyme assisted extraction of biomolecules as an approach to novel extraction technology: A review. *Food Research International, 108*, 309–330.

Nemerow, N. L., & Dasgupta, A. (1991). *Industrial and hazardous waste treatment*. N. p.

Noor, T., Javid, A., Hussain, A., Bukhari, S. M., Ali, W., Akmal, M., & Hussain, S. M. (2020). Types, sources and management of urban wastes. In *Urban ecology* (pp. 239–263). Elsevier.

Obi, F. O., Ugwuishiwu, B. O., & Nwakaire, J. N. (2016). Agricultural waste concept, generation, utilization and management. *Nigerian Journal of Technology, 35*(4), 957–964.

Ongondo, F. O., Williams, I. D., & Cherrett, T. J. (2011). How are WEEE doing? A global review of the management of electrical and electronic wastes. *Waste Management, 31*(4), 714–730.

Patwardhan, A. D. (2013). *Industrial solid wastes*. The Energy and Resources Institute (TERI).

Paul, S., Choudhury, M., Deb, U., Pegu, R., Das, S., & Bhattacharya, S. S. (2019). Assessing the ecological impacts of ageing on hazard potential of solid waste landfills: A green approach through vermitechnology. *Journal of Cleaner Production, 236*, 117643.

Pepper, D. A., Lada, H., Thomson, J. R., Bakar, K. S., Lake, P. S., & Mac Nally, R. (2017). A method to identify drivers of societal change likely to affect natural assets in the future, illustrated with Australia's native biodiversity. *Science of The Total Environment, 581*, 80–86.

Pham Van, D., Hoang, M. G., Pham Phu, S. T., & Fujiwara, T. (2018). Kinetics of carbon dioxide, methane and hydrolysis in co-digestion of food and vegetable wastes. *Global Journal of Environmental Science and Management, 4*(4), 401–412.

Polyakova, V., Degaev, E., & El Haddad, P. (2018). Reduction of ecological and economic risks in utilization of solid domestic wastes and construction waste. In *MATEC Web of Conferences* (Vol. 251, p. 06017). EDP Sciences.

Qi, M., Yang, Y., Zhang, X., Zhang, X., Wang, M., Zhang, W., Lu, X., & Tong, Y. (2020). Pollution reduction and operating cost analysis of municipal wastewater treatment in China and implication for future wastewater management. *Journal of Cleaner Production, 253*, 120003.

Qian, H., Zhang, M., Liu, G., Lu, T., Qu, Q., Du, B., & Pan, X. (2018). Effects of soil residual plastic film on soil microbial community structure and fertility. *Water, Air, & Soil Pollution, 229*(8), 1–11.

Rahman, M. M., Bodrud-Doza, M., Griffiths, M. D., & Mamun, M. A. (2020). Biomedical waste amid COVID-19: Perspectives from Bangladesh. *The Lancet. Global Health, 8*(10), e1262.

Rajaram T., & Das A. (2008). Water pollution by industrial effluents in India: Discharge scenarios and case for participatory ecosystem specific local regulation. *Futures*, 40(1), 56–69.

Rajendran, K., Lin, R., Wall, D. M., & Murphy, J. D. (2019). Influential aspects in waste management practices. In *Sustainable Resource recovery and zero waste approaches* (pp. 65–78). Elsevier.

Rajpal, A., Ali, M., Choudhury, M., Almohana, A. I., Alali, A. F., Munshi, F. M. A., Khursheed, A., & Kazmi, A. A. (2022). Abattoir wastewater treatment plants in India: Understanding and performance evaluation. *Frontiers in Environmental Science, 10*, 881623. https://doi.org/10.3389/fenvs.2022.881623

Sabourin-Provost, G., & Hallenbeck, P. C. (2009). High yield conversion of a crude glycerol fraction from biodiesel production to hydrogen by photo fermentation. *Bioresource Technology, 100*(14), 3513–3517.

Sana, S. S., Dogiparthi, L. K., Gangadhar, L., Chakravorty, A., & Abhishek, N. (2020). Effects of microplastics and nanoplastics on marine environment and human health. *Environmental Science and Pollution Research*, 1–14.

Sharma, A. K. (1998). *Bio medical waste (management and handling) rules*. Suvidha Law House.

Sharma, H. B., Vanapalli, K. R., Cheela, V. S., Ranjan, V. P., Jaglan, A. K., Dubey, B., Goel, S., & Bhattacharya, J. (2020). Challenges, opportunities, and innovations for effective solid waste management during and post COVID-19 pandemic. *Resources, Conservation and Recycling*, *162*, 105052.

Simatele, D. M., Dlamini, S., & Kubanza, N. S. (2017). From informality to formality: Perspectives on the challenges of integrating solid waste management into the urban development and planning policy in Johannesburg, South Africa. *Habitat International*, *63*, 122–130.

Singh, A. K., Gandhi, S., Siddiqui, N. A., Mondal, P., Nandan, A., & Haq, E. (2020). Solid waste problem due to packaging material and its management options. In *Advances in air pollution profiling and control* (pp. 133–141). Springer.

Skenderovic, I., Kalac, B., & Becirovic, S. (2015). Environmental pollution and waste management. *Balkan Journal of Health Science*, *3*(1), 2–10.

Sonak, S., Sonak, M., & Giriyan, A. (2008). Shipping hazardous waste: Implications for economically developing countries. *International Environmental Agreements: Politics, Law and Economics*, *8*(2), 143–159.

Soni, A., Patil, D., & Argade, K. (2016). Municipal solid waste management. *Procedia Environmental Sciences*, *35*, 119–126.

Stefanovsky, S. V., Yudintsev, S. V., Vinokurov, S. E., & Myasoedov, B. F. (2016). Chemical-technological and mineralogical-geochemical aspects of the radioactive waste management. *Geochemistry International*, *54*(13), 1136–1155.

Stokke, O. S., & Thommessen, O. B. (2013). *Yearbook of international cooperation on environment and development 2002–03*. Routledge.

Su, S., Li, S., Ju, J., Wang, Q., & Xu, Z. (2021). A building information modeling-based tool for estimating building demolition waste and evaluating its environmental impacts. *Waste Management*, *134*, 159–169.

Sydney, E. B., de Carvalho, J. C., Letti, L. A. J., Magalhaes Jr, A. I., Karp, S. G., Martinez-Burgos, W. J., de Souza Candeo, E., Rodrigues, C., de Souza Vandenberghe, L. P., Neto, C. J. D., & Torres, L. A. Z. (2021). Current developments and challenges of green technologies for the valorization of liquid, solid, and gaseous wastes from sugarcane ethanol production. *Journal of Hazardous Materials*, *404*, 124059.

Thines, K. R., Abdullah, E. C., Mubarak, N. M., & Ruthiraan, M. (2017). Synthesis of magnetic biochar from agricultural waste biomass to enhancing route for waste water and polymer application: A review. *Renewable and Sustainable Energy Reviews*, *67*, 257–276.

Unies, N. (1998). *European agreement concerning the international carriage of dangerous goods by road (ADR) and protocol of signature: done in Geneva on 30 September 1957*. UN.

United Nations. Economic Commission for Europe. Inland Transport Committee. (2014). *European agreement concerning the international carriage of dangerous goods by inland waterways (ADN): Including the annexed regulations, applicable as from 1 January 2015* (Vol. 2). United Nations Publications.

United States. Environmental Protection Agency. Office of Policy. (1999). *Inventory of US greenhouse gas emissions and sinks: 1990–1997*. The Agency.

Velvizhi, G., Shanthakumar, S., Das, B., Pugazhendhi, A., Priya, T. S., Ashok, B., Nanthagopal, K., Vignesh, R., & Karthick, C. (2020). Biodegradable and non-biodegradable fraction of municipal solid waste for multifaceted applications through a closed loop integrated refinery platform: Paving a path towards circular economy. *Science of the Total Environment*, *731*, 138049.

Wang, Q., Sang, H., Chen, L., Wu, Y., & Wei, Y. (2020). Selective separation of Pd (II) through ion exchange and oxidation-reduction with hexacyanoferrates from high-level liquid waste. *Separation and Purification Technology*, *231*, 115932.

Wath, S. B., Vaidya, A. N., Dutt, P. S., & Chakrabarti, T. (2010). A roadmap for development of sustainable E-waste management system in India. *Science of the Total Environment*, *409*(1), 19–32.

Wei, L., Li, X., Yang, W., Dai, Y., & Wang, C. H. (2020). Optimization of operation strategies of a syngas-fueled engine in a distributed gasifier-generator system driven by horticulture waste. *Energy Conversion and Management*, *208*, 112580.

Willis, K., Maureaud, C., Wilcox, C., & Hardesty, B. D. (2018). How successful are waste abatement campaigns and government policies at reducing plastic waste into the marine environment? *Marine Policy*, *96*, 243–249.

Xu, J., Krietemeyer, E. F., Boddu, V. M., Liu, S. X., & Liu, W. C. (2018). Production and characterization of cellulose nanofibril (CNF) from agricultural waste corn stover. *Carbohydrate Polymers*, *192*, 202–207.

Yang, F., Li, Y., Han, Y., Qian, W., Li, G., & Luo, W. (2019). Performance of mature compost to control gaseous emissions in kitchen waste composting. *Science of the Total Environment*, *657*, 262–269.

Yang, Z., Tang, S., Zhang, Z., Liu, C., & Ge, X. (2018). Characterization of PM10 surrounding a cement plant with integrated facilities for co-processing of hazardous wastes. *Journal of Cleaner Production*, *186*, 831–839.

Yong, Y. S., Lim, Y. A., & Ilankoon, I. M. S. K. (2019). An analysis of electronic waste management strategies and recycling operations in Malaysia: Challenges and future prospects. *Journal of Cleaner Production*, *224*, 151–166.

Yuan, P., Wang, J., Pan, Y., Shen, B., & Wu, C. (2019). Review of biochar for the management of contaminated soil: Preparation, application and prospect. *Science of the total environment*, *659*, 473–490.

Zhang, M., Wang, Y., Song, Y., Zhang, T., & Wang, J. (2016). Manifest system for management of non-hazardous industrial solid wastes: Results from a Tianjin industrial park. *Journal of Cleaner Production*, *133*, 252–261.

Zhefeng, L., Yueyan, L., & Gaofeng, Z. (2020, October). Analysis of greening ecology in landscape reconstruction of construction waste dump in wind-sand area. In *IOP conference series: Earth and environmental science* (vol. 585, No. 1, 012057). IOP Publishing.

2 Effect of Improper Waste Disposal on Environmental, Physical, and Emotional Well-Being of Human Beings

Jaya Bharti

2.1 INTRODUCTION

According to the World Health Organization (WHO), improper waste disposal can have disastrous consequences on human health. Inadequately disposed waste can trigger mortality, cancer and even reproductive health issues. These negative impacts may be due to the mismanagement of waste activities in our cities and suburban areas. Safe disposal of waste is important. If waste is not properly decomposed and recycled, the waste accumulates slowly and becomes a place that appears dirty and bad, where there are germs, flies, odors, etc.; it affects drinking water quality; it can affect the health of feed-seeking cattle and other domestic animals. The Environmental Protection Agency (EPA) defines waste products as "hazardous" if they are flammable, explosive, corrosive or toxic. The purpose of the present research in this chapter was to assess or discuss the effect of improper waste disposal on physical, environmental and emotional well-being among the general population across a locality (near and far *nagarpalika*). Data was collected using an interview schedule. The sample consisted of 100 persons, 50 of whom lived near *nagarpalika* trash-houses and 50 of whom lived in neat and clean areas. Their age range varied from 20 years to 50 years. Ex-post facto research with exploratory orientation was used. Incidental sampling was used to collect the data. The interviews were recorded after taking participants' prior consent. The language of the schedule was Hindi and English, both. There were 16 questions to understand the effect of improper waste disposal on physical, environmental and emotional well-being among the general population across a locality (near and far *nagarpalika*). The responses were recorded verbatim. With reference to results, each question is discussed in detail about the various responses given by the interviewer to explore three dimensions of well-being—i.e., physical, emotional and environmental.

DOI: 10.1201/9781003258377-2

2.2 WELL-BEING

Well-being is the state of an individual or group that can be called "good" in some sense. It includes many dimensions, such as the physical, mental, social, economic, psychological, spiritual or medical state.

2.2.1 Perceived Well-Being—Environmental

Environmental wellness is a process that involves learning about and contributing to the health of the planet. This includes establishing a sustainable lifestyle, protecting natural resources and eliminating pollutants and excessive waste.

2.2.2 Perceived Well-Being—Physical

Physical well-being is the ability to maintain a healthy quality of life that allows us to make the most of our daily activities without any fatigue or physical stress. According to Bharti (2020), "Physical well-being is an important component for the people living in and around the municipality. Physical well-being can be influenced by the health status and health-related behaviors of the people living in and around the municipality."

2.2.3 Perceived Well-Being—Emotional

Emotional wellness is the ability to generate positive emotions, moods, thoughts and feelings and to adapt when faced with adversity and stressful situations. Emotional well-being describes one's psychological and social well-being. According to Bharti (2020), "Emotional well-being refers to a state of mental well-being in which a person realizes one's own abilities, can cope with life's normal stresses, work productively and fruitfully, and enjoy the future in the making of family and society."

2.2.4 Narratives with Data Presentation

A narrative is a way of presenting connected events to tell a real story of persons life, and it's a way for understanding others' viewpoint.

Figure 2.1—Theme 1

The first question tried to explore the conceptualization of the word "Waste Disposal." For the majority of people belonging to Far *Nagarpalika* Trash-Houses (80%), they responded that the word included the Garbage. On the other hand, an 80% majority used words like "dirty places, smelly areas, hell," etc.; people belonging to Near *Nagarpalika* Trash-Houses consider it in terms of "illness, disorder, health problems and death."

Narrative of Far *Nagarpalika* Trash-Houses: "*As soon as I hear the name of the waste materials, it comes to my mind that at which places this garbage is thrown, where it is brought again for use, all these have a very harmful effect on our health.*"

Narrative of Near *Nagarpalika* Trash-Houses: "*The place of disposal of waste material should be established in such a way that they do not affect the* *general public, but who thinks that today many people like us or like me live around these places, due to which these things are affected on their health."*

Figure 2.1—Theme 2

In the second question, frequency of passing or visiting these waste-disposable areas was explored, and all the person belonging to Far *Nagarpalika* Trash-Houses reported that they visit "occasionally/sometimes." On the other side, people belonging to Near *Nagarpalika* Trash-Houses accepted that they lived in bad conditions; they have no choice to avoid these places, so they "daily/everyday" visit waste disposable areas.

Narrative of Far *Nagarpalika* Trash-Houses: *"My home and my place of work are far away from the places where the waste is disposed of. It is very rare that I have to leave there. Sometimes when I have to go to meet someone and if my friends live in such a place, then I cannot refuse even if I want to and I have to go there."*

Narrative of Near *Nagarpalika* Trash-Houses: *"Do not leave the places where waste materials are disposed of, we live next to them. Requested many times this place should be removed from the vicinity of our house but no hearing was held. Every day we have to face these places many times a day. Go on the terrace or come out of the house, this place is visible from the house itself, they smell very bad till the house."*

Categories	Far *Nagarpalika* Trash-Houses (50)	Near *Nagarpalika* Trash-Houses (50)
Garbage/dirty places/smelly areas/Hell	40 (80%)	5 (10%)
Compost	10 (20%)	1 (2%)
Waste disposal unit	–	2 (4%)
Gases	–	2 (4%)
Illness/disorders/health problems/death	–	40 (80%)

Theme 1: Question—What comes in your mind after listening to the term "waste disposal"?

Categories	Far *Nagarpalika* Trash-Houses (50)	Near *Nagarpalika* Trash-Houses (50)
Sometimes/occasionally	50 (100%)	–
Every day/daily	–	50 (100%)

Theme 2: Question—"How often do you pass or visit waste disposal areas?"

FIGURE 2.1 Theme 1—Semantics after listening to the term "waste disposal"; Theme 2—Visits near to waste disposal areas; Theme 3—Biggest worry with regards to improper waste disposal; Theme 4—Features of waste disposal technology.

Categories	Far *Nagarpalika* Trash-Houses (50)	Near *Nagarpalika* Trash-Houses (50)
Concern for future generations	50 (100%)	–
Concern for present status		50 (100%)

Theme 3: Question—"What are your biggest worries with regards to improper waste disposal?"

Categories	Far *Nagarpalika* Trash-Houses (50)	Near *Nagarpalika* Trash-Houses (50)
Turning waste into energy	40 (80%)	10 (20%)
Advances in routes/make closed and hidden areas	10 (20%)	40 (80%)

Theme 4: Question—"What features of waste disposal technology are most important to you?"

FIGURE 2.1 (Continued)

Figure 2.1—Theme 3

With reference to the biggest worry with regards to improper waste disposal, the people belonging to Far *Nagarpalika* Trash-Houses respond in the form of "concern for future generations." This, again, emphasizes the optimistic and healthy thinking of Far *Nagarpalika* Trash-House people. All the Near *Nagarpalika* Trash-House peoples responded, in this regard, that their "concern for present status" is that they feel ashamed, unsafe and unhealthy in these areas. In 2009, Porta et al. concluded that evidence of an association between living close to a landfill and adverse health effects is inconclusive. Ancona et al. 2014 stated that most of the published studies have methodological problems, including poor exposure assessment based only on distance from the source, use of health data at the aggregate level and limited possibility of adjusting for socioeconomic status.

Narrative of Far *Nagarpalika* Trash-Houses: *"Due to improper disposal of waste materials, my biggest concern is that it can be very harmful for the coming generation. This can have a very bad effect on their health I mean their physical and mental health. In many researches, we have seen that children become mentally ill because pollution is not clean, the place around us is not clean and if we do not take any steps for it today, then tomorrow it will affect our future generation. It can have very bad effects."*

Narrative of Near *Nagarpalika* Trash-Houses: *"My biggest concern is that today our children and our families are facing the brunt of the improper disposal of waste materials. 40 to 50% of the salary they get is being spent on their diseases and these are all the same diseases which they are getting due to living in such a place ie such a polluted place where waste materials* *are disposed then if today their If it is not right then tomorrow what should we talk about."*

Figure 2.1—Theme 4

Another question focuses on the features that people focus on the waste disposable technologies that they see. While 80% of the Far *Nagarpalika* Trash-House people focus on the "turning waste into energy" features of the waste disposal technology that they possess, 80% of the Near *Nagarpalika* Trash-House people focus on the waste disposal needs of "advances in routes/make these areas closed and hidden."

Narrative of Far *Nagarpalika* Trash-Houses: "*Whatever technology is adopted for proper disposal of waste materials, what I like the most is that they should be converted from waste into new energy, so that its very good use can be used in many other works i.e. we use garbage. Electricity can also be generated from waste. Garbage can also make new things from waste like just we heard that in China we have made bricks using garbage waste, in China we have built a new building using garbage waste, so we can do this too.*"

Narrative of Near *Nagarpalika* Trash-Houses: "*Whatever technology is adopted for the disposal of waste materials, I would like to say that there should be one in it that such places should be kept completely away from the houses or they should be covered completely so that they have a bad effect on our families and our families. The way soundproof rooms are made to prevent the sound from coming out on the children, similarly, in order not to let the stench of these garbage dumps come out, these garbage houses should be made in such a way that the stench can come out.*"

Figure 2.2—Theme 5

When the participants were asked about the type of waste disposal materials that they are mostly interested in, 50% of Far *Nagarpalika* Trash-Houses people responded 'platinum,' while the remaining 50% responded 'iridium' about waste disposal materials. In contrast, 60% of Near *Nagarpalika* Trash-Houses people want to remove 'smell and gases' and convert to fragrance.

Narrative of Far *Nagarpalika* Trash-Houses: "*While manufacturing waste materials, we get platinum and iridium from waste materials. It is composed of 80% platinum and 20% iridium. Platinum and iridium are used to make jewelry and make surgical princes. That's why I'm most interested in the platinum and iridium materials that come from the disposal of waste materials.*"

Narrative of Near *Nagarpalika* Trash-Houses: "*While disposing of waste materials, the gas coming out of them and the foul smell from them is* *converted into fragrance, then I will be most interested in that because of that all the problems will be removed and there is no harmful effect on the health of our children and our family. There will be no impact and living in such places will not be a reason for us to be ashamed."*

Figure 2.2—Theme 6

When participants were questioned about the reason/causes of improper waste disposal, all of the Near *Nagarpalika* Trash-House respondents responded that "ignorance and laziness" is the main cause of improper waste disposal. Far *Nagarpalika* Trash-House respondents responded that "greed" is the main cause of improper waste disposal.

Narrative of Far *Nagarpalika* Trash-Houses: *"The main reason for not disposing of waste properly is ignorance of people about proper disposal of waste, people are unaware of the consequences of their mindless acts and laziness: also improper waste disposal can be the reason because people do not dispose of proper waste. Don't follow the right rules of the game, they always throw it where they want and they don't care about its effect."*

Narrative of Near *Nagarpalika* Trash-Houses: *"Greed is the main reason for not disposing of waste properly. This can lead to improper waste disposal for example burning the wheel and plastic tires instead of keeping it or trading in extra automobile car tires to maximize on it."*

Categories	**Far *Nagarpalika* Trash-Houses (50)**	**Near *Nagarpalika* Trash-Houses (50)**
Platinum	25 (50%)	5 (10%)
Palladium	–	10 (20%)
Iridium	25 (50%)	5 (10%)
Smell and gases convert in fragrance	–	30 (60%)

Theme 5: Question—"Which type of waste disposal materials are you most interested in?"

Categories	**Far *Nagarpalika* Trash-Houses (50)**	**Near *Nagarpalika* Trash-Houses (50)**
Ignorance/laziness	–	50 (100%)
Greed	50 (100%)	–

Theme 6: Question—"What are the reasons/causes of improper waste disposal?"

FIGURE 2.2 Theme 5—Important waste disposal materials; Theme 6—Reason/causes of improper waste disposal; Theme 7—Effects of improper waste disposal; Theme 8—Way to fix this problem of waste disposal.

Categories	**Far *Nagarpalika* Trash-Houses (50)**	**Near *Nagarpalika* Trash-Houses (50)**
Affects our climate	30 (60%)	–
Air pollution	20 (40%)	–
Affects our climate	–	30 (60%)
Infectious to humans	–	20 (40%)

Theme 7: Question—"What are the main effects of improper waste disposal?"

Categories	**Far *Nagarpalika* Trash-Houses (50)**	**Near *Nagarpalika* Trash-Houses (50)**
Recycling	50 (100%)	–
Decomposed	–	50 (100%)

Theme 8: Question—"How to fix this problem according to you?"

FIGURE 2.2 (Continued)

Figure 2.2—Theme 7

For exploring the effects of improper waste disposal, Far *Nagarpalika* Trash-House respondents responded (60%) in the category of "effects of climate," and 40% responded in "soil contamination." On the other side, Near *Nagarpalika* Trash-House people responded (60%) to prefer to 'effects on health,' and 40% responded in the "infectious for humans" category. People who live close to municipal solid waste (MSW) landfills could be exposed to air pollutants emitted by the plants (landfill gas containing methane, carbon dioxide, hydrogen sulfide and other contaminants including volatile organic compounds, particulate matter and bioaerosols) or to contaminated soil and water (Mondal et al., 2023). The possible health effects related to residences close to these sites have been assessed in several original papers and evaluated in systematic reviews (Gouveia & Prado, 2010).

Narrative of Far *Nagarpalika* Trash-Houses: "*Improper waste disposal has a major impact on our climate. As some waste decomposes, it releases greenhouse gases into the atmosphere. Like Earth's trapping here which can be an effect of weather abnormality like more storms or typhoons coming every year till now and also we have an effect of this in the form of soil contamination, presence of man-made chemicals, soil erosion can see. The soft, smelly soil is caused by vapors and other changes happening day by day in the natural soil environment that could be harmful in the future*".

Narrative of Near *Nagarpalika* Trash-Houses: "*Improper waste disposal has major impacts on our health—such as lung disease, heart problems, skin irritation, breathing problems or abnormalities as well as transmission to humans, skin irritation and blood infections due to waste which occurs by direct contact, and also involves transmission of bacteria to wounds infected with waste.*"

Figure 2.2—Theme 8

In exploring the mindsets of people asked the way to fix this problem, it was identified that all the Far *Nagarpalika* Trash-House respondents responded with "recycling" of waste disposal. Near *Nagarpalika* Trash-House respondents mentioned that "decomposition" is the only way.

Narrative of Near *Nagarpalika* Trash-Houses: "*Decomposing your trash is an option to get rid of your household waste. You can compost a lot of items: food waste, animal waste, yard waste, and more. Decomposition is a type of process in which elements are broken down into different microscopic parts. In living beings this process starts after their death. An important contribution in this is made by micro-organisms, which gradually divide the corpse of any organism into very small pieces and make it a part of the ecosystem*".

Narrative of Far *Nagarpalika* Trash-Houses: "*Recycling waste is another option for getting rid of waste. There are many different things that are recyclable: paper, tin, aluminum, plastic and much more. Scientists have developed a new process that will allow precious metals to be recovered from electronic waste within seconds. This technology is not only better from the environmental point of view but it also consumes 500 times less energy than the method currently being used for this work. The process, developed by Rice University, is based on the Flash Joule heating method discovered last year. The advantage of this method is that the by-products that are left after recovering precious metals from electronic waste are completely safe, which can be dumped in landfills. Research related to this has been published in the journal Nature Communications.*"

Figure 2.3—Theme 9

Despite strict measures to stop it, we keep seeing the continual dumping of garbage and raw or untreated sewage. Any animal or marine life coming into contact gets impacted in the worst of ways. The inevitable formation of algal bloom and clusters contaminates and eventually suffocates marine life such as coral and fish (Paul et al., 2019). Furthermore, it is a vicious cycle which feeds itself. With reference to the question of harm towards animal and marine life by improper waste disposal, 100% of Far *Nagarpalika* Trash-House respondents opted for "the diseases carrying mosquitoes now spread sickness and death among the living population," while 80% of Near *Nagarpalika* Trash-House respondents opted for "waste contaminates the land on which we grow food and provides water for us and animals." Empirical evident by The National Institute of Environmental Health Sciences argues that an estimated 40+ tons of electric-waste (e-waste) is produced each year, and when this e-waste is illegally dumped, its chemicals mix. The sources cause harm to wildlife and human health, especially children who play and live near these fly-tipping sites.

Narrative of Far *Nagarpalika* Trash-Houses: "*The negative effects of improper waste management not only end in disgusting outlook but also* *affect the overall economy of the country. The state has to spend a lot of money to counter the effects of improper waste management. In addition, animals dependent on the environment are also at a great risk due to oil spills and leaching of chemicals that directly lead to soil and water contamination. Burning of any waste and plastic material cause's air and environment pollution."*

Narrative of Near *Nagarpalika* Trash-Houses*:* "*Land pollution is the result of dumping of garbage, garbage and other toxic substances causing the land to become contaminated or polluted. The source of land pollution comes from human elements such as dustbins and waste washed ashore from boats, oil rigs and sewage outlets."*

Figure 2.3—Theme 10

In contrast of question, 50% of Far *Nagarpalika* Trash-House respondents responded "employees whose workplaces manufacture or come into contact with waste materials." 40% of Near *Nagarpalika* Trash-House respondents preferred to respond "children and those who live near such facilities."

Narrative of Far *Nagarpalika* Trash-Houses: "*The people who have a place to work near the garbage houses, whoever pass from there or go to their house or go to their place of work, they are the people who get the most from the gas pollution coming out of the tortoise house and their.*"

Narrative of Near *Nagarpalika* Trash-House: "*The people whose houses are built near or little away from their lives around the garbage houses are not untouched by the diseases caused by the smell of gas coming out of the garbage houses so dangerous that you can tell us stay even at a distance of 1–2 km from him."*

Figure 2.3—Theme 11

60% of Far *Nagarpalika* Trash-House respondents opted for the category that 'soil' contamination is the deadly effect of improper waste disposal. On the other side, 60% and 40% of Near *Nagarpalika* Trash-House respondents responded that "water and air" is the deadly effect of improper waste disposal.

Narrative of Far *Nagarpalika* Trash-Houses: "*All the garbage that accumulates in the garbage houses is being disposed of very wrongly, due to which it is having a harmful effect on the health of human beings today, but the biggest effect is on our soil. Its fertility is decreasing. Today, the plastics found in the garbage houses are getting mixed in our soil, due to which we are not able to get pure and clean crop, it is directly affecting the crops."*

Narrative of Near *Nagarpalika* Trash-Houses: "*Due to improper disposal of waste in the garbage houses or due to not using the right technology, the very harmful effect of the smell and gases coming out of it is affecting our drinking water and breathing air.*"

Categories	Far *Nagarpalika* Trash-Houses (50)	Near *Nagarpalika* Trash-Houses (50)
The diseases carrying mosquitoes now spread sickness and death among the living population	50 (100%)	10 (20%)
Waste contaminates the land on which we grow food and provides water for us and animals	–	40 (80%)

Theme 9: Question—"Which type of harm is faced by animals and marine life by improper waste disposal?"

Categories	Far *Nagarpalika* Trash-Houses (50)	Near *Nagarpalika* Trash-Houses (50)
Employees whose workplaces manufacture or come into contact with waste materials	50 (100%)	10 (20%)
Children and those who live near such facilities	–	40 (80%)

Theme 10: Question—"What is the most significant human damage by improper waste disposal?"

Categories	Far *Nagarpalika* Trash-Houses (50)	Near *Nagarpalika* Trash-Houses (50)
Soil	30 (60%)	–
Nothing	20 (40%)	–
Water	–	30 (60%)
Air	–	20 (40%)

Theme 11: Question—"What is a deadly effect of improper waste disposal?"

FIGURE 2.3 Theme 9—Harm towards animal and marine life by improper waste disposal; Theme 10—Most significant human damage by improper waste disposal; Theme 11—Deadly effects of improper waste disposal.

Figure 2.4—Theme 12

Most of the Far *Nagarpalika* Trash-House respondents reported (60% and 40%) "obstacle towards modernization and health problems," while 80% and 20% of Near *Nagarpalika* Trash-House respondents reported "health problems as well as obstacles towards modernization."

> **Narrative of Far *Nagarpalika* Trash-Houses:** *"Today there has been a decrease in modernization due to garbage houses, it looks bad to see. It also seems very wasteful to pass around them, if a garbage house is built in a very good area, then it does not look very good to see, people do not want to take home there nor do they want to do any other kind of development there."*

Categories	Far *Nagarpalika* Trash-Houses (50)	Near *Nagarpalika* Trash-Houses (50)
Obstacle towards modernization	30 (60%)	10(20%)
Health problems	20 (40%)	40 (80%)

Theme 12: Question—"Which type of problem is faced by the public due to improper waste disposal?"

Categories	Far *Nagarpalika* Trash-Houses (50)	Near *Nagarpalika* Trash-Houses (50)
Moisture production from waste	50 (100%)	–
Mosquitoes and rats	–	50 (100%)

Theme 13: Question—"Which type of disease-carrying pests are there by improper waste disposal?"

Categories	Far *Nagarpalika* Trash-Houses (50)	Near *Nagarpalika* Trash-Houses (50)
Effect on economy of country	50 (100%)	–
Can't say	–	50 (100%)

Theme 14: Question—"Does improper waste disposal adversely affect the local economy or not?"

Categories	Far *Nagarpalika* Trash-Houses (50)	Near *Nagarpalika* Trash-Houses (50)
Positive and optimistic	30 (60%)	10 (20%)
Negative and pessimistic	20 (40%)	40 (80%)

Theme 15: Question—"In what way can proper waste disposal affect your mental and physical performance?"

FIGURE 2.4 Theme 12—Problem faced by the public due to the improper waste disposal; Theme 13—Carrying pests by improper waste disposal; Theme 14—Adverse effect on the local economy of improper waste disposal; Theme 15—Effect of proper waste disposal on mental and physical performance.

Narrative of Near *Nagarpalika* Trash-Houses: "*Garbage houses have the biggest impact on our health. Every day we and the children of our family get affected by diseases caused by garbage houses, due to which more than half the money is spent in it.*"

Figure 2.4—Theme 13

For exploring the disease-carrying pests, 100% of Far *Nagarpalika* Trash-House respondents reported "moisture production from waste," and their counterparts reported "mosquitoes and rats" are the big problem.

Narrative of Far *Nagarpalika* Trash-Houses: *"Moreover, moisture production from waste is a breeding ground from appliances and food scraps."*

Narrative of Near *Nagarpalika* Trash-Houses: *"Mosquitoes and rats are known to live and breed in sewage areas, and both are known to carry life-threatening diseases."*

Figure 2.4—Theme 14

With reference to first impression of the person, 100% of Far *Nagarpalika* Trash-House respondents responded "Yes, there is an effect on economy of country." On the other side, Near *Nagarpalika* Trash-House respondents responded "Can't say" regarding improper waste disposal effects the local economy or not.

Narrative of Far *Nagarpalika* Trash-Houses: *"More than a direct impact on our environment, improper waste disposal also affects our economy at both a local and national level."*

Narrative of Near *Nagarpalika* Trash-Houses: *"People living near the garbage houses said that we do not know how the garbage houses are affecting our economy today, all we know is that it is affecting our child's health, how our economy is affected."*

Figure 2.4—Theme 15

This question focuses on the mental and physical health of the respondents. The result of this is to induce a stress reaction, which could take forms of impaired well-being, physiological dysregulation and the ten observed factors which result in diseases. There is no way management and control of inappropriate waste disposal will be discussed without mentioning the cost involved in undertaking the processes and steps to be followed, with the objective of disease prevention and environmental protection (Calijuri et al., 2004).

Narrative of Far *Nagarpalika* Trash-Houses: *"The people living around the garbage houses told that if the waste is disposed very properly then it has a very good effect on our health and on our society, because of this we are seeing many types of problems today."*

Narrative of Near *Nagarpalika* Trash-Houses: *"We have never seen the proper disposal of the waste coming out of the garbage houses, we have only seen this, what problems does it cause."*

2.3 IMPLICATIONS AND SUGGESTIONS

Most of the waste generated by most households is organic. Therefore, the most suitable, sustainable and environmentally friendly method for organic waste is to make compost from them by recycling and reusing them. Composting at home will encourage organic farming and reduce the need for chemical fertilizers. The kitchen garden is the most preferred option at the household level, as the family receives some edible items such as fruits or vegetables from the garden. However, this would be possible only where open land is available near the house. Garden irrigation can be done.

2.4 CONCLUSION

The result clearly reflects the effect of improper waste disposal on physical, environmental and emotional well-being among the general population across a locality (near and far *nagarpalika*). Poor waste disposal activities cause serious environmental problems in a place. Improper waste disposal is the disposal of waste in a way that has negative consequences for the environment. Examples include garbage, hazardous waste that is dumped in the ground and non-recyclable items that must be recycled.

The people living near the garbage houses—the people whose houses are built near the garbage houses—they have to face many problems due to the garbage houses; many harmful effects fall on their health because of these garbage houses. They face both mental and physical problems, they have health-related problems, they have to live in those places even if they do not want to, but on the other hand, people living away from garbage houses face fewer such problems, and it has little effect on their mental and physical health. Waste-disposal management has less impact, as they live away from such places and do not want to pass around these places.

REFERENCES

Ancona, C., Mataloni, F., Badaloni, C., et al. (2014). Residential cohort approach in industrial contaminated sites: The ERAS Lazio project. *Journal of Preventive Epidemiology, 38*(Suppl 1), 158–161.

Bharti, J. (2020). Psychological health among caregivers of person with mental illness. *Mukta Shabd Journal, 9*(12), 1141–1146.

Calijuri, M. L., Marques, E. T., Lorentz, J. F., Azeredo, R. F., & Carvalho, C. A. B. (2004). Multi-criteria analysis for the identification of waste disposal areas. *Geotechnical and Geological Engineering, 22*(2), 299–312.

Gouveia, N., & Prado, R. R. (2010). Health risks in areas close to urban solid waste landfill sites. *Rev Saude Publica, 44*, 859–866. www.speedyclearances.com/post/effects-of-improper-waste-disposal-on-human-health

Mondal, T., Choudhury, M., Kundu, D., Dutta, D., & Samanta, P. (2023). Landfill: An eclectic review on structure, reactions and remediation approach. *Waste Management, 164*, 127–142. https://doi.org/10.1016/j.wasman.2023.03.034

Paul, S., Choudhury, M., Deb, U., Pegu, R., Das, S., & Bhattacharya, S. S. (2019). Assessing the ecological impacts of ageing on hazard potential of solid waste landfills: A green approach through vermitechnology. *Journal of Cleaner Production, 236*, 117643. https://doi.org/10.1016/j.jclepro.2019.117643

3 An Integrated Approach Towards Sustainable Waste Management

Decentralized and Community-Based Practices

Emilda K. Joseph, Bindi Varghese, Tomy K. Kallarakal, and Jose K. Antony

3.1 INTRODUCTION

One of the most important problems faced by many natural and vulnerable destinations around the world is waste management. This problem primarily faced revolves around the amount of waste and the nature of the waste being generated. All human activities create leftovers, and people living together create even more waste. The waste that is created from every activity that we do is broadly divided into three categories: it can be organic, inorganic or toxic waste. These can be further classified into solid, liquid and gaseous waste, and each type of waste will require a special kind of treatment.

Numerous studies have found that, in many economically progressive countries, resource improvement is highly common, and it is generally undertaken at the local community levels. Sustainable materials are frequently and continuously used as feedstock for reuse. The rapid depletion of scarce resources can be controlled, and waste management problems can be avoided to a great extent if natural materials can be more resourcefully and professionally used (Wang et al., 2011; Paul et al., 2019). Moreover, by efficiently using the resources from our ecosystem, the rate of exhaustion of natural resources can be reduced, and the waste management problems can be regulated, as we will be better able to reuse the waste that will help us to safeguard our income (Joseph & John, 2017). The following are two effective methods to solve waste management problems in a natural destination: decentralized waste management and community-based waste management.

3.1.1 Decentralized Waste Management

Decentralized waste management involves the collection, treatment, disposal and reuse of the waste materials from individual homes, local communities and houseboats. Decentralized waste management systems are highly appropriate for

DOI: 10.1201/9781003258377-3

semi-urban, rural and remote areas, as they are generally localized and more controlled waste management systems. It can be said that decentralized waste management systems deliver services nearer to the points of origin or generation of waste (Joseph & John, 2017). It is an onsite waste management system that is used to treat and recycle relatively small quantities of waste materials, mainly originating from individuals, groups of dwellings and businesses that are located relatively close to each other (Figure 3.1-A). The decentralized approach to waste management treatment is seen as advantageous and highly recommended for a number of reasons:

i. It saves money: By deciding on anticipatory strategies such as assessing a community's needs and conditions, in order to manage waste materials before a crisis occurs, thereby avoiding unnecessary costs
ii. Easy to handle: Managing the decentralized waste is easier, viable and offers lasting alternatives to centralized waste management treatment facilities, particularly in small and rural communities where such practices can turn more money-spinning

On the whole, the decentralized waste management system is a small-scale and eco-friendly practice which requires only little infrastructure and fewer costs. It protects the health of the individuals as well as the natural environment, and it also enhances the property value and lives of the local population.

3.1.2 Community-Based Waste Management

The community-based waste management approach revolves around a cooperative concept that leads to the common goal of making transformations in the community solid waste management practices in the areas of source segregation, recovery of recyclable materials and storage processes prior to collection. These initiatives are aimed at involving the local residents in the implementation and management of waste management practices in a bid to protect their natural environment. According to Indrianti (2016), "Community-based waste management offers an alternative income source for community members". Community based waste management focuses on "long term sustainability, community development, community involvement, and local benefits". It is an important initiative designed by the environmental management teams to promote sustainable tourism exercises related to an increased sense of socio-cultural and environmental responsibilities in tourism and sustainability (Indrianti, 2016). From "an environmental and economic perspective, unless the local people are also involved in such practices, it is likely that over time the resources on which tourism depends will be destroyed and the investments will be lost" (Joseph, 2018). Community-based waste management practices require advanced management of the total waste generated through sorting, composting and implementing the 3R's; namely, reuse, recycle and reduce. The 3R's concept will assist to reduce the amount of waste generated at the community levels and it will positively impact the amount of waste going into landfills (Mondal, 2023). However, lack of wholehearted community participation, absence of a robust social structure and lack of support from the community leaders and stakeholders can act as major roadblocks in the realization of community based waste management initiatives (Joseph, 2018). Indrianti (2016), in her article "Community-based Solid Waste Bank Model for Sustainable Education," reiterates that community-based waste management can create a sense of togetherness, and it can help to work towards the eradication of many of the common problems in a destination. Community members are the most important stakeholders in waste management activities, and therefore, they have to take up active participation in solving the problems in waste management by modifying their perceptions and attitudes for the successful execution of solid waste management programs. Seminal contributions have been made by a number of researchers to prove that community-based waste management is always oriented towards long-term sustainability, and the literatures on community-based waste management focus on community involvement, community support and perception of stakeholders on the benefits and costs of community involvement in solid waste management activities and their local-level responsibilities and participation in tourism development and waste management (Figure 3.1-B).

Community-based waste management is recognized the world over as the single most important solution to waste management problems in many tourism destinations around the world (Ridho & Nasution, 2017). It is an integrated approach (Februati & Dayana, 2018) that identifies the local community's involvement in conserving the local environment in a sustainable manner. It offers immense opportunities to people to exercise control over their environment and to involve, support and enhance the aesthetic value of their environment (Suburb, 2019). This strategy has helped the community to become more conscious of the waste management issues around their environment, and in this context, the local community plays a significant role, and their participation is very much required to solve the waste management problems (Satori et al., 2018). The main purpose of this concept is to bring about changes in the waste management system amongst the local community. It can create a sense of belonging in the local community (Satori et al., 2018), as it involves the participation of different stakeholders. Perception, policy planning, involvement, awareness, participation and willingness of stakeholders towards waste management are essential for successful community-based waste management practices (Shatnawi, 2018). The success of community-based waste management depends on the participation of all the stakeholders in the destination (Suburb, 2019). The outcomes will help the community to understand the real situation and problems of waste management (Joseph, 2018). The main objective of community-based waste management is to empower the local community members in every aspect of tourism and destination management and to ensure the environmental, social and economic sustainability of the region. Training, support facilities and financial assistance are prerequisites to ensure the sustainability of community-based waste management practices (Kaur et al., 2013). Therefore, social preparedness, stakeholder identifications, social mobilizations, community organizations, community development, community monitoring and evaluation are all highly recommended for the proper implementation of community-based waste management initiatives (Joseph et al., 2017). A recent

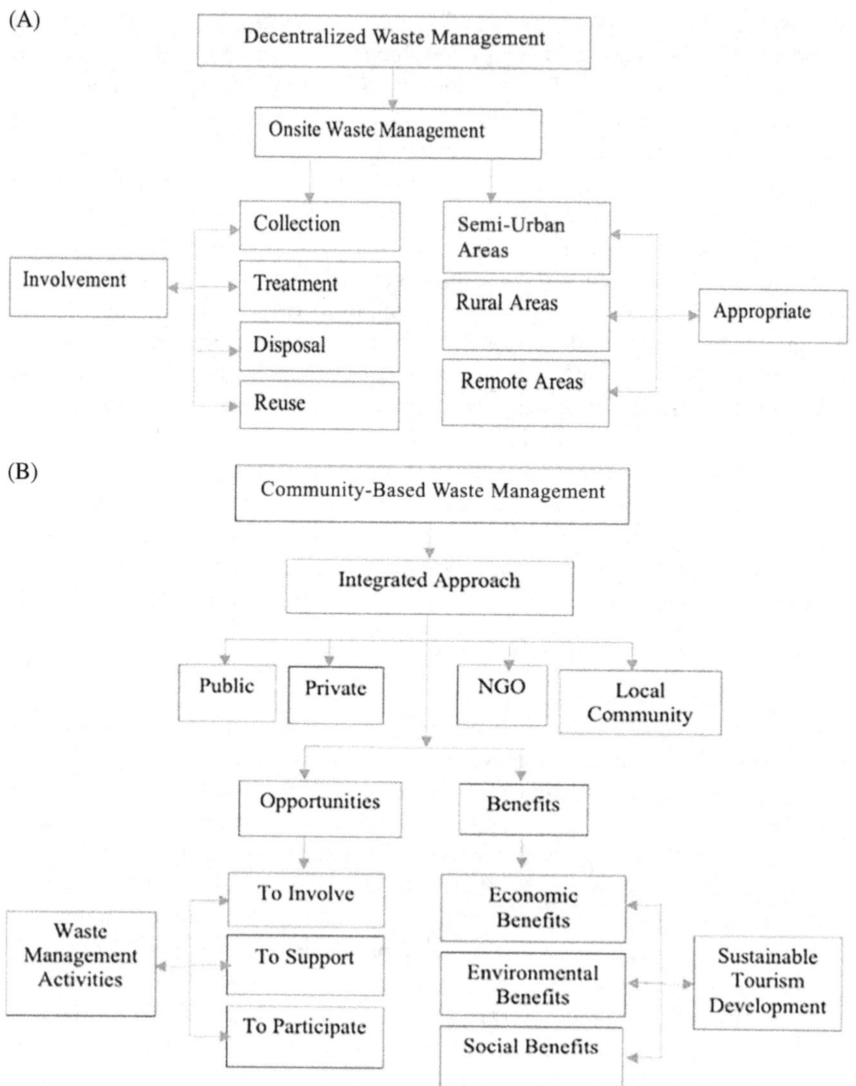

FIGURE 3.1 (A)—A conceptual framework for decentralized waste management; (B)—A conceptual framework for community-based waste management.

study (Dickella et al., 2017) explains the key steps for community based solid waste management plans:

- Preliminary Meetings: Preliminary meetings include the planning sessions and creating awareness among the stakeholders about the problems in the destinations
- Waste Profile Determination: A survey needs to be conducted to understand the nature and characteristics of the waste being produced

- Partnerships: Ensure the support of the public, private and local community members
- Awareness Programme: Conduct awareness programmes for all the stakeholders
- New System of Waste Management: Introduce a new system of waste management with active participation of the local community members
- Share the Experience: Finally, share the results and experience of the initiative with other similar destinations

Likewise, many researchers recommended community-based waste management as the ideal and most appropriate solution to waste management problems (Chengula, 2015). Community involvement and community support are imperative for this solution (Ridho & Nasution, 2017). Abdullahi and Polytechnic (2017) said "community based waste management involves various forms of community involvement, including awareness, behaviour and participation. It also requires numerous institutional support and recognitions in order to be successful". These activities are normally carried out by members of the local community and management (Joseph et al., 2017). However, the major problems of community-based waste management include financial and operational challenges, low participation and inadequate involvement from community members.

In order to achieve sustainability, it is necessary to overcome these hindrances in any tourism destination (Abdullahi & Polytechnic, 2017). The community-based waste management model is successful only when there is wholesome and active support from all the local community members and all the other stakeholders involved in the programme (Mubaiwa, 2006). Consultations with community members indicate that a positive and favourable behaviour can be created only through strong public involvement in various clean-up activities (Joseph et al., 2017). The main factors that affect community-based waste management include economic affordability, environmental effectiveness and social acceptability. These three factors are most important for sustainable waste management systems (Satori et al., 2018). Joseph et al. (2019) further explained that the participation of the local community members is very much a requirement in solid waste management programmes, as they are continuous maintenance systems and the experience earned can be a learning phase for everyone involved in the solid waste management process. It should also be clearly noted that it is not always necessary that the same approach could be fruitful and successful for all the communities. Hence, an integrated waste management approach can be adopted in all the major destinations through the active participation of all stakeholders. Moreover, partnerships with women's self-help groups, household associations, local non-governmental organizations, academic institutions and private ventures can further strengthen the community-based waste management programmes (Premakumara, 2012). However, this cannot be treated as a complete solution to solid waste management problems in any tourism destination because solid waste management initiatives are constantly facing a lot of financial challenges and lack of corporation from different stakeholders (Kasala, 2014).

Community-based waste management can be used as a powerful concept to reduce waste management problems in tourism destinations (Joseph et al., 2019). This approach can be implemented to create awareness among community members

TABLE 3.1
Elements of Community-Based Waste Management

	Elements of Community-Based Waste Management	Authors
Community Involvement	Promotion of re-usable materials Reduction and recycling of waste materials Development of waste segregate mechanisms Providing facilities for public litter bins Facilitating environmentally aware service providers Preparing an environmental impact assessment Offering environmental protection	(Mongkolnchaiarunya, 2005); (Furqan, 2013); (Nicholas et al., 2009); (Malik et al., 2015); (Prabhakaran et al., 2013)
Community Support	Local government support Providing support to minimize waste problems Empowering the local community in waste management Implementing the new waste-treatment methods Designing plans for waste disposal Providing a coordinator for waste management	(Lee, 2013; Muresan et al., 2016)
Perceived Costs	Water pollution Solid waste generation Destruction of flora and fauna Explosive growth of tourism Threat to the environment	(Dangi & Jamal, 2016; Muresan et al., 2016)
Perceived Benefits	Local community participation Income Employment opportunities Preserve natural and cultural resources Improve environmental awareness Opportunities for women	(Dangi & Jamal, 2016; Muresan et al., 2016)

and to identify and solve the problems by themselves (Joseph et al., 2020). Thus, members of the communities realize the real problems revolving around improper waste management and their consequences, and they always try to develop their own waste management techniques. Such processes help the communities to become deeply aware of the real situations being faced during waste management. The elements of community-based waste management can be discussed by using the dimensions listed in Table 3.1.

3.2 DISCUSSION

People are often unaware of the numerous benefits of decentralized waste management and community-based waste management systems. If these practices are to be accepted for sustainable development, people need to be instructed and trained on the advantages and values of these programmes, and at the same time, all forms of waste should be considered as anti-environmental and anti-social and not just as mere "waste". The waste should be recycled and reused so that it can help the local communities in pursuing economic growth and in promoting healthier environments. Previous studies had found that the local community could be influenced by the perceived benefits of waste management in three major areas; namely, social, economic and environmental factors. But, the perceived costs involved in waste management often negatively affect the local community and their local environment (Gursoy et al., 2002).

There are statistical evidences to conclude that, the higher the perceived benefits, the higher would be the sustainable development initiatives through community-based waste management initiatives. However, the higher the perceived costs, the higher would be the reluctance to employ sustainable development initiatives through community-based waste management practices. Therefore, the study substantiates that local community involvement and support are always interlinked to the perceived benefits and to the perceived costs. Perceived benefits positively support and involve local communities in such programmes, whereas perceived costs negatively affect the involvement of local communities in these initiatives. Community support, community involvement and the perception of community-based waste management practices are crucial because community support and involvement can have a direct impact on the solid waste management issues in a destination, as studied by Kumar and Nandini (2013). Similarly, strict efforts should be enforced to ban all unplanned construction activities that can affect the sustainability of destinations. Every place, be it tourism destinations, villages, towns, *panchayats* or municipalities need to have their own appropriate waste management methods in place at or near the areas of waste generation to ensure sustainable development. Efforts need to be ensured to restrict and ban the use of plastic and other non-degradable materials at the destinations and the use of plastic bags below the stipulated microns, as it is mandated that plastic carry bags must have a minimum thickness of 40 microns. The use of such plastics should therefore be strictly banned by the local administration and government.

3.3 THE STUDY SETTING: A CASE STUDY ON THE BACKWATER DESTINATIONS IN SOUTH KERALA, INDIA

Kerala has always been one among the most exclusive destinations for travel lovers. Lakes, beaches, backwaters and houseboats are unique selling attractions of Kerala's tourism industry. But in Kerala, waste materials pose major challenges and troubles to the backwater regions in the state, having such high tourist footfalls and activities. The inappropriate disposal of waste materials, solid waste and littering can damage the physical appearance of the destinations. Increased generation of solid waste and loss of the natural environment can definitely present the upcoming generations with an irretrievable future. Tourism has a prominent role to play in the preservation of the natural resources, and therefore, community-based, decentralized waste management is the new initiative in which local communities can participate in various forms of tourism activities and thereby help to minimize the waste management issues in the backwater regions, supporting the sustainability of tourism.

Community-based waste management practices can offer numerous economic benefits to the host community, and at the same time, it can also result in the conservation of the natural environment (Joseph et al., 2020). Development of community-based waste management practices are highly worthwhile as the four variables—namely, community support, community involvement, perceived benefits and perceived costs—can be employed as measuring variables to determine the effectiveness of sustainable tourism development initiatives in the tourism destinations. Stakeholders of the backwater regions need to develop programmes to foster community support and community involvement that will enhance perceived benefits and reduce perceived costs of sustainable tourism development practices. Similarly, stakeholders will have to emphasize activities that will enhance the perceived benefits of sustainable tourism. By expanding the perceived benefits, stakeholders guarantee that the local community members are also more likely to support and involve in tourism activities that will lead to sustainable tourism development. Different stakeholders can use the findings and outcomes of such initiatives to evaluate and enhance policy formulations on community-based waste management proposals. To achieve sustainability of the waste management initiatives, action plans need to be highlighted and implemented so that wholesome involvement from all the stakeholders can be elicited (Joseph et al., 2020).

3.4 ENVIRONMENT REGENERATION AND CHALLENGES

The strategic context for environmental regeneration has not been well developed in the past. The prime concern among the regulators and stakeholders has always been the policies that have lacked strategic vision and long-term perspectives. The study setting for the chapter explores the lack of broader considerations and poorer attempts to design strategies in order to handle waste management issues (Joseph et al., 2020). Institutional arrangements with a focus on region-wide partnerships can definitely aid improved amenities so as to create centres of attention on environmental quality. Drawing on the expertise of a number of novel environmental partnerships and local initiatives in the state of Kerala has always led to the introduction of new ideas and new ways of engaging all stakeholders in the processes (Figure 3.2).

3.5 ENVIRONMENTAL CRISIS MANAGEMENT AND SCOPE FOR FURTHER RESEARCH

To understand the specific distribution of the research paradigm on environmental crisis and management, the chapter has successfully indicated the macro- and micro-perspectives to deal with strategic issues and diverse stakeholders while assessing environmental crisis due to improper waste management practices. Past researches on crisis management have lacked adequate integration, whereas, in this chapter, sincere efforts have been made to integrate and build upon the current and available knowledge so as to create a multidisciplinary approach to research in crisis management. The study has used both social-engagements and structural-research perspectives on decentralized and community-based waste management approaches. The chapter has managed to offer a robust framework depicting waste management processes and researchable propositions for the integration of various perspectives. Suggestions on the implications of the research have also been offered. The chapter has been concluded by proposing some likely avenues for future research and by discussing many of the strengths and weaknesses evident in the present areas of waste management. Future

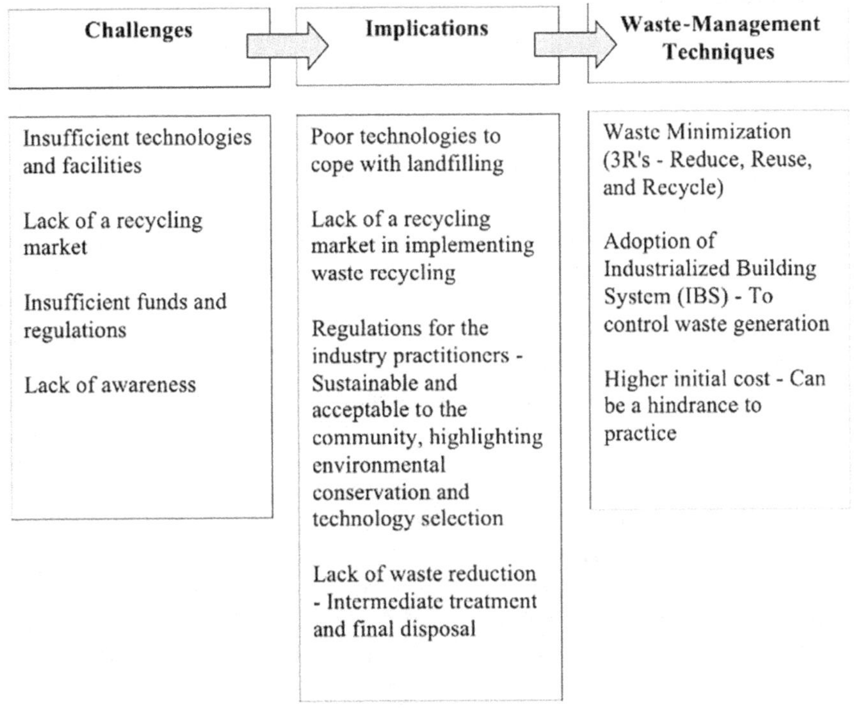

FIGURE 3.2 Change in achieving sustainable waste management.

research should consider the potential effects of similar studies across a wider spectrum of different waste management methods. In the future, more research is definitely required and invited to study in-depth about the perceptions of the local communities on different waste management techniques. Future studies can also be indicative of the major issues that can challenge different waste management practices and methods.

3.6 CONCLUSION

Sustainable approaches towards waste management are definitely a challenge and they can incur a high cost towards efficient practices. This can be successfully achieved and practiced only with the wholehearted support and participation of the local community members. Such sustainable initiatives will not only offer higher income and better employment opportunities to the local communities, but they will also lead to the conservation of the natural environment, thereby leading to the safeguarding of the less renewable and scarce resources. Waste minimization is an innovative approach towards practicing the sustainable model through the 3R's; namely, reduce, reuse and recycle the waste. Adoption of the Industrialized Building System (IBS) is yet another effective mechanism to control the unwanted generation of waste. IBS can be defined as a system in which all the building parts are produced in mass either in a factory or at a site under very strict quality control conditions. The system focuses on investment in equipment, facilities and technologies with the main objectives of maximizing production and output and minimizing wastage, while optimally utilizing the labour resources. This process calls for a high initial cost, and the high implementation cost also incurred can be a hindrance to practice this method. But, with less dependence on labour and with less production time, a lot of valuable time can be saved, and this can also reduce the monetary losses. Development of tourism in the backwater regions of Kerala has made the destinations more convenient. But, with more convenience has come the generation of more waste. All stakeholders and the host communities have to work with an integrative approach in order to handle the waste being produced in these regions. Adopting effective waste management practices should be preferred over disposing off or carelessly releasing waste into the environment. A scientific mechanism can make the practitioners more conscious about the significance of keeping their environments clean, tidy and eco-friendly. A localized approach by integrating the right technologies can also be ideal in materializing a sound system in place. More regulations and guidelines with a sustainable outlook will support the industry practitioners. In an ideal context, an acceptable ideology will be mutually beneficial. A sensitized approach towards the local community, with emphasis on environmental conservation and sustainable technology, is ideal. Involving members of the local community in waste management initiatives will help to offer them more employment opportunities and allow them to earn more income. It will also improve the cleanliness in the destinations and will encourage the local community members to improve discipline and minimize dumping of waste from their households. Households in the region will also cooperate while handling the waste, and this can reduce the cost incurred in managing a large amount of waste. Every individual needs to develop the growth mindset that it is each one's duty and responsibility to keep their surroundings clean and handle the waste generated from every homes judiciously. Local community members are, without doubt, the ambassadors of their destinations, and hence, they have a prominent role to play in promoting their destinations through sustainable development. Sustainability should therefore be made a part of everyday life, and, as is well known, truly investing in sustainability will require steadfast dedication to finding solutions and also involving the people in implementing these solutions. Today, if waste management is handled judiciously, it will offer huge potential to transform problems into solutions and to lead the way forward towards sustainable development.

REFERENCES

Abdullahi, Y. R., & Polytechnic, K. (2017, December). Community based solid-waste management strategy: A case study of Kaduna metropolis. *WIT Transactions on Ecology and the Environment*, *8*, 761–772.

Chengula, A. A., Lucas, B. K., & Mzula, A. (2015). Assessing the awareness, knowledge, attitude and practice of the community towards solid-waste disposal and identifying the threats and extent of bacteria in the solid-waste disposal sites in Morogoro municipality in Tanzania. *Journal of Biology, Agriculture and Healthcare*, *5*(3), 54–65. https://iiste.org/Journals/index.php/JBAH/article/download/20188/20663

Dangi, T. B., & Jamal, T. (2016). An integrated approach to "sustainable community-based tourism". *Sustainability (Switzerland)*, *8*(5). https://doi.org/10.3390/su8050475

Dickella, G., Premakumara, J., Chatura, W., Nadeesha, G., Nadeeka, S. A., & Masaki, T. (2017). *Community-based solid-waste management in Galle City – A pilot project in China Town*. CITYNET Yokohama Project Office. https://www.iges.or.jp/en/pub/community-based-solid-waste-management-galle/en

Februati, T., & Dayana. (2018). *The participation of community-based organizations on waste management in the city municipal of Medan. The participation of community-based organizations on waste management in the city municipal of Medan*. IOP Conference Series : Earth and Environmental Science. https://doi.org/10.1088/1755-1315

Furqan, M. (2013). Supporting indicators for the successful solid waste management based on community at Rawajati, South Jakarta. *Jurnal Wilayah Dan Lingkungan*, *1*(3), 245. https://doi.org/10.14710/jwl.1.3.245-250.

Gürsoy, D., Jurowski, C., & Uysal, M. (2002). Resident attitudes. *Annals of Tourism Research*, *29*(1), 79–105. https://doi.org/10.1016/s0160-7383(01)00028-7

Indrianti, N. (2016). Community-based Solid Waste Bank model for sustainable education. *Procedia – Social and Behavioral Sciences*, *224*, 158–166. https://doi.org/10.1016/j.sbspro.2016.05.431

Joseph, E. K. (2018). Community-based waste management for sustainable tourism in backwater regions of Kerala, India. In B. Varghese (Ed.), *Evolving paradigms in tourism and hospitality in developing countries a case study of India* (pp. 120–142). Apple Academic Press.

Joseph, E. K., & John, S. (2017). *Community-based decentralized waste management for sustainable tourism* (Vol. 1). Lambert Publications.

Joseph, E. K., Kallarackal, T. K., & Varghese, B. (2017). Community-based waste management: Backwater tourism as a case example. *Atna*, *2*(12), 85–94. https://doi.org/10.12727/ajts.18.5

Joseph, E. K., Kallarackal, T. K., & Varghese, B. (2019). A sustainable approach to Community Based Waste Management in the Backwaters of South Kerala. *Journal of Tourism*, *XX*(2), 21–37.

Joseph, E. K., Kallarakal, T. K., Varghese, B., & Antony, J. K. (2020). Sustainable tourism development in the backwaters of south Kerala, India: The local government perspective. *Geojournal of Tourism and Geosites*, *33*(4), 1532–1537. https://doi.org/10.30892/gtg.3

Kasala, S. E. (2014, September). *Critical analysis of the challenges of solid waste management initiatives* (pp. 1064–1074). Keko Machungwa Informal Settlement, Dar Es Salaam.

Kaur, A., Jawaid, A., & Othman, N. B. A. (2013). The impact of community-based tourism community development in Sarawak. *Journal of Borneo Kalimantan Institute of Borneo Studies, UNIMAS*, 15–26.

Kumar, M., & Nandini, N. (2013). Community attitude, perception and willingness towards solid waste management in Bangalore city, Karnataka, India. *Internal Journal of Environmental Sciences*, *4*(1), 87–95. https://doi.org/10.6088/ijes.2013040100009

Lee, T. H. (2013). Influence analysis of community resident support for sustainable tourism development. *Tourism Management*, *34*, 37–46. https://doi.org/10.1016/j.tourman.2012.03.007

Malik, N. K. A., Abdullah, S. H., & Manaf, L. A. (2015). Community Participation on Solid Waste Segregation Through Recycling Programmes in Putrajaya. *Procedia Environmental Sciences*, *30*(June 2016), 10–14. https://doi.org/10.1016/j.proenv.2015.10.002

Mondal, T., Choudhury, M., Kundu, D., Dutta, D., & Samanta, P. (2023). Landfill: An eclectic review on structure, reactions and remediation approach. *Waste Management*, *164*, 127–142. https://doi.org/10.1016/j.wasman.2023.03.034

Mongkolnchaiarunya, J. (2005). Promoting a community-based solid waste management initiative in local government: Yala municipality, Thailand. *Habitat International*, *29*(1), 27–40. https://doi.org/10.1016/S0197-3975(03)00060-2

Mubaiwa, A. (2006). Community based waste management in urban areas. In *Proceedings from the 2nd International Conference on Appropriate Technology*. National University of Science and Technology (NUST).

Mureşan, I. C., Oroian, C., Harun, R., Arion, F., Poruţiu, A., Chiciudean, G. O., Todea, A., & Lile, R. (2016). Local residents' attitude toward sustainable rural tourism development. *Sustainability*, *8*(1), 100. https://doi.org/10.3390/su8010100.

Nicholas, L. N., Thapa, B., & Ko, Y. J. (2009). Residents' perspectives of a world heritage site. The Pitons Management Area, St. Lucia. *Annals of Tourism Research*, *36*(3), 390–412. https://doi.org/10.1016/j.annals.2009.03.005

Paul, S., Choudhury, M., Deb, U., Pegu, R., Das, S., & Bhattacharya, S. S. (2019). Assessing the ecological impacts of ageing on hazard potential of solid waste landfills: A green approach through vermitechnology. *Journal of Cleaner Production, 236*, 117643.

Prabhakaran, S., Nair, V., & Ramachandran, S. (2013). Marine waste management indicators in a tourism environment: Exploring possibilities for Semporna District, Sabah. *Worldwide Hospitality and Tourism Themes*, *5*(4), 365–376.

Premakumara, D. G. J., (2012). Establishment of the community-based solid waste management system in Metro Cebu, the Philippines. In KITA & IGES (Eds.), *The report for the establishment of the waste management system in Metro Cebu, Philippines, Kitakyushu, Japan*.

Ridho, H., & Nasution, M. A. (2017). Analysis of community-based waste management in Medan. *Advances in Social Science, Education and Humanities Research, 81*(ICOSOP 2016), 513–516.

Satori, M., Megantara, E. N., Gunawan, B., Sciences, N., Padjadjaran, U. (2018). Review of the Influencing Factors of Integrated. *International Journal of Geomate*, *15*(48), 34–40.

Shatnawi, R. S. (2018). Solid waste management: Classification and public perception on management options at applied science university. *Jordan Journal of Civil Engineering*, *12*(3), 379–388.

Sinthumule, N. I., & Mkumbuzi, S. H. (2019). Participation in community-based solid waste management in Nkulumane suburb, Bulawayo, Zimbabwe. *Resources, 8*(1), 30.

Wang, H., He, J., Kim, Y., & Kamata, T. (2011). Municipal solid waste management in small towns: An economic analysis conducted in Yunnan, China. *World Bank Policy Research Working Paper, 5767*. https://doi.org/10.1596/1813-9450-5767

4 Solid Waste Management

Thamilmaraiselvi B, Bhuvaneswari P, Steffi PF, and Sangeetha K

4.1 INTRODUCTION

Solid waste (the global challenge) is made up of organic and inorganic waste materials that are generated as a result of a community's various lifestyle activities (1, 2). Solid waste materials are generated in a variety of ways, such as industrial, residential, and commercial activities of human beings (3). Unplanned urbanisation and economic activities are the major contributing factors in the exponential rise in solid wastes (4). India is facing the same issue as a result of an increasing population that has generated a significant amount of municipal solid waste (MSW) in recent years (5). The solid wastes can be divided into three major categories, like Physical Characteristics, Sources, and Toxicity. Based on Physical Characteristics, it is divided into three categories: Solid-State Wastes, Liquid-State Wastes, and Gaseous-State Wastes. Based on sources, it is categorized as Municipal (Urban) Wastes, Industrial Waste, and Commercial Wastes. Based on toxicity, it is categorized as Hazardous Wastes and Non-hazardous Wastes (3, 6).

Solid waste management lacks a holistic approach to encompass the entire cycle of garbage collection, reduction, and recycling from a global perspective. Solid waste management is a worldwide issue due to environmental contamination (7). Open dumping and burning are the most common waste-treatment and final-disposal methods in poor countries (8). The generation of solid waste is difficult to manage on a global scale due to the rapid rise in population, economic expansion, urbanisation, and industrialisation advancement. Where plastic, glass, and metal have a developed market in developing nations, the informal sector plays a major role in their utilisation (9).

4.2 DIFFERENT TYPES OF WASTES—BRIEF OVERVIEW

4.2.1 Based on Physical Characteristics

4.2.1.1 Solid-State Waste Materials

Any solid-state material which is expected to be unusable and is cleared out from factories, businesses, and households that turn out completely different types of solid waste that includes plastics, paper, metals, and solid forms of chemicals, cloth pieces, food, and animal remains, including faecal materials (10). Solid waste does not have a moisture level of greater than 20%. Solid waste densities vary, depending on location, storage time, season of the year, and other factors. If a specific amount

DOI: 10.1201/9781003258377-4

of waste must be disposed of in a landfill, it must first be compacted or lowered to an acceptable density. Any standard compaction can reduce the volume of trash by 75%, increasing density by a factor of ten. Compaction reduces the density of 100 to 400 kgm^3. Note that the compaction ratio should not exceed 1.5:1. Disposal, scavenging, wetting, drying, handling, and other factors might cause density changes.

The size of waste varies from particle size to small, medium, large, and huge. Solid wastes can be opaque (like bricks, wood, and batteries), translucent (like tissues, butter paper, and frosted glass), or transparent (like glass and plastic sheets). Solid wastes are containing metallic materials having magnetic qualities that can be easily separated by magnetic separators, such as batteries and iron rods. Combustible waste is one that has ignitability as one of its qualities—paper, leaves, matchboxes, and other wastes, for example. Glass, ceramic, metal, dust, and ash are examples of non-combustible solid waste. Carbon, hydrogen, nitrogen, sulphur, and oxygen are found in the majority of dry solid wastes. Some solid waste may also contain metals. The percentage of waste that is transformed into vapour when it is burned. The volatility rate of typical municipal solid waste is between 40% and 60%. Lipids, carbohydrates, proteins, natural fibres, and plastics all contribute to the chemical composition of solid wastes; hence, their percentage change affects the chemical composition.

4.2.1.2 Liquid-State Waste

Industrial waste-water is generated from the activities associated with washing during the manufacturing process, cleaning of objects, and mixing of chemicals (11). Both faeces and urine (human excreta) and different kinds of water-using activities are placed underneath liquid waste. The mixture of human excretory product from water-flushed bogs and different wastes from home and business are referred to as sewage (12). Rain water cleanses land surfaces having different kinds of wastes into water bodies such as rivers, ponds, and lakes (13). Another kind of liquid waste is urban run-off which will be the source of pollution. Urban run-off will contain a significant amount of organic material that may have been released as a result of open evacuation or improper treatment of organic waste generated by houses and businesses (14).

4.2.1.3 Gaseous-State Waste

Gaseous waste is described as abnormally high concentrations of chemical elements in the air that can harm living organisms (15). Clean air is composed of 78% nitrogen, 21% oxygen, and 1% trace gases. Particulate materials such as black soot, as well as various incompatible gaseous substances such as carbon monoxide, carbon dioxide, sulphur oxides, nitrogen oxides, nitrates, ozone, sulphates, and organic hydrocarbons, will be present in impure air (16). The majority of these are also found in clean air as trace gases. Abnormal quantities, on the other hand, create contaminants. Intense pollution is indicated by the emission of black smoke and invisible and odourless gases like carbon monoxide and carbon dioxide (17). Among them, carbon monoxide is very dangerous to humans (18). Inefficient burning of fuel can produce carbon monoxide (inadequate air supply to a charcoal stove in a home) and, if breathed in large quantities, is often deadly (19). Carbon dioxide is a vital pollutant which is involved in global climate-change.

4.2.2 Based on Source of Generation

4.2.2.1 Urban or Municipal Solid Waste

Municipal entities dispose of municipal solid waste gathered from various locations, such as residential residences, marketplaces, streets, and urban areas, which are referred to as "refuse" (20). Municipal solid trash is a mixture of paper, clothing, plastic, glass, metals, and organic matter generated by households and business facilities such as hotels, small cafes, boarding houses, and marketplaces (21). The degree of commercial and industrial activity, as well as the standard of living, lifestyle, and food habits, all contribute to vastly varying MSW proportions, which will change based on climatic conditions and location (22). Municipal solid-trash is collected on a local level, and the amount collected is governed by the size and consumption of the population (23).

4.2.2.2 Industrial Waste

Industrial waste is generated by chemical plants, paint factories, mining operations, cement factories, metallurgic plants, textile industries, food-processing industries, petroleum industries, and thermal-power plants (24). These industries turn out different kinds of waste products at the time of production and processing.

4.2.2.3 Commercial Waste

With the expansion of modern cities, industries, and automobiles, a massive amount of garbage is generated every day, including marketplaces, highways, buildings, hotels, business complexes, automobile workshops, machineries, and so on (24). The tremendous amount of waste discharged from hospitals, nursing homes, and medical institutes is very dangerous in nature (25). Several chemicals and discarded objects released by these machines endanger human health and spread a wide range of infectious diseases (26).

4.2.2.4 Agricultural Waste

Plant and animal waste are created from agricultural space. For agriculture, fertiliser, pesticides, and alternative chemicals are used. Excess usage can cause land and water pollution (27). Pesticides such as chlorinated hydrocarbons, DDT, BHC, endrin, dieldrin, lindane, parathion, malathion, and endosulfan pollute the soil, which is then absorbed by the crops (28). Sugar factories, tobacco processing units, slaughter homes, placentals, and poultry jointly manufacture agricultural waste.

4.2.3 Based on Toxicity

4.2.3.1 Hazardous Waste

Hospitals and industries generate hazardous waste like metals, chemicals, drugs, lather, pulp, electroplating, dye, and rubber, etc. The nature of these wastes could be corrosive, inflammable, explosive, or react when exposed to other materials (26).

4.2.3.2 Non-Hazardous Waste

Textile mills, food processing companies, cotton, paper, and sugar mills all produce non-hazardous waste that poses no risk to human or environmental health (27). This type of garbage is usually recycled.

4.3 METHODS OF SOLID WASTE DISPOSAL

The methods of solid waste disposal and management include the following:

4.3.1 Solid Waste Open-Burning

Open burning of municipal solid waste (MSW) is a significant non-point source of emissions in developing nations like India. Open burning of municipal solid waste is a potential source of 10 primary pollutants that pollute the environment, including hydrocarbons, furans, particulate matter, carbon dioxide, sulphur oxides, nitrogen oxides, benzene, toluene, ethyl group aromatic hydrocarbons, and 1-hexene (29). Agricultural areas and rural areas are more likely to burn waste openly, so there is a greater risk of dioxins and furans contaminating crops, streams, and lakes. Dioxins and furans deposited on plants by open burning are sometimes toxic. As a result of the exposure to smoke, health risks like headaches, nausea, and alternative metabolism symptoms were developed (30).

4.3.2 Sea-Dumping Process

The process of dumping solid garbage into the sea is typically done in coastal cities. The majority of the solid garbage was dumped in the north-east Atlantic and the Arctic, and nearly all of the radioactive waste was dumped in the sea as well (31). When heavy rains or floods wash garbage and debris into the waters, untreated or partially treated human waste and sewage water end up in the oceans. This is referred to as "trash dumping," and it is one of the most common sources of ocean contamination. This is not environmentally friendly. The "Ocean Dumping Act" regulated this process, and the Marine Protection, Research, and Sanctuaries Act of 1972 prohibited dumping materials into the ocean that have a negative impact on human health as well as marine habitat (32).

4.3.3 Solid Wastes Sanitary Landfills

Sanitary landfills are a scientific technique of disposing of solid waste on property that does not pose a health risk to the public, yet they are no better than open dumps. The sanitary landfill method, according to standard procedures, provides a satisfactory and nuisance-free solid waste disposal operation (33, 34). Landfilling is not suitable for extremely toxic wastes that include hazardous fumes or dust emissions and require greater isolation during unloading and disposal. Sanitary landfills ensure waste safety by allowing safe decomposition, alternative layers of debris, and soil speed decomposition, all of which reduce the harm caused by accumulated waste. The trash generated at the landfill has the potential to contaminate the environment, including the air, water, soil, and groundwater (35).

4.3.4 Incineration Method

Incineration is a waste disposal procedure that involves the combustion of waste items. The garbage is converted into ash (inorganic ingredients of the waste) and solid lumps carried by flue gas and heat during the incineration process. This method has significant advantages for the treatment of wastes such as clinical wastes and some hazardous wastes, in which germs and poisons are eliminated by high temperatures (36). Generally, hospitals and research centres burn biological tissue, blood-contaminated materials, and other medical wastes such as hypodermic needles and tubing in incinerators. Countries such as Japan, Singapore, Denmark, and the Netherlands use waste-combustion processes to generate energy. In 2005, incinerators reduced original waste by 80–85% and volume by 95–96%, depending on the composition and degree of recovery materials that included metals from the requested recycling, accounting for 4.8% of total electricity consumed and 13.7% of total domestic heat consumed in Denmark (37).

4.3.5 Composting Process

Composting is comparable to sanitary landfilling in that it produces a stable end-product that works as a soil conditioner and serves as a fertiliser base. Composting can be aided by the use of biodegradable solid waste. It is extremely popular in developing nations, such as India (38). Decomposing materials are separated from glass, metal, and other inorganic elements during sorting and separating activities.

Hammer mills and rotary shredders are used. The composting process for pulverised waste can either be carried out by open windows or in an enclosed facility. A windrow is a pile of low, long refuse. Turning or mixing the organics every few days will ensure that the microbes can digest the organics. In a dry environment, complete digestion of waste might take five to eight weeks. Due to the metabolism of aerobic bacteria, an active compost pile reaches around 65 °C (150 °F), killing pathogenic organisms that may be present in the waste material. The two ways are open-window composting and mechanical composting.

4.3.5.1 Open-Window Composting

Refuse is placed in the constructed piles (1.5 m in height, 2.5 m in width) at about 60% moisture content. The temperature will rise by about 70 °C due to biological activity (microbial actions). The pile is turned up for cooling and aeration to an anaerobic condition (39). Moisture content is set to 60%.

4.3.5.2 Mechanical Composting

One or more closed tanks or digesters with rotating vanes can be used to mix and aerate the solid waste in a mechanical-composting system. It takes roughly a week for the solid waste to be completely digested. Night soil and cow dung are added to the refuse for the enrichment of compost (40). The pit is filled with alternative layers of refuse (depth of 30–40 cm) and cow dung or night soil (laid over it in a thin layer). Excess moisture draining provisions are arranged at the bottom of the pit. To maintain the pH (neutralisation of acidity) in the compost, a layer of ash, ground limestone, or loamy soil is placed. In India, the Indore method, the Bangalore method, the NADEP method, and the Coimbatore method were developed for composting (41).

4.3.5.3 Vermicomposting

Vermicomposting is a process of composting using earthworms that increases familiarity with industrial as well as household settings. Earthworm species such as *Eisenia fetida* (red wigglers), *Eisenia hortensis* (European night crawlers), Eudrilus eugeniae (African night crawlers), and *Perionyx excavatus* (blue worms) are mostly preferred for the composting bed or pit method (42). Heaping the organic waste material along with cattle dung slurry for 20 days is known as "predigestion," which partially digests the organic material and prepares the environment for earthworm digestion. High-quality earthworms are introduced which eat rotting plant debris in the soil and swallow homes (43).

4.3.6 Disposal by Ploughing into the Fields

Organic waste provides numerous advantages in the agricultural field. Farmers employed this occurrence to improve the fertility of the soil in ancient times. Animal dung waste is added to the area, ploughed and mixed with soil adequately (44). It becomes part of the earth and provides essential nutrients required for the growth of crops. It lowers the cost of fertilisers. But this method is not commonly used for municipal solid waste and is not eco-friendly because of the nature of waste (45).

4.3.7 Disposal by Hog Feeding

Hog feeding is considered an effective method of disposal of waste material. Hogs eat various foods, including organic waste from kitchen yards and grocery stores (46). Using grinders, refuse is well grounded and then transferred into sewers. It requires a more significant amount of attention.

4.3.8 Salvaging Procedure

Few kinds of plastic, paper, rags, metals and glass are often retrieved, recycled and then reused. This is often thought about as an advanced approach of solid waste management. Salvaged material ought to be treated with care and should be avoided to mix with surrounding materials. Salvaged material is then reused by passing through different stages of recycling (47).

4.3.9 Fermentation

Fermentation, or biological digestion, is the chemical breakdown of a substance by using microbes. Complex materials are converted to easily utilised form because of the enzyme synthesised by the microorganisms. The process is classified into two methods, such as aerobic fermentation and anaerobic fermentation, based on the presence or absence of oxygen. At the end of fermentation, depending upon the nature of metabolic activity, microorganisms produce end products that may leads to digestion process (48).

4.4 ADVANTAGES AND DISADVANTAGES OF SOLID WASTE MANAGEMENT

4.4.1 Advantages

- **Reduces the quantity of waste**: Dumping, burning, and fermentation processes reduce the space occupied by the waste.
- **Reduces pollution in the environment**: The quantity of waste reduced means it is possible to create a disease-free, pollution-free environment.
- **Keeps the environment clean and fresh**: It is easy to keep the environment clean when solid waste management is successful because it can not only eliminate waste quantity but also reduce greenhouse gases such as methane and carbon dioxide.
- **Recycling helps to earn money and creates employment**: Solid waste management process requires more man power in every step, and they can produce recycled products. Using worms, vermicompost is prepared from the waste. That obviously increases the employability (49).
- **Conserves energy:** Resource generation with energy conversion.

4.4.2 Disadvantages

- **Processes are not always price-effective**: The process needs more spacious areas and money contributors to start a plant, and it consumes more time. Depending upon the nature of the waste, alternative management process in needed (50).
- **The final product has a very efficient shelf-life**: Recycled products having less quality and short shelf-life.
- **The locations are risky**: Microorganism such as bacteria and fungi are easily grown in the waste, and they may cause diseases to the person who works/handles the waste frequently.
- **Practices are not done regularly basis.**

The primary premise of solid waste management is to apply the 3Rs to reduce trash volume (reduce, reuse, and recycle). Even though we have several methodologies, depending upon the nature of waste, a suitable process will be selected to minimise the quantity of the waste, which should be properly recycled in an eco-friendly manner. Selection of a solid waste management process will vary country to country; it depends upon the composition of the waste, availability of space, cost effectiveness, and also their climatic condition. Success is only possible if we reduce the usage of non-degradable materials, and we can use solar energy (reduce the e-waste) eco-friendly things.

REFERENCES

(1) Abdallah, M., Hamdan, S., & Shabib, A. (2021). Development of a dynamic optimization framework for waste management systems. *MethodsX*, *8*, 101203.

(2) Akmal, T., & Jamil, F. (2021, April 15). Testing the role of waste management and emental quality on health indicators using structural equation modeling in Pakistan. *International Journal of Environmental Research and Public Health, 18*(8).
(3) Atabani, A. E., Ali, I., Naqvi, S. R., Badruddin, I. A., Aslam, M., Mahmoud, E., et al. (2021, July 30). A state-of-the-art review on spent coffee ground (SCG) pyrolysis for future biorefinery. *Chemosphere, 286*(Pt 2), 131730.
(4) Akram, S. V., Singh, R., AlZain, M. A., Gehlot, A., Rashid, M., Faragallah, O. S., et al. (2021, April 14). Performance analysis of IoT and long-range radio-based sensor node and gateway architecture for solid waste management. *Sensors (Basel), 21*(8).
(5) Alshehrei, F., & Ameen, F. (2021, June). Vermicomposting: A management tool to mitigate solid waste. *Saudi Journal of Biological Sciences, 28*(6), 3284–3293.
(6) Assuah, A., Sinclair, A. J. (2021, June 15). Solid waste management in western Canadian First Nations. *Waste Management, 129*, 54–61.
(7) Shi, Y., Wang, Y., Yue, Y., Zhao, J., Maraseni, T., & Qian, G. (2021, October). Unbalanced status and multidimensional influences of municipal solid waste management in Africa. *Chemosphere, 281*, 130884.
(8) Tong, Y. D., Huynh, T. D. X., Khong, T. D. (2021, April 1). Understanding the role of informal sector for sustainable development of municipal solid waste management system: A case study in Vietnam. *Waste Management, 124*, 118–127.
(9) Sharma, H. B., Vanapalli, K. R., Samal, B., Cheela, V. R. S., Dubey, B. K., & Bhattacharya, J. (2021, August 12). Circular economy approach in solid waste management system to achieve UN-SDGs: Solutions for post-COVID recovery. *Science of the Total Environment, 800*, 149605.
(10) Burns, C., Orttung, R. W., Shaiman, M., Silinsky, L., & Zhang, E. (2021, May 1). Solid waste management in the Arctic. *Waste Management, 126*, 340–350.
(11) Capoor, M. R., & Parida, A. (2021, June). Biomedical waste and solid waste management in the time of COVID-19: A comprehensive review of the national and international scenario and guidelines. *Journal of Laboratory Physicians, 13*(2), 175–182.
(12) De-la-Torre, G. E., Rakib, M. R. J., Pizarro-Ortega, C. I., Dioses-Salinas, D. C. (2021, June 20). Occurrence of personal protective equipment (PPE) associated with the COVID-19 pandemic along the coast of Lima, Peru. *Science of the Total Environment, 774*, 145774.
(13) Espuny, M., Faria, N. A., da Motta Reis, J. S., Dos Santos Neto, S. T., Nunhes, T. V., de Oliveira, O. J. (2021, June 24). Building new paths for responsible solid waste management. *Environmental Monitoring and Assessment, 193*(7), 442.
(14) Gibellini, S., Abu, Q. H., & Vaccari, M. (2021, March). Municipal solid waste management in refugee hosting communities: Analysis of a case study in northern Jordan. *Waste Management Research*, 734242X21994656.
(15) Kalvani, N., Mesdaghinia, A., Yaghmaeian, K., Abolli, S., Saadi, S., Rashidi, M. A., et al. (2021, June). Evaluation of iron and manganese removal effectiveness by treatment plant modules based on water pollution index; a comprehensive approach. *Journal of Environmental and Health Science Engineering, 19*(1), 1005–1013.
(16) Doaemo, W., Dhiman, S., Borovskis, A., Zhang, W., Bhat, S., Jaipuria, S., et al. (2021, April 29). Assessment of municipal solid waste management system in Lae City, Papua New Guinea in the context of sustainable development. *Environmental Development Sustainability*, 1–31.
(17) Espinoza, P. L., Ziegler-Rodriguez, K., Espinoza Perez, A. T., Vasquez, O. C., & Vazquez-Rowe, I. (2021, April 1). Closing the gap in the municipal solid waste management between metropolitan and regional cities from developing countries: A life cycle assessment approach. *Waste Management, 124*, 314–324.

(18) Fereja, W. M., Chemeda, D. D. (2021, April 28). Status, characterization, and quantification of municipal solid waste as a measure towards effective solid waste management: The case of Dilla Town, Southern Ethiopia. *Journal of the Air and Waste Management Association, 2,* 187–201.
(19) Haddad, M. B., De-la-Torre, G. E., Abelouah, M. R., Hajji, S., & Alla, A. A. (2021, July 27). Personal protective equipment (PPE) pollution associated with the COVID-19 pandemic along the coastline of Agadir, Morocco. *Science of the Total Environment, 798*, 149282.
(20) Kasemy, Z. A., Rohlman, D. S., & Abdel Latif, A. A. (2021, June). Health disorders among Egyptian municipal solid waste workers and assessment of their knowledge, attitude, and practice towards the hazardous exposure. *Environmental Science and Pollution Research International*, *28*(24), 30993–1002.
(21) Sanniyasi, E., Gopal, R. K., Gunasekar, D. K., Raj, P. P. (2021, August 26). Biodegradation of low-density polyethylene (LDPE) sheet by microalga, Uronema africanum Borge. *Science Reports, 11*(1), 17233.
(22) Liang, Y., Song, Q., Wu, N., Li, J., Zhong, Y., & Zeng, W. (2021). Repercussions of COVID-19 pandemic on solid waste generation and management strategies. *Frontiers of Environmental Science and Engineering*, *15*(6), 115.
(23) Marti, L., & Puertas, R. (2021, May 1). Influence of environmental policies on waste treatment. *Waste Management, 126*, 191–200.
(24) Murphy, E. L., Bernard, M., Iacona, G., Borrelle, S. B., Barnes, M., McGivern, A., et al. (2021, September 1). A decision framework for estimating the cost of marine plastic pollution interventions. *Conservation Biology*, *2*, e13827.
(25) Rafew, S. M., Rafizul, I. M. (2021, June 15). Application of system dynamics model for municipal solid waste management in Khulna city of Bangladesh. *Waste Management, 129*, 1–19.
(26) Senekane, M. F., Makhene, A., & Oelofse, S. (2021, May 18). Methodology to Investigate Indigenous Solid Waste Systems and Practices in the Rural Areas Surrounding Maseru (Kingdom of Lesotho). *International Journal of Environmental Research and Public Health, 18*(10).
(27) Sharma, B. K., Chandel, M. K. (2021, April 1). Life cycle cost analysis of municipal solid waste management scenarios for Mumbai, India. *Waste Management, 124*, 293–302.
(28) Rakib, M. R. J., De-la-Torre, G. E., Pizarro-Ortega, C. I., Dioses-Salinas, D. C., Al-Nahian, S. (2021, August). Personal protective equipment (PPE) pollution driven by the COVID-19 pandemic in Cox's Bazar, the longest natural beach in the world. *Marine Pollution Bulletin, 169*, 112497.
(29) Usman, M., Ho, Y. S. (2021, April). COVID-19 and the emerging research trends in environmental studies: A bibliometric evaluation. *Environmental Science Pollution Research International, 28*(14), 16913–16924.
(30) Varjani, S., Shah, A. V., Vyas, S., Srivastava, V. K. (2021, November). Processes and prospects on valorizing solid waste for the production of valuable products employing bio-routes: A systematic review. *Chemosphere*, *282*, 130954.
(31) Vinti, G., Bauza, V., Clasen, T., Medlicott, K., Tudor, T., Zurbrugg, C., et al. (2021, April 19). Municipal Solid waste management and adverse health outcomes: A systematic review. *International Journal of Environmental Research and Public Health, 18*(8).
(32) Overgaard, H. J., Dada, N., Lenhart, A., Stenstrom, T. A. B., & Alexander, N. (2021, August 1). Integrated disease management: Arboviral infections and waterborne diarrhoea. *Bulletin of World Health Organization, 99*(8), 583–592.
(33) Wang, Y., Levis, J. W., Barlaz, M. A. (2021, April 20). Development of streamlined life-cycle assessment for the solid waste management system. *Environmental Science Technology, 55*(8), 5475–5484.

(34) Choudhury, M., Jyethi, D. S., Dutta, J., Purkayastha, S. P., Deb, D., Das, R., . . . Bhattacharyya, K. G. (2022). Investigation of groundwater and soil quality near to a municipal waste disposal site in Silchar, Assam, India. *International Journal of Energy and Water Resources, 6*(1), 37–47. https://link.springer.com/article/10.1007/s42108-021-00117-5

(35) Paul, S., Choudhury, M., Deb, U., Pegu, R., Das, S., & Bhattacharya, S. S. (2019). Assessing the ecological impacts of ageing on hazard potential of solid waste landfills: A green approach through vermitechnology. *Journal of Cleaner Production, 236*, 117643. https://doi.org/10.1016/j.jclepro.2019.117643.

(36) Warmadewanthi, I., Wulandari, D., Cahyadi, M. N., Pandebesie, E. S., Anityasari, M., Dwipayanti, N. M. U., et al. (2021, August). Socio-economic impacts of the COVID-19 pandemic on waste bank closed-loop system in Surabaya, Indonesia. *Waste Management Research, 39*(8), 1039–1047.

(37) Xia, W., Jiang, Y., Chen, X., & Zhao, R. (2021, July 16). Application of machine learning algorithms in municipal solid waste management: A mini review. *Waste Management Research*, 734242X211033716.

(38) Yong, Z. J., Bashir, M. J. K., Hassan, M. S. (2021, July 1). Biogas and biofertilizer production from organic fraction municipal solid waste for sustainable circular economy and environmental protection in Malaysia. *Science Total Environment, 776*, 145961.

(39) Yousefi, M., Oskoei, V., Jonidi, J. A., Farzadkia, M., Hasham, F. M., Abdollahinejad, B., et al. (2021, May 3). Municipal solid waste management during COVID-19 pandemic: Effects and repercussions. *Environmental Science Pollution Research International, 28*, 32200–32209.

(40) Zainal, S. F. F. S., Aziz, H. A., Omar, F. M., Alazaiza, M. Y. D. (2021, July 10). Influence of Jatropha curcas seeds as a natural flocculant on reducing Tin (IV) tetrachloride in the treatment of concentrated stabilised landfill leachate. *Chemosphere, 285*, 131484.

(41) Zhang, J., Qin, Q., Li, G., Tseng, C. H. (2021, June 1). Sustainable municipal waste management strategies through life cycle assessment method: A review. *Journal of Environmental Management, 287*, 112238.

(42) Zhang, L., Zhao, T., Shi, E., Zhang, Z., Zhang, Y., & Chen, Y. (2021, April). The non-negligibility of greenhouse gas emission from a combined pre-composting and vermicomposting system with maize stover and cow dung. *Environmental Science Pollution Research International, 28*(15), 19412–19423.

(43) Zhong, H., Yang, S., Zhu, L., Liu, C., Zhang, Y., & Zhang, Y. (2021, April). Effect of microplastics in sludge impacts on the vermicomposting. *Bioresource Technology, 326*, 124777.

(44) Yuvaraj, A., Thangaraj, R., Ravindran, B., Chang, S. W., & Karmegam, N. (2021, January 1). Centrality of cattle solid wastes in vermicomposting technology—A cleaner resource recovery and biowaste recycling option for agricultural and environmental sustainability. *Environmental Pollution, 268*(Pt A), 115688.

(45) Sohal, B., Singh, S., Singh, S. I. K., Bhat, S. A., Kaur, J., Singh, J., et al. (2021, September). Comparing the nutrient changes, heavy metals, and genotoxicity assessment before and after vermicomposting of thermal fly ash using Eisenia Fetida. *Environmental Science Pollution Research International, 28*(35), 48154–48170.

(46) Rini, J., Deepthi, M. P., Saminathan, K., Narendhirakannan, R. T., Karmegam, N., & Kathireswari, P. (2020, October). Nutrient recovery and vermicompost production from livestock solid wastes with epigeic earthworms. *Bioresoure Technology, 313*, 123690.

(47) Zziwa, A., Jjagwe, J., Kizito, S., Kabenge, I., Komakech, A. J., & Kayondo, H. (2021, February 1). Nutrient recovery from pineapple waste through controlled batch and continuous vermicomposting systems. *Journal of Environmental Management, 279*, 111784.

(48) Zhong, H., Liu, X., Zhu, L., Yang, Y., Yan, S., & Zhang, X. (2019, October). Bioelectrochemically-assisted vermibiofilter process enhancing stabilization of sewage sludge with synchronous electricity generation. *Bioresoure Technology, 289*, 121740.

(49) Bavaghar, Z. M., Abbas, R. A. (2021, April 15). Designing an integrated municipal solid waste management system using a fuzzy chance-constrained programming model considering economic and environmental aspects under uncertainty. *Waste Management*, *125*, 268–279.
(50) Bertanza, G., Mazzotti, S., Gomez, F. H., Nenci, M., Vaccari, M., Zetera, S. F. (2021, May 1). Implementation of circular economy in the management of municipal solid waste in an Italian medium-sized city: A 30-years lasting history. *Waste Management*, *126*, 821–831.

5 Solid Wastes
Basic Knowledge on Sources and Types for Better Management

Tanmoy Jana, Subhrajeet Sahoo, Karthikeyan Ramesh, Shyamapada Ghosh, Vimala Raghavan, Rehab A. Rayan, Abhisek Nalluri, Preetam Bhardwaj, and Siva Sankar Sana

5.1 INTRODUCTION

Waste is an obvious by-product of anthropogenic activity. As the countries are hurtling towards urbanization, more and more by-products of an urban lifestyle, which is known as waste, are generated. According to Pullen et al., even the rate of producing waste materials has overcome the rate of urbanization. The glossary of the National Waste Report 2010 includes the definition for waste: "Any discarded, unwanted, redundant, neglected or excess material envisioned intended for re-processing, reuse, reclaim, recuperation, or cleansing by a detach procedure commencing that which created the material, or for trade, whether of whichever value" (Pullen et al., 2013). Distinct types of waste material are generated from variable sources. Most of the waste materials of total waste are solid. These types of waste are also known as solid waste, which achieved more attention. Solid desecrate is discussed to as several rebuff or rubbish, mire from a water supply healing lodge, wastewater healing plant, or else any air effluence organizes ability and former rejected matter, comprise of hard, semi-solid, fluidic, or restrain gaseous substances resultant from industrial, manufacturing, mining, farming process and from society actions (Gyimah et al., 2019). Misra and Panday provides the definition of waste material in 2005 as "*a material turns out to be waste once it is discarded devoid of hoping to be remunerated for its essential worth*" (Misra & Pandey, 2005). In the urbanization of the countries such as India, heaps of rubbish are common sights. Improper disposal of waste contaminates landscapes, air, pollutes our water sources like wells, lakes, rivers, and ultimately, affect the health of the population living nearby the polluted area (Fuss et al., 2018; Sharma et al., 2018; Liu et al., 2017; Sudibyo et al., 2017; Shams et al., 2017; Simatele et al., 2017; Paul et al., 2019). Admittance to a sanitation facility and hygienically sufficient water is consequently deemed as very important to the health and well-being of people (Prüss-Ustün et al., 2019). The inveterate crisis of waste administration

 DOI: 10.1201/9781003258377-5

increases with enlarged industrialization in towns. Economic and technological advancement is the main reason behind the diversity of solid waste and the complexity of their management. The multifaceted character of infection epidemics—that is, in the current time period, the increased number of cases of cholera and erstwhile diarrhoeal diseases—authenticate this. The varying financial drifts and quick urbanization obscure solid waste managing (SWM) in embryonic countries. Nowadays, solid desecrate is not only swelling in combination but also altering in amount from a little kilogram to tonnage quantity (Paes et al., 2020). Indecorous disposal is perceived through discarding of rebuff in sewers, strewing of streets, thereby making them pungent, or, in shaggy regions, is a source of eyesores, further significantly generating reproduction of grounds for pathogens and disease vectors (Colvero et al., 2020; Ko et al., 2020; Rajpal et al., 2022; Purkayastha et al., 2015). Deprived institutional structure, stumpy capability, and lack of assets; both human beings and resources have put sanitation circumstances and waste administration in many cities of the developing world in an extremely disgraceful condition (Gutberlet & Baeder, 2008; Choudhury et al., 2022). For the proper administration of waste, resources and the kind of solid waste must also be considered well with regard to disposal and organization of diverse types of waste in different ways (Yahalom et al., 2020).

5.2 REDUNDANCY WITH TYPES OF SOLID WASTES BY SOLID

Distinct landfill sites are widely used for the disposal of tons of garbage regularly. Such waste is generated through industries, workplaces, households, and many other farms. The fields of landfill emit pungent odours if garbage is not treated appropriately. It contaminates nearby air, which adversely affects the health, ecosystems, and the entire environment. The key sources of solid waste are:

5.2.1 Residential

Residential buildings are amongst the key contributors of inorganic wastes and organic wastes like plastic products, food scraps, fabrics, bottles, pulp, cardboard, asbestos, chemicals, and hazardous waste appliances, batteries, rubber, old bed frames, discarded oil, etc. Many households contain dumpsters in which they can discard their garbage, and afterwards, the dumpster is cleaned for processing by a sanitation worker or company or treatment plant (Yahalom et al., 2020; Choudhury & Choudhury, 2014; Vu et al., 2020).

5.2.2 Industrial

Factories are recognized as solid waste major suppliers. These are comprised of light and heavy manufacturing, industrial plants, building sites, food processing plants, fertilizer, and power stations. These industries produce solid sludge of wasted food, domestic sewage, packaging waste, ashes, building materials, hazardous wastes, sewage sludge, etc. (Papapetrou et al., 2018).

5.2.3 Commercial

Business infrastructures (such as malls, resorts, marts, supermarkets, cafes, and apartment complexes) are becoming further responsible sources of solid waste. Garbage, vegetables, wood, chemicals, wood, bottles, special waste, packaging products, and other toxic waste are produced from most of these amenities (Phu et al., 2019).

5.2.4 Institutions

Educational institutions such as schools, colleges, military barracks, universities, jails, as well as many other public institutions also generate garbage. A few common sources of household waste are latex wastes, bottles, food scraps, liquids, lumber, cardboard, ceramic materials, commodities, electrical, and other toxic waste (Hosetti, 2006; Duygan et al., 2021).

5.2.5 Destruction and Construction Regions

Destruction and building sites often generate waste products. Building projects comprise major highway construction locations and infrastructure construction, road maintenance or reconstruction sites. Some waste materials created here includes concrete, metal, iron, screws, rubber, coaxial cables, latex, polymers, dust soil, and ceramic waste (Lu & Yuan, 2012).

5.2.6 Municipal Services

Metropolitan hubs are also contributing to the solid waste problem in several nations today. Most solid pollutants emitted by municipalities include garbage pickup, shore and garden waste, site waste, sewage systems, and waste from sludge treatment plants (Jouhara et al., 2017).

5.2.7 Treatment Centres and Sites

Light and heavy treatment centres such as sugar refineries, processing plants, hydroelectric dams, marble mining plants, and composite or alloys manufacturing sites are also major sources of unwanted waste or garbage products such as electrical components, plastics, ceramics, alcoholic products, etc. (Sana et al., 2020; D'Adamo et al., 2019; Zhang & Klenosky, 2016).

5.2.8 Agriculture

Farming lands, dairy-products processing-units, crop fields, distilleries, and slaughterhouses have become also cause and source of hazardous waste. They are responsible for the production of rotten food, pesticide containers, farm waste, and other treacherous materials (Bhat et al., 2020; Zou et al., 2018).

5.2.9 Biomedical

This applies to hospital, biomedical appliances, and pharmaceutical companies. Hospitals generate a variety of solid waste such as bandages, syringes, used gloves, paper, medications, food waste, plastics, and chemicals. All these wastes would also require careful disposal, creating severe environmental issues along with the people in these facilities (Patil & Bohara, 2020; Rahman et al., 2020; Choudhury et al., 2022).

5.2.10 Rubbish

Rubbish is comprised of flammable and inflammable hard wastes coming from institutions, domestic, and industrial activities. This eradicates food desecrate and other highly rescinded materials. In general, flammable rubbish is comprised of hard objects such as cardboard, tabloid, fabric, equipment's, plastics, fleece, rubber, timber, and garden trimmings. Inflammable rubbish is comprised of metallic cans such as iron containers, aluminium canisters, tin containers, non-ferrous metals, etc. (Martinez & Beilmann, 2020; Tvedten & Candiracci, 2018).

5.2.11 Trash

Trash is described as waste that would generally be in housing or domestic refuse comprised of brush and dissipate from small domestic preservations that is ignitable. Mostly debarred, it is naturally poisonous, unsafe, and harmful wastes which shall be selected as "harmful wastes" by country and/or federal dictatorial authority, comprising suitable command, as well as important liquor waste or any farming waste, exhumed ground, slabs, pebbles, wreckage concrete, and waste components occurring by foremost devastation, fitting equipment and maintenance, bathrooms, sink, fathom parts, bath containers, vehicles or wagon parts, apparatus, car batteries, tree wood and boughs higher than 6 inches in length, as well as tree baffle (Russo et al., 2019; Pollans, 2017).

5.2.12 Food Waste

Food wastes are several unprocessed or roasted grocery materials, which are spare. Food wastes are the raw or natural residues provoked by the managing, organizing, storage, trading, and food preparation and dishing up of food (McClellan & Halden, 2010). Food waste is comprised of additional, unconsumed shares of meals and clippings from food-organizing commotion in kitchens, canteens, and cafés (Pollans, 2017; Schanes et al., 2018).

5.2.13 Yard Waste

Yard waste is hard vegetative dissipate resultant from courtyard preservation such as sniping and plummeting of plant ingredients such as tree branches, brush, and foliage, and similarly, vegetative objects that have a carbon and nitrogen proportion superior to 75 (Sharma & Dubey, 2020; Lee et al., 2019; Panigrahi & Dubey, 2019).

5.2.14 Ashes and Residues

MSW ignitions engender solid wastes describe as ash, which can include several of the ingredients which were previously existing in the waste. MSW authorities' plants diminish the prerequisite of landfill capability, as the discarding of MSW ash requires fewer territorial regions than does unprocessed MSW (Gomes et al., 2020; Leng et al., 2019).

5.2.15 Special Waste

Special waste is desecrated matter, since its chemical constituents, physical description, or natural characters necessitate either an extraordinary administration process or authorize or present an infrequent hazard to human well-being, possessions, apparatus, and the atmosphere (de Figueiredo, 2018; Pacheco et al., 2018; Cossu et al., 2012).

5.2.16 Hazardous Waste

Harmful wastes intimidate diverse groups of people, introducing waste creators, landfill employees, collectors, waste pickers, and nearby residents at risk. The leachate from a landfill may be hazardous as well; its level of toxicity is directly related to the toxicity and quantity of hazardous materials mixed with other solid waste. The variety and classes of materials and sources from households to medical and industrial facilities makes this particularly thought provoking in these materials (Fazzo et al., 2017; Torkashvand et al., 2020).

5.3 TAXONOMY OF SOLID WASTE

Hard waste may be characterized in different approaches, depending upon diverse factors. Source and vulnerability potentially are the most important amongst them. According to sources or causes, solid waste is categorized into numerous types. These are business-related, inhabited, organization, manufacturing, building or devastation, metropolitan services, procedure and farming waste.

5.3.1 Residential Waste

Housing wastes are several waste matters produced from housing actions. In easy words, these are the rejected commodities of a domestic place or household. Residential waste is also recognized as the household or domestic waste. This is engendered from the regular, everyday actions of household utilization (Yahalom et al., 2020; Vu et al., 2020).

- Waste producer: solo or multiple habitation, public house, and motels, picnic grounds, etc.
- Example of wastes: food dissipate, cardboard, synthetic, tumbler materials, leather, metals, enclosure waste, timber, fabric, cloth, ash, etc. It also includes unusual types of waste consisting of electronics, battery, tires, lubricant, etc., and domestically perilous waste.

5.3.2 Industrial Waste

Industrialized wastes are the dissipate commodities of diverse industries. In a few countries, because of the building materials of different industries, a considerable quantity of waste commodities is formed. Industrial waste production and also the conformation of waste fluctuate, dependent on the nature and characteristic of manufacturing units and the equipment or procedures utilized in those production units. Not only hard waste but waste from the manufacturing units—a massive quantity of compounds—pollute (in fluid form) and gases (source of air contamination) also pollute. Some perilous wastes are also produced in production or manufacturing units (Papapetrou et al., 2018; Woolley et al., 2018).

- Waste generators: light and heavy manufacturing industries, food industries, fabric industries, power and chemical plants, etc.
- Example of wastes: food waste, dirt, and gravel, scrap metals, rubbish, oil, solvents, chemicals, etc.

5.3.3 Commercial Waste

Commercial wastes are the waste materials generated from commercial activity. This type of waste is generated as a result of business sector activities. Commercial waste arises from the activities of wholesalers, shops, stores, markets, offices, etc. It is a type of industrial waste (Phu et al., 2019).

- Waste generators: commercial waste is generated as a result of such activities as manufacturing or industrial process, wholesale or retail trading, fishing, mining, different business activities including administrative services (Hupponen et al., 2018).
- Example of wastes: paper, cardboard, plastic, glass, cans, ash, etc.

5.3.4 Institutional Waste

Institutional wastes are the solid wastes generated by institutional enterprises such as social, educational, government services, and others. Institutional waste is similar in nature to household waste. In different countries, there are such many institutions. From those places, small amounts of waste are generated, as compared to others (Hosetti, 2006; Duygan et al., 2021).

- Waste generators: schools, prisons, government sectors, etc.
- Example of wastes: waste products generated from institutional enterprises are similar to the waste products of commercial waste.

5.3.5 Building and Destruction Waste

Building and destruction waste is produced from building or destruction actions. In countries generally, developmental activities consist of construction and demolition. Such activity produces mostly inert and non-bio-degradable materials. This waste is a noteworthy constituent of the entire solid waste measures. It is not considered to be a part of municipal solid waste (Lu et al., 2017).

- Waste generators: any kind of construction or demolition activities of buildings, roads, bridges, flyovers, subway, etc.
- Example of wastes: such waste contains mainly static and non-eco-friendly objects such as concrete, swathe, metal, synthetic, gypsum, etc. Building dissipate may include lead, asbestos, or other such perilous materials.

5.3.6 Municipal-Service Waste

Municipal waste is the waste materials collected by the municipalities of that area or by the local authorities. In most countries, most areas are under the municipality. Due to the standard lifestyle, an enormous amount of waste is produced in the municipality area (Choudhury & Dutta, 2017). Waste of these types is the same as of the residential wastes (Kumar & Samadder, 2017).

5.3.7 Agricultural Waste

Agricultural waste is the waste materials produced by agricultural activities. In such countries as India, many people live their lives by depending on agriculture. As a result, enormous amounts of such waste products are generated. It includes waste materials from farms, poultry houses, cultivated land, etc. The quantity of agricultural waste is significantly low compared with other waste products of different sources. Agricultural wastes can be both natural (organic) and non-natural wastes. Throughout many developing and developed countries, large quantities of food and crop processing, forestry, and animal waste are generated each year. However, most of these wastes are biodegradable (Bhat et al., 2020; Zou et al., 2018).

- Waste generators: cultivated land, poultry farms, Cote, etc.
- Example of wastes: these wastes are of two types—
    - Natural waste—food waste, rice straw, manure, animal excreta, animal carcasses, etc.
    - Non-natural waste—pesticide, insecticide, discarded pesticide containers, plastics, bags and sheets, packaging waste, etc.

5.3.8 Biomedical Waste

Biomedical, or hospital, waste is the waste materials which are infectious (or potentially infectious) in nature and is generated from hospitals and such places incorporated with medical activities. These are generated from medical and biological sources and activities such as the treatment of disease, diagnosis, etc. Also, it includes the waste of research laboratories. Research laboratory waste contains organisms or bio-molecules that are restricted from environmental release. Radioactive materials and hazardous chemicals are also generated from medical facilities (Patil & Bohara, 2020; Rahman et al., 2020).

- Waste generators: hospitals, health clinics, nursing homes, emergency medical services, medical research laboratories, etc.
- Example of wastes: discarded blood, pus, body fluids, human excreta, microbiological culture, used bandages, dressings, discarded gloves, discarded medicines, etc.

5.3.9 Municipal Solid Waste

Municipal solid waste, or MSW, is generally defined as the non-hazardous solid waste materials which are collected from a city, town, or village that requires routine collection and proper management by the municipality. MSW is also known as trash or rubbish. It consists of household waste (private home), commercial establishments,

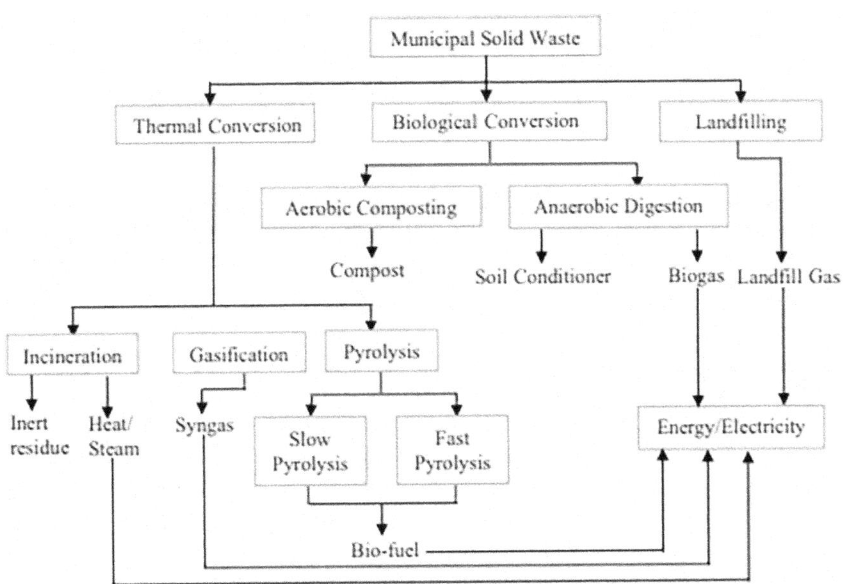

FIGURE 5.1 Municipal solid waste treatment techniques and their products [adapted with permission from (Malinauskaite et al., 2017), (2017, Elsevier)].

TABLE 5.1
Typical reaction conditions and products from thermal treatment processes [adapted with permission from (Malinauskaite et al., 2017), (2017, Elsevier)]

Parameters		Incineration	Pyrolysis	Gasification
Principle		Full oxidative combustion	Thermal degradation of organic material in the absence of oxygen	Partial oxidation
Operating temperature ('C)		850–1200	400–800	800–1600
Atmosphere		Presence of sufficient oxygen	Absence of oxygen	Controlled supply of oxygen
Reaction products	Solid	Bottom ash, fly ash, slag, other non-combustible	Ash. char (combination of non-combustible and carbon)	Ash, stag
	Liquid	subsumes like metals and glass	Condensate of pyrolysis gas (pyrolysis od, wax, tar)	
	Gas	CO_2, H_2O, O_2, N_2,	Pyrolysis gas (H_2, CO. hydrocarbons. H_2O, N_2)	Syngas (H_2, CO, CO_2, CH_2, H_2O, N_2)
Pre-treatment		Not necessary	Required	Required
Raw MSW		Usually preferred	Usually not preferred	Usually not preferred

institutional waste, and waste from streets, although street wastes are generated from commercial and residential activities. Generally, MSW and domestic waste are synonyms. However, MSW does not include waste materials from the construction, industry, and sewage sludge, demolition, or agricultural waste. In many countries, especially in urban areas, industrial solid wastes are managed as a detailed stream. Actually, the waste generated from different sources, which can be included in MSW, varies from country to country (Choudhury & Choudhury, 2014; Jouhara et al., 2017; Malinauskaite et al., 2017).

5.3.10 Hazard Potential

Waste materials in this category are commonly known as hazardous waste. The Hazardous Waste Act 1989 (Regulation of Exports and Imports) covers only hazardous waste. This type of waste possesses an improved risk to human health along with the environment. Hazardous waste materials show the following characteristics: toxicity, reactivity, ignitability, and corrosivity. Hazardous waste materials are of different types. They can be toxic, infectious, and radioactive materials. Waste materials (not all) from different sources are hazardous in nature. All of these do not have the same characteristics. Some of them are toxic, some are radioactive, and others are infectious (Ilankoon et al., 2018; Saeidi-Mobarakeh et al., 2020).

5.3.11 Poisonous or Toxic Wastes

Poisonous wastes are venomous in trace quantity. Several may have an immediate or discriminating consequence on humans as well as animals. Mutagenic or carcinogenic wastes instigate genetic transformations in the children of exposed people and animals. Examples: weighty metals, pesticides (Krupskaya et al., 2019; Ndanguza et al., 2020; Marfe & Di Stefano, 2016).

5.3.12 Reactive Wastes

This type of waste has a propensity to respond vigorously to water and atmosphere, is unhinged to heat or shock, produces poisonous gases, or disintegrates during routine administration. Examples: nitro-glycerine, gun powder (Luo et al., 2020; Wilson et al., 2018).

5.3.13 Ignitable Waste

Ignitable wastes are those which burn at moderately low temperatures (< 60 °C) and are capable of natural ignition during transport or disposal. Examples: paint thinners, petrol, and alcohol (Patil & Bohara, 2020).

5.3.14 Corrosive wastes

Corrosive wastes destroy living tissues by chemical reactions. Examples: different acids and bases (Marzorati et al., 2019).

5.3.15 Infectious Wastes

This type wastes include used bandages, human tissue from surgery, and hypodermic needles found in hospital wastes (Diaz et al., 2008).

5.4 SOLID WASTE MANAGEMENT

Waste creation is an inescapable characteristic of living; it becomes a subject of critical inevitability to scrutinize the procedure from production to dumping in direct ways, to control the region of significant concern, and obstruct them in command to evade toxic waste in huge amounts and also offer superior healthiness. Solid Waste Management (SWM) is defined as the organization, production, storage, gathering, transportation and relocation, dispensation, and dumping of hard waste by trustworthy methods with best practices of public health as well as financial and economic, legal, administrative, and environmental considerations. SWM can be broken into five main elements: production, storage, gathering, transport, and dumping.

5.4.1 Production

Production of solid waste is the step when materials turn insignificant as well as worthless, and they are necessitated no longer. Substances which are rubbish to one individual may be precious to another. For example, waste objects like tins and containers may be thought extremely useful by children (Sokka et al., 2007).

5.4.2 Storage

Storage is a method for carrying on unwanted materials, proceeding to gathering and concluding with dumping. Where on-site dumping systems are functional, people throw away items unswervingly into family pits, the storage which is dispensable. In case of crisis, mostly in premature stages, it is expected that the ostentatious residents will lob away domestic waste in inadequately defined stacks close to residence areas. In this casing, discarding storage amenities has to be afforded reasonably, and these should be positioned near people where they can utilize them without difficulty. Improved storage accommodations include: large containers—i.e., oil drums, communal bins, etc.; small containers—i.e., plastic bins, household containers, etc.; shallow pits (Erfani et al., 2018).

5.4.3 Collection

Waste is gathered in containers for shipping purposes and is taken to the ending dumping location. Any gathering organization should be wisely designed to make sure that storage amenities are not over-encumbered. Collection volumes and intervals of waste gathering should be approximated carefully (Das & Bhattacharyya, 2015).

5.4.4 Transportation

In this phase, hard waste is transferred to the finishing dumping location. There are numerous transportation modes that may be implemented, and the chosen method depends upon local availability as well as the volume of waste to be transported (Putri et al., 2019). Types of transportation are divided into three classes:

- Animal-powered: cart pulled by donkey.
- Human-powered: open hand-cart, hand-cart with bins, tricycle, and wheelbarrow.
- Motor: standard wagon, tractor and clip, tipper-truck.

5.4.5 Disposal

This is the finishing phase of solid waste administration, where desecrate is predisposed with least risks (Singh, 2019; Wang et al., 2018). There are four foremost techniques for solid waste dumping:

- Land application: land-filling or burial, composting, incineration or burning, reprocessing.

The most common of these is indisputably land function, even though all four are frequently functional in crisis cases. On-site and off-site discarding are illustrated later.

5.5 ON-SITE DISPOSAL OPTIONS

The technology choices described in the following sections are common guidelines for on-site disposal and storage of waste; these may be adapted for each particular site and situation (Eregno & Heistad, 2019).

5.5.1 Communal Pit-Disposal

It is the simplest solid waste management system, where consumers dispose of the waste in a communal pit directly. The size of this pit will depend upon the quantity of people it serves. A 6-cubic-meter communal pit is required per 50 people. The pit should be fenced off to prevent small children from falling in, and distance should not be more than 100 m from the dwellings to be served. Ideally, the waste should be covered at least weekly with a thin soil layer to minimize the breeding of flies and other pests (Rimélé et al., 2018).

- Advantages: It requires very little action and maintenance.
- Disadvantages: The distance to the pit may cause indiscriminate disposal; and regularity of waste workers is required to manage the pits.

5.5.2 Family Pit-Disposal

Family pits provide a better long-term option where there is enough space. These should be shallow—up to 1 m deep. The family pits should be fortified to daily cover the waste with soil from sweeping or ash which comes from fires used for cooking. This method is best suited where the families have large plots and organic food wastes are the major constituent of domestic refuse (Rimélé et al., 2018; Lu & Cai, 2012).

- Advantages: Here, families are responsible for managing their own waste (no external waste workers are required).
- Disadvantages: Involves significant community mobilization for the construction, processing, and maintenance of pits; and reasonable space is very much needed.

5.5.3 Communal Bins

Communal bins, or containers, are planned to collect waste where wastes will not be spread by wind or animals and where it can easily be removed for transport to the targeted place for disposal. Plastic containers are inappropriate, as they can be blown over by the wind and easily be removed, and they may be desirable for alternate uses. A widely accepted solution is to provide oil drums cut in half. The bases of the drums should be perforated to allow the passing out of liquid and to prevent their use for other purposes. A lid and handles can be used for its betterment (Zen & Siwar, 2015).

Generally, a 100-liter bin should be provided for every 100 people at feeding centres, every 50 people in domestic areas, and every 10 market stalls, and the bins should be emptied daily.

- Advantages: It is a highly hygienic management method; and final disposal of waste is well away from the locality.
- Disadvantages: Appropriate gathering and transport are needed; the method requires time to execute, and proficient administration is very much required.

5.5.4 Family Bins

Family bins are utilized in crisis circumstances, as they require a meticulous compilation and shipping arrangement, and there is an enormous number of dustbins or containers necessary. Later on, society members can be optimistic about creating their individual rebuff creel or containers and to seize responsibility to vacant those in common pits (Khale & Worku, 2013; Metcalfe et al., 2012).

- Advantages: Families are accountable for preserving and compiling pits; and potentially, a helpful waste administration process.
- Disadvantages: Here, a large number of bins are needed; proper collection, transportation, and human resources are required; it takes time to implement, and well-organized management is vital.

5.5.5 Common Dumping without Bins

For various community organizations, marketplaces, or allocation centers, solid waste administration methods exclusive of bins or baskets can be utilized, whereby consumers dispose the desecrate unswervingly on the land. This is able to simplify effort if cleaners are working to remove recurrently around market booths, congregate the waste, and transfer it to a separate, off- site discarding location. This is probably appropriate for vegetable waste, but slaughterhouse waste should be disposed of in liquid-tight containers and buried at a different place (Duygan et al., 2021; Tibu et al., 2019).

- Advantages: Method is quick to apply and execute; there is negligible dependence on abuser action, and it can be in queue with every day exercise.
- Disadvantages: Efficient and effective management is needed, and full-time waste workers are required.

5.6 OFF-SITE DUMPING PREFERENCES

The equipment options planned underneath—that is the overall option for the final disposal of waste off-site.

5.6.1 Land-Filling

The most widely accepted off-site disposal process is land-filling. Once the solid desecrate is disposed off-site, it is generally engaged to a landfill site, where solid desecrate is positioned in a big ditch or dugout in the land, which is back-crammed with exhumed soil every day desecrate is poured. Preferably, about 0.5 m of soil must wrap the deposit rebuff to prevent animals from excavating the waste and flies, along with

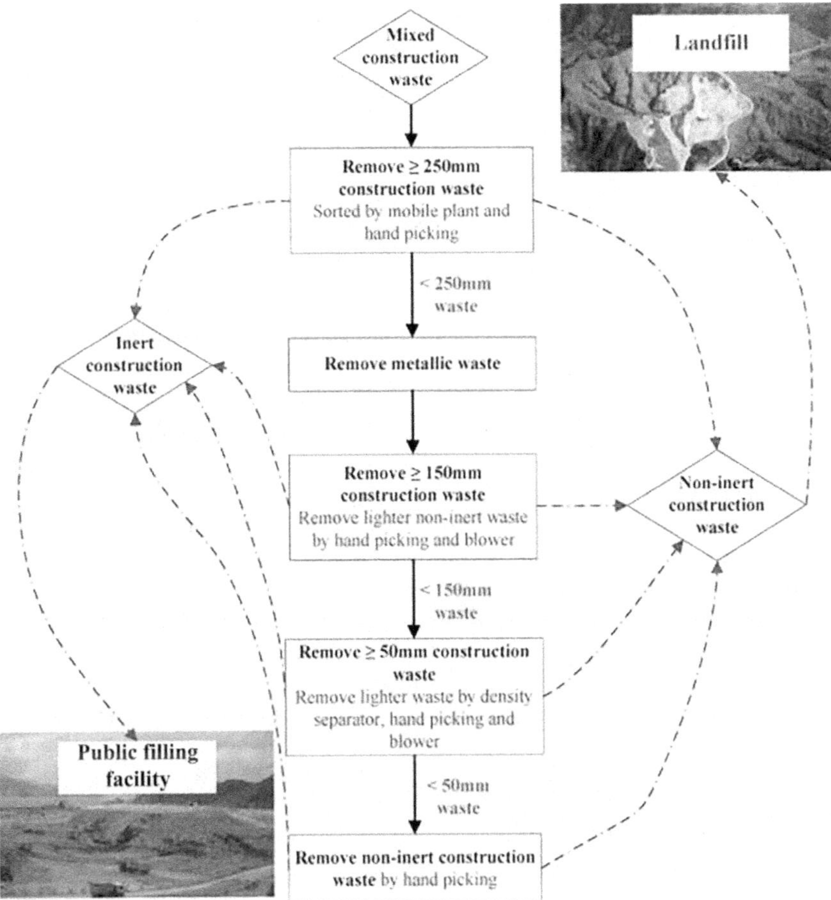

FIGURE 5.2 Flowchart of off-site construction waste sorting in Hong Kong, China [adapted with permission from (Lu & Yuan, 2012), (2012, Elsevier)].

the erstwhile disease-producing vector from breeding. The landfill site's location is determined through the meeting with the local establishment as well as the neighboring people. The sites are properly enclosures, and at slightest, 1 kilometer downwind from the nearest locality (Lee et al., 2017; Kamaruddin et al., 2017; Paul et al., 2019).

- Advantages: It is a vigorous dumping technique if it is efficiently administered.
- Disadvantages: An enormous region is required for land filling.

5.6.2 Incineration

Incineration or burning is frequently utilized for flammable waste removal; this must be consigned off-site or at a rational distance from the locality. Blazing wastes within the residential region or in the locality may produce noteworthy flame exposure and smoke. Smouldering might be utilized to diminish the waste quantity and might be helpful, where space is inadequate, in favour of landfill or burial. Waste must be burnt up within pits that are the equivalent method to land-filling. Similar shortcomings as sitting landfill sites would pertain here also (Makarichi et al., 2018; Dong et al., 2018).

- Advantages: Huge amount of flammable waste is condensed significantly throughout blazing, and it is appropriate in offsite depths to shrink scavenge.
- Disadvantages: There will be bonfire vulnerability.

5.6.3 Composting

Composting of vegetable and diverse natural wastes can be pertaining under numerous circumstances. As individuals contain their personal vegetable plots or gardens, natural waste can be excavated from the earth to adjoin fibre and humus. These compose the waste securely and also support in the rising progression. This should be fortified anywhere potentially, predominantly in the concluding phase of an urgent situation or programme (Awasthi et al., 2020; Akdeniz, 2019). Proper management of composting requires watchful monitoring for waste decomposition to organize chemical and humidity level and encourage movement of microbes. This is intended to produce fertilizer which is very much secure to use as well as well manure. Significant knowledge as well as experience is very much needed for such systems. Generally, they are not suitable in emergency cases (Guo et al., 2019; Chen et al., 2020).

- Advantages: It is beneficial for crops and environment friendly also.
- Disadvantages: Exhaustive administration and skilled workers should be mandatory for major procedures.

5.6.4 Reprocess and Recycle

Reprocessing and recycling methods are questionable to appropriate, although the reprocessing of various waste substances may be achievable intermittently. Plastic luggage, tins, boxes, and glass will frequently be repeatedly recycled because they are probably remarkable products in several circumstances. The strong tradition of reprocessing of waste was primarily to diminish severe quantities of waste in the majority of developing countries (Nomura et al., 2020; Eriksen et al., 2018).

- Advantages: Recycling is environment friendly.
- Disadvantages: There is limited probability; and it is very expensive.

5.7 CONSEQUENCES OF DEPRIVED SOLID WASTE ORGANIZATION

Because of inappropriate waste dumping schemes predominantly by MSW groups, ravages pile up and develop into a serious crisis. Persons wash their houses and place their debris all around, which affects the atmosphere as well as the neighborhood. Due to this kind of discarding of ravage materials, eco-friendly resources decay and crumble under unhygienic, inappropriate, and abandoned circumstances. After decomposition, an awful odour is created, and it turns into a breeding ground for many infectious organisms and disease-causing insects (Sharma, 2018). During the solid waste-collection process, perilous wastes frequently blend with the usual refuse alongside other combustible waste; that is why this dumping procedure is even more dangerous and harder. When harmful dissipate from batteries include lead (Pb), pesticides, cleaning solvents, zinc (Zn) or mercury (Hg), radioactive resources, and electronic dissipate as well as plastics blended up among papers and former scraps that burn up generate carbon dioxins and harmful gasses. These gases are very much toxic and have the potential to cause several diseases, including eye problems, breathing problem, and cancer (Sana et al., 2020).

5.8 CONCLUSION AND FUTURE PROSPECTS

Over the forthcoming 10–15 years, regions of different countries have to face waste management challenges. We know that the region's capacity for proper disposal or management of waste is limited. So, waste management requires efficient and effective strategies and long-term planning of waste disposal. Most countries face some issues regarding waste management—lack of public awareness, proper planning, and strategies, economic problems, insufficient government priority, etc.

REFERENCES

Akdeniz, N. (2019). A systematic review of biochar use in animal waste composting. *Waste Management, 88*, 291–300. https://doi.org/10.1016/j.wasman.2019.03.054

Awasthi, S. K., Sarsaiya, S., Awasthi, M. K., Liu, T., Zhao, J., Kumar, S., & Zhang, Z. (2020). Changes in global trends in food waste composting: Research challenges and opportunities. *Bioresource Technology, 299*, 122555. https://doi.org/10.1016/j.biortech.2019.122555

Bhat, V. S., Kanagavalli, P., Sriram, G., John, N. S., Veerapandian, M., Kurkuri, M., & Hegde, G. (2020). Low cost, catalyst free, high-performance supercapacitors based on porous nano carbon derived from agriculture waste. *Journal of Energy Storage, 32*, 101829. https://doi.org/10.1016/j.est.2020.101829

Chen, T., Zhang, S., & Yuan, Z. (2020). Adoption of solid organic waste composting products: A critical review. *Journal of Cleaner Production, 272*, 122712. https://doi.org/10.1016/j.jclepro.2020.122712

Choudhury, M., & Choudhury, M. (2014). Trends of Urban Solid Waste Management in Agartala City, Tripura, India. *Universal Journal of Environmental Research & Technology*, *4*(4).

Choudhury, M., & Dutta, J. (2017). A comparative study of municipal solid waste management status for three major towns of Upper Assam-India. *International Journal of Waste Resources*, *7*(291), 2.

Choudhury, M., Jyethi, D. S., Dutta, J., Purkayastha, S. P., Deb, D., Das, R., . . . Bhattacharyya, K. G. (2022). Investigation of groundwater and soil quality near to a municipal waste disposal site in Silchar, Assam, India. *International Journal of Energy and Water Resources, 6*(1), 37–47. https://link.springer.com/article/10.1007/s42108-021-00117-5

Colvero, D. A., Ramalho, J., Gomes, A. P. D., de Matos, M. A. A., & da Cruz Tarelho, L. A. (2020). Economic analysis of a shared municipal solid waste management facility in a metropolitan region. *Waste Management, 102*, 823–837. https://doi.org/10.1016/j.wasman.2019.11.033

Cossu, R., Lai, T., & Pivnenko, K. (2012). Waste washing pre-treatment of municipal and special waste. *Journal of Hazardous Materials, 207*, 65–72. https://doi.org/10.1016/j.jhazmat.2011.07.121

D'Adamo, I., Ferella, F., & Rosa, P. (2019). Wasted liquid crystal displays as a source of value for e-waste treatment centers: A techno-economic analysis. *Current Opinion in Green and Sustainable Chemistry, 19*, 37–44. https://doi.org/10.1016/j.cogsc.2019.05.002

Das, S., & Bhattacharyya, B. K. (2015). Optimization of municipal solid waste collection and transportation routes. *Waste Management, 43*, 9–18. https://doi.org/10.1016/j.wasman.2015.06.033.

de Figueiredo, G. M. P. (2018). Campuses in the Global South: Is sustainability possible without considering social and territorial dimensions? In *Towards green campus operations* (pp. 93–103). Springer.

Diaz, L. F., Eggerth, L. L., Enkhtsetseg, S. H., & Savage, G. M. (2008). Characteristics of healthcare wastes. *Waste Management, 28*(7), 1219–1226. https://doi.org/10.1016/j.wasman.2007.04.010

Dong, J., Tang, Y., Nzihou, A., Chi, Y., Weiss-Hortala, E., & Ni, M. (2018). Life cycle assessment of pyrolysis, gasification and incineration waste-to-energy technologies: Theoretical analysis and case study of commercial plants. *Science of the Total Environment, 626*, 744–753. https://doi.org/10.1016/j.scitotenv.2018.01.151

Duygan, M., Stauffacher, M., & Meylan, G. (2021). What constitutes agency? Determinants of actors' influence on formal institutions in Swiss waste management. *Technological Forecasting and Social Change, 162*, 120413. https://doi.org/10.1016/j.techfore.2020.120413

Eregno, F. E., & Heistad, A. (2019). On-site treated wastewater disposal systems–The role of stratified filter medias for reducing the risk of pollution. *Environment International, 124*, 302–311. https://doi.org/10.1016/j.envint.2019.01.008

Erfani, S. M. H., Danesh, S., Karrabi, S. M., Shad, R., & Nemati, S. (2018). Using applied operations research and geographical information systems to evaluate effective factors in storage service of municipal solid waste management systems. *Waste Management, 79*, 346–355. https://doi.org/10.1016/j.wasman.2018.08.003

Eriksen, M. K., Pivnenko, K., Olsson, M. E., & Astrup, T. F. (2018). Contamination in plastic recycling: Influence of metals on the quality of reprocessed plastic. *Waste Management, 79*, 595–606. https://doi.org/10.1016/j.wasman.2018.08.007

Fazzo, L., Minichilli, F., Santoro, M., Ceccarini, A., Della Seta, M., Bianchi, F., Comba, P., & Martuzzi, M. (2017). Hazardous waste and health impact: A systematic review of the scientific literature. *Environmental Health, 16*(1), 1–11. https://doi.org/10.1186/s12940-017-0311-8

Fuss, M., Barros, R. T. V., & Poganietz, W. R. (2018). Designing a framework for municipal solid waste management towards sustainability in emerging economy countries-An application to a case study in Belo Horizonte (Brazil). *Journal of Cleaner Production, 178*, 655–664.

Gomes, H. I., Funari, V., Dinelli, E., & Soavi, F. (2020). Enhanced electrodialytic bioleaching of fly ashes of municipal solid waste incineration for metal recovery. *Electrochimica Acta, 345*, 136188. https://doi.org/10.1016/j.electacta.2020.136188

Guo, X. X., Liu, H. T., & Wu, S. B. (2019). Humic substances developed during organic waste composting: Formation mechanisms, structural properties, and agronomic functions. *Science of the Total Environment, 662*, 501–510. https://doi.org/10.1016/j.scitotenv.2019.01.137

Gutberlet, J., & Baeder, A. M. (2008). Informal recycling and occupational health in Santo André, Brazil. *International Journal of Environmental Health Research, 18*(1), 1–15. https://doi.org/10.1080/09603120701844258.

Gyimah, P., Mariwah, S., Antwi, K. B., & Ansah-Mensah, K. (2019). Households' solid waste separation practices in the Cape Coast Metropolitan area, Ghana. *GeoJournal*, 1–17.

Hosetti, B. B. (2006). *Prospects and perspective of solid waste management*. New Age International.

Hupponen, M., Grönman, K., & Horttanainen, M. (2018). Areas on which to focus when seeking to reduce the greenhouse gas emissions of commercial waste management. A case study of a hypermarket, Finland. *Waste Management, 76*, 1–18. https://doi.org/10.1016/j.wasman.2018.03.024

Ilankoon, I. M. S. K., Ghorbani, Y., Chong, M. N., Herath, G., Moyo, T., & Petersen, J. (2018). E-waste in the international context–A review of trade flows, regulations, hazards, waste management strategies and technologies for value recovery. *Waste Management, 82*, 258–275. https://doi.org/10.1016/j.wasman.2018.10.018

Jouhara, H., Czajczyńska, D., Ghazal, H., Krzyżyńska, R., Anguilano, L., Reynolds, A. J., & Spencer, N. (2017). Municipal waste management systems for domestic use. *Energy, 139*, 485–506. https://doi.org/10.1016/j.energy.2017.07.162

Kamaruddin, M. A., Yusoff, M. S., Rui, L. M., Isa, A. M., Zawawi, M. H., & Alrozi, R. (2017). An overview of municipal solid waste management and landfill leachate treatment: Malaysia and Asian perspectives. *Environmental Science and Pollution Research, 24*(35), 26988–27020. https://doi.org/10.1007/s11356-017-0303-9

Khale, S., & Worku, Z. (2013). Factors that affect municipal service delivery in Gauteng and North West Provinces of South Africa. *African Journal of Science, Technology, Innovation and Development, 5*(1), 61–70. https://doi.org/10.1080/20421338.2013.782143

Ko, S., Kim, W., Shin, S. C., & Shin, J. (2020). The economic value of sustainable recycling and waste management policies: The case of a waste management crisis in South Korea. *Waste Management, 104*, 220–227. https://doi.org/10.1016/j.wasman.2020.01.020

Krupskaya, L. T., Mazur, E. A., & Filatova, M. Y. (2019, June). The technological solution for management of toxic wastes accumulated in the last century (on the example of the closed mining enterprise in the Primorsky Krai). In *IOP conference series: Earth and environmental science* (Vol. 272, No. 3, 032014). IOP Publishing. https://doi.org/10.1088/1755-1315/272/3/032014

Kumar, A., & Samadder, S. R. (2017). A review on technological options of waste to energy for effective management of municipal solid waste. *Waste Management, 69*, 407–422. https://doi.org/10.1016/j.wasman.2017.08.046

Lee, E., Bittencourt, P., Casimir, L., Jimenez, E., Wang, M., Zhang, Q., & Ergas, S. J. (2019). Biogas production from high solids anaerobic co-digestion of food waste, yard waste and waste activated sludge. *Waste Management, 95*, 432–439. https://doi.org/10.1016/j.wasman.2019.06.033

Lee, U., Han, J., & Wang, M. (2017). Evaluation of landfill gas emissions from municipal solid waste landfills for the life-cycle analysis of waste-to-energy pathways. *Journal of Cleaner Production, 166*, 335–342. https://doi.org/10.1016/j.jclepro.2017.08.016

Leng, L., Bogush, A. A., Roy, A., & Stegemann, J. A. (2019). Characterisation of ashes from waste biomass power plants and phosphorus recovery. *Science of the Total Environment, 690*, 573–583. https://doi.org/10.1016/j.scitotenv.2019.06.312

Liu, Y., Xing, P., & Liu, J. (2017). Environmental performance evaluation of different municipal solid waste management scenarios in China. *Resources, Conservation and Recycling, 125*, 98–106.

Lu, W., Webster, C., Chen, K., Zhang, X., & Chen, X. (2017). Computational Building Information Modelling for construction waste management: Moving from rhetoric to reality. *Renewable and Sustainable Energy Reviews, 68*, 587–595. https://doi.org/10.1016/j.rser.2016.10.029

Lu, W., & Yuan, H. (2012). Off-site sorting of construction waste: What can we learn from Hong Kong? *Resources, Conservation and Recycling, 69*, 100–108.

Lu, Z., & Cai, M. (2012). Disposal methods on solid wastes from mines in transition from open-pit to underground mining. *Procedia Environmental Sciences, 16*, 715–721. https://doi.org/10.1016/j.proenv.2012.10.098

Luo, Z., Lam, S. K., Hu, S., & Chen, D. (2020). From generation to treatment: A systematic reactive nitrogen flow assessment of solid waste in China. *Journal of Cleaner Production, 259*, 121127. https://doi.org/10.1016/j.jclepro.2020.121127

Makarichi, L., Jutidamrongphan, W., & Techato, K. A. (2018). The evolution of waste-to-energy incineration: A review. *Renewable and Sustainable Energy Reviews, 91*, 812–821. https://doi.org/10.1016/j.rser.2018.04.088

Malinauskaite, J., Jouhara, H., Czajczyńska, D., Stanchev, P., Katsou, E., Rostkowski, P., Thorne, R. J., Colon, J., Ponsá, S., Al-Mansour, F., & Anguilano, L. (2017). Municipal solid waste management and waste-to-energy in the context of a circular economy and energy recycling in Europe. *Energy, 141*, 2013–2044. https://doi.org/10.1016/j.energy.2017.11.128

Marfe, G., & Di Stefano, C. (2016). The evidence of toxic wastes dumping in Campania, Italy. *Critical Reviews in Oncology/Hematology, 105*, 84–91. https://doi.org/10.1016/j.critrevonc.2016.05.007

Martinez, F., & Beilmann, K. (2020). Waste and postsocialism in Estonia: Becoming European through the management of rubbish. *Environment and Planning C: Politics and Space, 38*(7–8), 1348–1366. https://doi.org/10.1177/2399654420925083

Marzorati, S., Verotta, L., & Trasatti, S. P. (2019). Green corrosion inhibitors from natural sources and biomass wastes. *Molecules, 24*(1), 48. https://doi.org/10.3390/molecules24010048

McClellan, K., & Halden, R. U. (2010). Pharmaceuticals and personal care products in archived US biosolids from the 2001 EPA national sewage sludge survey. *Water Research, 44*(2), 658–668. https://doi.org/10.1016/j.watres.2009.12.032

Metcalfe, A., Riley, M., Barr, S., Tudor, T., Robinson, G., & Guilbert, S. (2012). Food waste bins: Bridging infrastructures and practices. *The Sociological Review, 60*, 135–155. https://doi.org/10.1111/1467-954x.12042

Misra, V., & Pandey, S. D. (2005). Hazardous waste, impact on health and environment for development of better waste management strategies in future in India. *Environment International, 31*(3), 417–431.

Ndanguza, D., Nyirahabinshuti, A., & Sibosiko, C. (2020). Modeling the effects of toxic wastes on population dynamics. *Alexandria Engineering Journal, 59*(4), 2713–2723. https://doi.org/10.1016/j.aej.2020.05.013

Nomura, K., Peng, X., Kim, H., Jin, K., Kim, H. J., Bratton, A. F., Bond, C. R., Broman, A. E., Miller, K. M., & Ellison, C. J. (2020). Multiblock copolymers for recycling polyethylene–poly (ethylene terephthalate) mixed waste. *ACS Applied Materials & Interfaces, 12*(8), 9726–9735. https://doi.org/10.1021/acsami.9b20242

Pacheco, R. M., Laurenti, A., dos Santos Silva, B. E., Mattes, I., & Meireles, S. (2018). Diagnosis of chemical and special waste management in a higher education institution: A methodology for data acquisition and processing. In *Towards green campus operations* (pp. 205–217). Springer.

Paes, M. X., de Medeiros, G. A., Mancini, S. D., Bortoleto, A. P., de Oliveira, J. A. P., & Kulay, L. A. (2020). Municipal solid waste management: Integrated analysis of environmental and economic indicators based on life cycle assessment. *Journal of Cleaner Production, 254*, 119848.

Panigrahi, S., & Dubey, B. K. (2019). Electrochemical pretreatment of yard waste to improve biogas production: Understanding the mechanism of delignification, and energy balance. *Bioresource Technology, 292*, 121958. https://doi.org/10.1016/j.biortech.2019.121958

Papapetrou, M., Kosmadakis, G., Cipollina, A., La Commare, U., & Micale, G. (2018). Industrial waste heat: Estimation of the technically available resource in the EU per industrial sector, temperature level and country. *Applied Thermal Engineering, 138*, 207–216. https://doi.org/10.1016/j.applthermaleng.2018.04.043

Patil, P. M., & Bohara, R. A. (2020). Nanoparticles impact in biomedical waste management. *Waste Management & Research, 38*(11), 1189–1203. https://doi.org/10.1177/0734242x20936761

Paul, S., Choudhury, M., Deb, U., Pegu, R., Das, S., & Bhattacharya, S. S. (2019). Assessing the ecological impacts of ageing on hazard potential of solid waste landfills: A green approach through vermitechnology. *Journal of Cleaner Production, 236*, 117643. https://doi.org/10.1016/j.jclepro.2019.117643

Phu, S. T. P., Takeshi, F., Giang, H. M., & Dinh, P. V. (2019). An analysis of the commercial waste characterization in a tourism city in Vietnam. *International Journal of Environment and Waste Management, 23*(3), 319–335. https://doi.org/10.1504/ijewm.2019.099015

Pollans, L. B. (2017). Trapped in trash: 'Modes of governing' and barriers to transitioning to sustainable waste management. *Environment and Planning A, 49*(10), 2300–2323. https://doi.org/10.1177/0308518x17719461

Prüss-Ustün, A., Wolf, J., Bartram, J., Clasen, T., Cumming, O., Freeman, M. C., Gordon, B., Hunter, P. R., Medlicott, K., & Johnston, R. (2019). Burden of disease from inadequate water, sanitation and hygiene for selected adverse health outcomes: An updated analysis with a focus on low-and middle-income countries. *International Journal of Hygiene and Environmental Health, 222*(5), 765–777.

Pullen, S., Palmer, J., Chiveralls, K., Zillante, G., Zuo, J., & Wilson, L. (2013). Exploring factors leading to the holistic reduction of waste in construction. In M. A. Schnabel (Ed.), *Cutting edge: 47th International conference of The Architectural Science Association* (pp. 355–364). The Architectural Science Association (ANZAScA).

Purkayastha, S. P., Choudhury, M., Deb, D., & Paul, C. (2015). Arsenic contamination in ground water is a serious threat in the North Karimganj block of Karimganj district, Southern part of Assam, India. *Journal of Chemical and Pharmaceutical Research, 7*(8), 371–378.

Putri, S. R., Muda, K. B., & Alia, F. (2019, September). Mapping of municipal solid waste transportation system, case studies Seberang Ulu region Palembang City. In *IOP conference series: Materials science and engineering* (Vol. 620, No. 1, 012051). IOP Publishing. https://doi.org/10.1088/1757-899x/620/1/012051

Rahman, M. M., Bodrud-Doza, M., Griffiths, M. D., & Mamun, M. A. (2020). Biomedical waste amid COVID-19: Perspectives from Bangladesh. *The Lancet. Global Health, 8*(10), e1262. https://doi.org/10.1016/s2214-109x(20)30349-1

Rajpal, A., Ali, M., Choudhury, M., Almohana, A. I., Alali, A. F., Munshi, F. M. A., Khursheed, A., & Kazmi, A. A. (2022). Abattoir wastewater treatment plants in India: Understanding and performance evaluation. *Front. Environ. Sci., 10*, 881623. https://doi.org/10.3389/fenvs.2022.881623.

Rimélé, M. A., Dimitrakopoulos, R., & Gamache, M. (2018). A stochastic optimization method with in-pit waste and tailings disposal for open pit life-of-mine production planning. *Resources Policy, 57*, 112–121. https://doi.org/10.1016/j.resourpol.2018.02.006

Russo, I., Confente, I., Scarpi, D., & Hazen, B. T. (2019). From trash to treasure: The impact of consumer perception of bio-waste products in closed-loop supply chains. *Journal of Cleaner Production, 218*, 966–974. https://doi.org/10.1016/j.jclepro.2019.02.044

Saeidi-Mobarakeh, Z., Tavakkoli-Moghaddam, R., Navabakhsh, M., & Amoozad-Khalili, H. (2020). A bi-level and robust optimization-based framework for a hazardous waste management problem: A real-world application. *Journal of Cleaner Production, 252*, 119830. https://doi.org/10.1016/j.jclepro.2019.119830

Sana, S. S., Dogiparthi, L. K., Gangadhar, L., Chakravorty, A., & Abhishek, N. (2020). Effects of microplastics and nanoplastics on marine environment and human health. Environmental Science and Pollution Research: 1–14. https://doi.org/10.1007/s11356-020-10573-x

Schanes, K., Dobernig, K., & Gözet, B. (2018). Food waste matters-A systematic review of household food waste practices and their policy implications. *Journal of Cleaner Production, 182*, 978–991. https://doi.org/10.1016/j.jclepro.2018.02.030

Shams, S., Sahu, J. N., Rahman, S. S., & Ahsan, A. (2017). Sustainable waste management policy in Bangladesh for reduction of greenhouse gases. *Sustainable Cities and Society, 33*, 18–26.

Sharma, A., Ganguly, R., & Gupta, A. K. (2018). Matrix method for evaluation of existing solid waste management system in Himachal Pradesh, India. *Journal of Material Cycles and Waste Management, 20*(3), 1813–1831.

Sharma, H. B., & Dubey, B. K. (2020). Co-hydrothermal carbonization of food waste with yard waste for solid biofuel production: Hydrochar characterization and its pelletization. *Waste Management, 118*, 521–533. https://doi.org/10.1016/j.wasman.2020.09.009

Sharma, V. D. (2018). *Solid waste management: Challenges & opportunities for Bharat* (p. 63). PK Chaubey.

Simatele, D. M., Dlamini, S., & Kubanza, N. S. (2017). From informality to formality: Perspectives on the challenges of integrating solid waste management into the urban development and planning policy in Johannesburg, South Africa. *Habitat International, 63*, 122–130.

Singh, A. (2019). Managing the uncertainty problems of municipal solid waste disposal. *Journal of Environmental Management, 240*, 259–265. https://doi.org/10.1016/j.jenvman.2019.03.025

Sokka, L., Antikainen, R., & Kauppi, P. E. (2007). Municipal solid waste production and composition in Finland—Changes in the period 1960–2002 and prospects until 2020. *Resources, Conservation and Recycling, 50*(4), 475–488. https://doi.org/10.1016/j.resconrec.2007.01.011

Sudibyo, H., Pradana, Y. S., Budiman, A., & Budhijanto, W. (2017). Municipal solid waste management in Indonesia-A study about selection of proper solid waste reduction method in DI Yogyakarta Province. *Energy Procedia, 143*, 494–499.

Tibu, C., Annang, T. Y., Solomon, N., & Yirenya-Tawiah, D. (2019). Effect of the composting process on physicochemical properties and concentration of heavy metals in market waste with additive materials in the Ga West Municipality, Ghana. *International Journal of Recycling of Organic Waste in Agriculture, 8*(4), 393–403. https://doi.org/10.1007/s40093-019-0266-6

Torkashvand, J., Farzadkia, M., Sobhi, H. R., & Esrafili, A. (2020). Littered cigarette butt as a well-known hazardous waste: A comprehensive systematic review. *Journal of Hazardous Materials, 383*, 121242. https://doi.org/10.1016/j.jhazmat.2019.121242

Tvedten, I., & Candiracci, S. (2018). "Flooding our eyes with rubbish": Urban waste management in Maputo, Mozambique. *Environment and Urbanization, 30*(2), 631–646. https://doi.org/10.1177/0956247818780090

Vu, H. L., Ng, K. T. W., Fallah, B., Richter, A., & Kabir, G. (2020). Interactions of residential waste composition and collection truck compartment design on GIS route optimization. *Waste Management, 102*, 613–623. https://doi.org/10.1016/j.wasman.2019.11.028

Wang, F., Cheng, Z., Reisner, A., & Liu, Y. (2018). Compliance with household solid waste management in rural villages in developing countries. *Journal of Cleaner Production, 202*, 293–298. https://doi.org/10.1016/j.jclepro.2018.08.135

Wilson, J. C., Benbow, S., & Metcalfe, R. (2018). Reactive transport modelling of a cement backfills for radioactive waste disposal. *Cement and Concrete Research, 111*, 81–93. https://doi.org/10.1016/j.cemconres.2018.06.007

Woolley, E., Luo, Y., & Simeone, A. (2018). Industrial waste heat recovery: A systematic approach. *Sustainable Energy Technologies and Assessments, 29*, 50–59. https://doi.org/10.1016/j.seta.2018.07.001

Yahalom, S., Guan, C., & Johansson, E. (2020). An innovative intermodal solution to urban residential waste disposal in large cities: A marine highway solution to a growing environmental problem. *Maritime Economics & Logistics*, 1–13. https://doi.org/10.1057/s41278-020-00164-5

Zen, I. S., & Siwar, C. (2015). An analysis of household acceptance of curbside recycling scheme in Kuala Lumpur, Malaysia. *Habitat International, 47*, 248–255. https://doi.org/10.1016/j.habitatint.2015.01.014

Zhang, L., & Klenosky, D. B. (2016). Residents' perceptions and attitudes toward waste treatment facility sites and their possible conversion: A literature review. *Urban Forestry & Urban Greening, 20*, 32–42. https://doi.org/10.1016/j.ufug.2016.07.016

Zou, K., Deng, Y., Chen, J., Qian, Y., Yang, Y., Li, Y., & Chen, G. (2018). Hierarchically porous nitrogen-doped carbon derived from the activation of agriculture waste by potassium hydroxide and urea for high-performance supercapacitors. *Journal of Power Sources, 378*, 579–588. https://doi.org/10.1016/j.jpowsour.2017.12.081

6 Municipal Solid Waste Management of Different Socioeconomic Groups Toward Sustainable Development Goals

Pankaj Kanti Jodder, Rabeya Sultana Leya, Md. Sohel Rana, and Bristi Sarkar

6.1 INTRODUCTION

Nowadays, waste management is becoming an acute problem in nature. Construction and excavation waste as well as domestic, industrial, institutional, commercial, municipal, and manufacturing waste all require serious attention (Hoornweg et al., 2015). In the 18th century, poor waste management led to outbreaks of diseases such as the plague and cholera, which killed over 380,000 people in Hamburg. Every year, an estimated 590 to 880 million tons of methane (CH_4) are released into the atmosphere (Saidan et al., 2019; Bashar, 2021). For several developing countries, notably Asia, the rising amount of waste generated by production and consuming processes poses a significant difficulty (Afroz et al., 2010; Ahsan et al., 2014). In 2025, Asia's municipal solid waste generation rate is expected to be 1.8 million metric tons per day, up from the current rate of 1 million tons per day (Bashar, 2021). The overall system for managing municipal solid waste (MSW) is strongly influenced by the quantity of garbage generated on a daily basis in the home (Dangi et al., 2011). For developing and implementing a feasible management system of solid waste for any particular location, it is essential to measure the waste-generation rate and categorize its composition (Gomez et al., 2008). In addition, to establish a garbage-collection and treatment scheme for urban residential units, household waste should be quantified and characterized. Because it varies in both amount and content at different income levels, household garbage is a complex substance that contains a variety of heterogeneous materials (Ogwueleka, 2013). During the previous several decades, the amount of MSW generated in cities of developing countries has grown many times. Most of the countries which are developing have a substantial rate of MSW production (Jadoon et al., 2013). The expansion in waste generation has become one of the obvious consequences of overpopulation. Urban sprawl and slum growth result through migration, and these regions create large amounts of uncontrolled

DOI: 10.1201/9781003258377-6

solid waste generation in Bangladesh's main cities (Sujauddin et al., 2008). Due to rural migration, there has been a significant increase in population growth in metropolitan areas, changing living conditions of city dwellers, growth in the urban economy, as well as social advancements in metropolitan areas, and other factors are all contributing to this massive increase in the MSW amount in Asia and Africa's emerging cities (Suthar & Singh, 2015). Developing countries like Bangladesh are thinking on waste production caused by urbanization, population development, and also for improved lifestyles (Bhuiyan, 2010). Food waste makes up a large amount (68–81%) of Bangladesh's urban waste (Jodder et al., 2022; Ahsan et al., 2014; Bhuiyan, 2010), which produces CH_4 when it decomposes anaerobic respiration. Waste generation is ascending to 0.1343 million tons yearly (Ahsan et al., 2014), according to current trends. According to Ahsan et al. (2014), waste generation rates in Bangladeshi cities average 0.2–0.5 kg/person/day; 0.56 kg/person/day in Dhaka and 0.2 kg/person/day in Barisal are the highest the lowest (Hoornweg & Bhada-Tata, 2012). It is predicted to rise by 0.6 kg/person/day in Bangladesh's by 2025 (Haque & Islam, 2021). The rate of garbage generation has grown from 1.1 million tons in 1970 to 5.2 million tons in 2015 (Islam, 2016; Hoornweg & Bhada-Tata, 2012). About 78% is created in the household branch, 20% in the the commercial branch, 1% in manufacturing, and so on (Ahsan et al., 2014). The management of produced municipal solid waste in most of the emerging countries is inadequate, resulting in major challenges with garbage storage, collection, and final disposal (Al-Khatib et al., 2010; Paul et al., 2019; Batool & Ch, 2009).

Khulna City is Bangladesh's third biggest metropolis, commercial center, and major port, with a population of 1.13 million people and a land area of about 45 square kilometers (Rahman et al., 2008). Between 1991 and 2002, the city's population grew dramatically, from 0.62 million to 1.13 million, due to the expansion of industry and commercial business. Due to various changes in consumption levels and the affordability of products available, the quantity of solid waste has maximized in addition to the growth in population (Rahman et al., 2008). The overall amount of MSW produced in Khulna City is 420 to 520 tons per day (Ahsan et al., 2014). MSW is often dumped in secondary-disposal sites (SDS) by the inhabitants individually, community-based groups, or non-government organizations using a door-to-door collection process (Alam & Mosharraf, 2020). Khulna City Corporation (KCC) implements MSWM by transporting MSW from SDS to the disposal site (FDS) at Rajbandh, roughly 7 kilometers from the city center (Moniruzzaman et al., 2011). MSWM has resulted in numerous greenhouse gas (GHG) concentrations, such as CO_2 from the manufacture of new products and CH_4 from the breakdown of organic garbage in landfill spaces (Bari et al., 2012; Mondal et al., 2023).

SWM is a holistic and integrated exposure that has an impact on sustainable development (Rodić & Wilson, 2017a). SDGs' guidelines are largely similar to the aims that have led to the development of SWM practices and strategies, especially public healthcare (SDG 3), environmental challenges (SDG 6 and 13), and resources valuation (SDG 11), with climatic changes (SDG 13) being added more recently (Wilson, 2007). For example, SDG 12.3 (reduce worldwide for each food waste across consumer levels) and SDG 14 (prevent uncontrolled usage-disposal of hazardous waste that causes marine pollution and plastic-particles issues) cannot be achieved without achieving sustainable SWM targets. Uncollected waste, improperly discarded, is having serious health and environmental consequences (SDG 6 and 13). The overall cost of building and maintaining basic, appropriate waste management methods is many times greater than the amount of mitigating these consequences (Kaza et al., 2018). As the primary utility system that more than 2 billion people now lack, SWM can be connected to 12 of the 17 UN-SDGs (Rodić & Wilson, 2017a). By 2050, global garbage is anticipated to increase to two times more than previously. Unless no modifications are introduced in this sector, solid waste emission levels are expected to rise to 2.38 billion metric tons of carbon dioxide equivalents each year by 2050 (Kaza et al., 2018). People in developing countries leave their villages in search of job opportunities and a better standard of living for themselves and their family members in city areas; they often begin their urban lives in slums or nearly identical poor neighborhoods, where they face extremely difficult living standards and inadequate basic infrastructural facilities (Hande, 2019). SDGs Target 11.1 addresses special needs: "Make sure access to sufficient, secure, and economical dwellings and basic facilities, and rehabilitate settlements as a whole by 2030." Management of solid waste could be one of those fundamental services in the livelihoods of people around the world, especially those who are most in need of progress (Rodić & Wilson, 2017a; Wilson, 2007).

To address these issues, this study primarily focuses on solid waste generation and administration in Khulna City to reach specific sustainable development goals. Since there are few solid waste management studies conducted on Khulna City in Bangladesh, the local community, in collaboration with NGOs and the municipal corporation, has devised a system for managing solid waste in the city using limited recovery and land-fill approaches. This research will also attempt to make policy recommendations for community-based solid waste management approaches that are linked with the SDGs. The findings of this study could support decision-makers in characterizing municipal solid waste in various socioeconomic groups to achieve long-term development goals. Because cities are the right size for transforming policies into actual ways of responding to geographical constraints and are extensively involved in the execution of municipal waste monitoring plans, this study has been undertaken at city scale. The significance of management at city scale, particularly to achieve sustainable development goals, has also been explored in this study. Additionally, it should be noted that this is brand-new work in the context of Khulna City, Bangladesh.

6.2 METHODS

6.2.1 Geographical Information of Research Area

Khulna City is the third most populous city in the south-western territory of Bangladesh at 22°49′0″N and 89°33′0″E (Sikder et al., 2015). It covers a total area of 45.65 km^2 (Rahman & Kabir, 2019) and has about 1.30 million populations (*BBS*, 2015). As a result of growing resettlement from nearby districts and urbanization, waste production in the city has progressively grown. Waste management is an overarching factor in managing communities that may be directly related to SDGs, particularly SDG 11.6 (Rodić & Wilson, 2017b). The management of MSW is intimately linked to the amount of solid waste produced. Khulna City is becoming more contaminated by rubbish, since there isn't a good waste-controlling mechanism. Rapid population expansion and major migration from rural to urban regions have put a pressure on the city of Khulna's present solid waste management infrastructure in recent years (Haque et al., 2018). MSW output is expected to reach 520 Mg (megagram) per day, with the bulk of kitchen waste (food and vegetable wastes) and residential areas serving as the primary source of MSW. Approximately half of the rubbish generated at the home level gets disposed of daily at a dumping site, leaving the other waste uncollected and unregulated (Ahsan et al., 2015). So, in Khulna City, the waste management system is alarming and disturbing to lessen the negative effects of pollution to SDGs. That's why Khulna City has been selected as a study area. The study has been conducted at ward level. Three representatives' wards (17, 23 and 24) are being considered here (Figure 6.1).

6.2.2 Respondent Selection and Measurement Techniques

In order to select three representatives of higher, middle and lower socioeconomic groups in Khulna City, the study has considered wards 17, 23 and 24; these wards have been selected purposely for the research because these three wards are mixed up with three different socioeconomic groups of people together. Mostly in context of sampling, it can be observed that, in a major metro area of 10 to 10,000 thousand, populations required an average of 380 samples for attaining typically suggested confidence levels and margin of error. But sometimes, collecting wastes samples from 384 residences for 7 days for a city is impractical and expensive; hence, this study gathered 144 sample residences (48 from each socioeconomic group; households from the selected three wards) (UN HABITAT, 2019). A stratified random-sampling technique has been used for selecting households (Manly, 2008). On the basis of household monthly income, the entire research has considered three socioeconomic groups:

1. Lower socioeconomic group: Households' monthly income is less than 20,000 BDT
2. Middle socioeconomic group: Households' monthly income is 20,000–40,000 BDT
3. Higher socioeconomic group: Households monthly income is more than 40,000 BDT

In order to collect demographic and financial information, a survey questionnaire was delivered to 144 randomly selected representative households. All the data have been collected in between July 2021. One of the most exact ways for determining solid waste production rate and composition (waste collection from chosen residences using the waste-segregation method) have been followed to investigate the waste-production rate and its characteristic (Gu et al., 2015). Following this method, household waste has been collected for consecutively 7 days and weighted using an

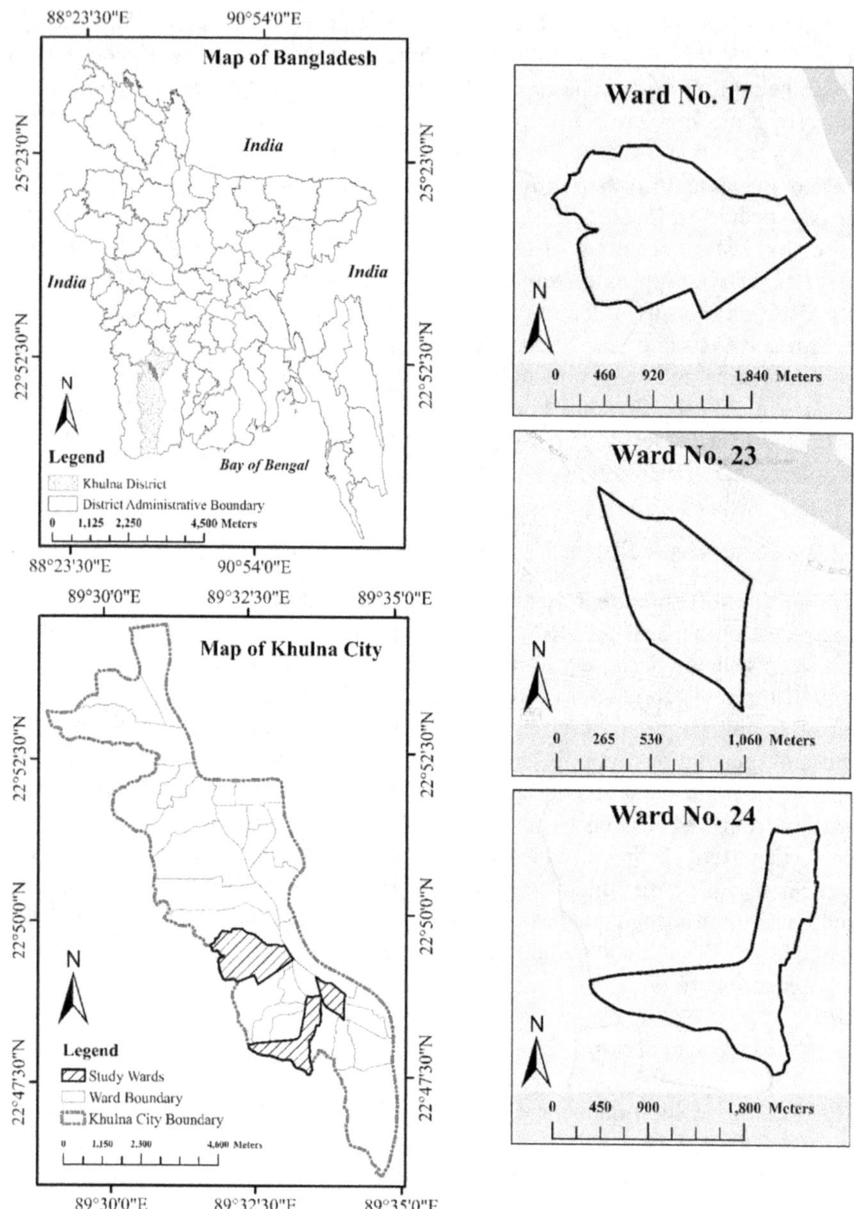

FIGURE 6.1 Geographical location of research area.

electronic weighing machine. The amount of garbage collected from each household and their categories are measured and recorded separately. Trash bags labeled with socioeconomic group and house number were supplied to each household, and residences were instructed to dump each and every single piece of garbage to that provided bag for that particular day. For calculating the average waste creation per residence and per person generation rate, equation 1 and equation 2 have been followed, respectively.

$$Wh = \sum_{i=1}^{n} Wi / 7 \tag{6.1}$$

$$Wpc = \Sigma_{i=1}^{n} \frac{wi}{7\, x \ number \ of \ HH \ members} \tag{6.2}$$

Where ***Wh*** is the average household waste; ***i*** is the individual household waste categories; ***n*** is the total number of household waste categories; ***Wi*** is the individual households generated waste; ***Wpc*** is the average per-person waste.

In the next stage, the waste cleaner segregated the dumped waste according to the specified 12 categories, accordingly, with direct supervision. Each of the categories were weighted individually using an electronic weighing-machine (Harun & Rashid, 2019). The waste composition categories for this study are given as follows with a brief description: kitchen, garden, and park wastes—paper and cardboard, film plastics, dense plastics, metals, glass, textiles and shoes, wood; special wastes—composite products; and other.

6.3 DISCUSSION OF THE FINDINGS

6.3.1 Socioeconomic Features of Households

Household socioeconomic characteristics influence waste production rate; for example, age of members, numbers of members, educational background, employed persons, land-occupying status, duration of stay, and environmental concern of that household (Table 6.1). Average age of members has been found to be 32 years. Most of the selected families have household members in between 4 to 6, which is around 56.3%. A majority of the respondents have primary- and secondary-level educational backgrounds. Around 75% of households have a single earning member. Land-occupying status shows that 46.5% of respondents have their own land, and for the rest, 53.56% are tenants. In the case of the living duration, 39% of residences have been occupied over the last 5 years; and 30%, for the last 6 to 10 years. Most importantly, in recent times among the different socioeconomic groups of households, 61.1% are concerned about the environment. A few (29.86%) have no concern, and the rest (9.03%) have no response on this.

6.3.2 Generation of Household Waste

Table 6.2 shows that the average waste-producing rates, both for household and per person, are on a daily basis, varying with their socioeconomic status. Here, findings show that middle and higher socioeconomic groups of households are producing

TABLE 6.1
Socioeconomic Features of Households

Variable name	Frequency	Percentage	Cumulative percentage
1. Age of household members			
< 20 years	97	15.4	15.4
21 to 30 years	182	28.9	44.3
31 to 40 years	217	34.4	78.7
41 to 50 years	95	15.1	93.8
> 50 years	39	6.2	100
Mean = 32.21 (years), Standard deviation = 14.25 (years)			
2. Number of household members			
< 4	51	35.4	35.4
4 to 6	81	56.3	91.7
> 6	11	7.6	100
Mean = 4.23, Standard deviation = 1.91			
3. Respondent's education level			
< 5	28	19.4	19.4
5 to 10	56	38.9	58.3
10 to 12	51	35.4	93.7
< 12	9	6.3	100
Mean = 8.59, Standard deviation = 2.99			
4. Number of employed persons in household			
1	108	75.0	75
2	33	22.9	97.9
3	3	2.1	100
Mean = 1.27, Standard deviation = 0.49			
5. Land-occupying status			
Land owner	67	46.5	46.5
Tenant	77	53.5	100
6. Duration of stay			
< 1 years	9	6.3	6.3
1 to 5 years	56	38.9	45.2
6 to 10 years	43	29.9	75.1
11 to 15 years	18	12.5	87.6
16 to 20 years	11	7.6	95.2
> 20 years	7	4.9	100
Mean = 7.88 years, Standard deviation = 5.70 years			
7. Environmental concern			
No concern	43	29.9	29.9
Have concern	88	61.1	91
No response	13	9.0	100

TABLE 6.2
Rate of Average Waste Generation Per Household Per Day

Categories of socioeconomic group	Average [kg/HH/day]	Average [kg/person/day]
Lower socioeconomic group	1.472	0.329
Middle socioeconomic group	2.143	0.501
Higher socioeconomic group	2.138	0.562

TABLE 6.3
The Proportion of Domestic Waste with Socioeconomic Characteristics

Variables		HH earnings	Average (kg) [daily per person]	Average (kg) [daily per HH]
HH earnings	Pearson correlation	1	.912**	.768*
Average (kg) [daily per person]		.912**	1	.940**
Average (kg) [daily per HH]		.768*	.940**	1

** At the 0.01 level, correlation is significant (2-tailed)
* At the 0.05 level, the correlation is significant (2-tailed)

nearly same amount of waste per day (2.1 kg/day), whereas the lower socioeconomic group is producing around 1.5 kg per day. On the other hand, the maximum per-person waste-producing rate has been found to be 0.56 kg daily for the affluent social class, whereas the middle socioeconomic group is also standing nearly with the amount of 0.50 kg per person daily. Last, the lower socioeconomic groups of households have a value of 0.33 kg per person daily, which is the lowest among socioeconomic categories.

6.3.3 The Proportion of Domestic Waste with Socioeconomic Characteristics

The degree of relationship between different parameters (monthly earnings, number of members, per-person waste creation, and household's waste production) have been determined using correlation analysis. From Table 6.3, it is found that there is a strong, positive correlation between socioeconomic group and average per-capita waste-generation daily ($r = 0.912$), indicating that a person belonging to the higher

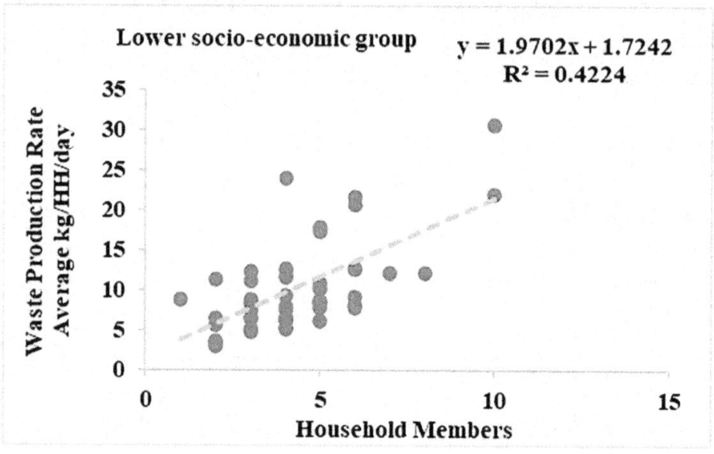

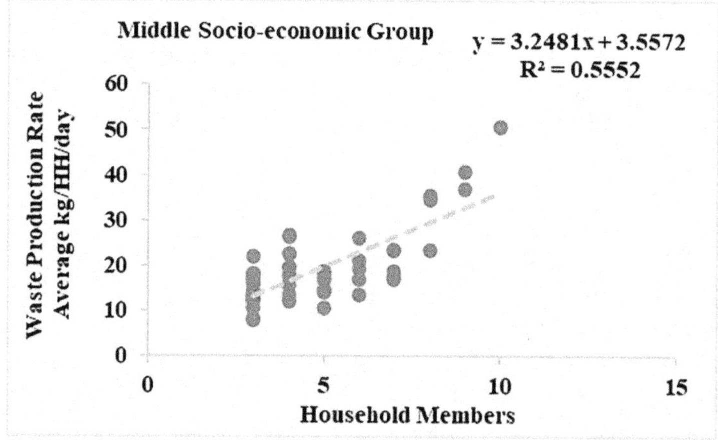

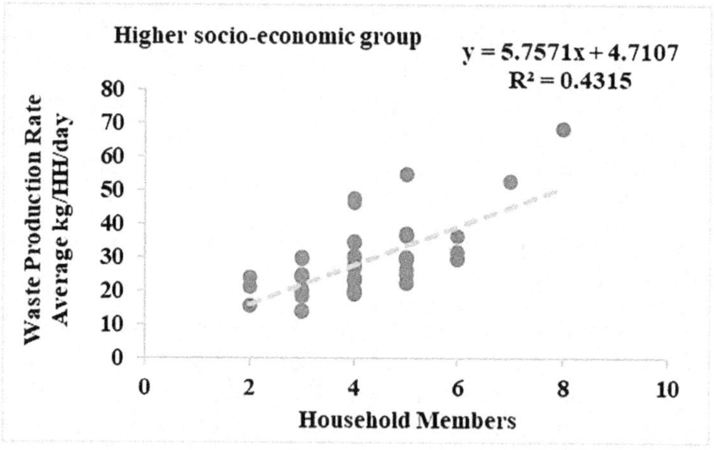

FIGURE 6.2 Proportion of trash generation based on the occupants' number.

socioeconomic group household produces a higher volume of solid trash every day. And again, the socioeconomic group has a significant, positive relationship with the average amount of garbage produced per household (r = 0.768), indicating that the higher socioeconomic household group produces a greater volume of solid garbage each day. Furthermore, the correlation value between average per-capita trash-generation and average waste-generation per home reveals a very high, positive association (r = 0.940), which indicates that the per-person waste is larger at those households that produce a larger quantity of solid waste every day. On the contrary, Figure 6.2 depicts the link between number of household occupants and daily average waste-generation. In this case, a linear regression model has been performed, and the value of R squared has been assessed. Though there is a moderate relationship found for both of these socioeconomic groups (lower, middle, and higher socioeconomic categories received 0.42, 0.55, and 0.43, respectively), it has a clear meaning that the greater the household size, the larger the amount of daily waste generation.

6.3.4 Composition of Domestic Waste Across Diversified Socioeconomic Groups

Table 6.4 illustrates the waste-composition characteristics among the diversified socioeconomic groups. The characteristics of household solid waste generated from the selected 144 households shows that the kitchen waste (240.47 kg/day) has the largest share (87.37%) among the other categories. Garden/park waste, film plastics, paper and cardboard, other, and dense plastics have the share of 2.40%, 1.99%, 1.99%, 1.58%, 1.29%, respectively (Table 6.4), whereas the remaining categories contain the very lowest share in the generation of daily household-waste. Figure 6.4 indicates that medium and upper socioeconomic classes produce nearly same amount of waste: 101.40 and 102.37 kg per day, respectively. The lower socioeconomic group has a lower amount of waste production per day (70.47 kg). The result also shows that household income-level is a determinant of characteristics of household waste. The finding shows that the higher socioeconomic level generates a substantially greater quantity of kitchen waste (90.37 kg), garden/park waste (2.72 kg), and paper and cardboard (2.27 kg) per day.

Whereas the other compositions of wastes—film plastics (2.27 kg), textiles and shoes (1.22 kg), processed wood (1.11 kg)—are greater in amount per day for middle income household levels. This occurs as a result of disparities in household consumption habits caused by socioeconomic status. In other words, households with a higher socioeconomic status generate a variety of kitchen waste because they consume a variety of foods, and these different kinds of foods are packaged in plastics, whereas households with a middle socioeconomic grouping are nearly identical to households with a high income-level. It also explains the observed increased use of film plastics and dense plastics for all of these socioeconomic groups (Figure 6.3).

6.4 WASTE MANAGEMENT TARGETING SDGs

This study emphasizes the creation and management of solid waste in Khulna City in order to meet specified sustainable development goals. For each of the specific

TABLE 6.4
Composition of Domestic Waste across Diversified Socioeconomic Groups

Waste categories	Lower socioeconomic group (kg/day)	Middle socioeconomic group (kg/day)	Higher socioeconomic group (kg/day)	Total (kg/day)
1. Kitchen	60.78	89.32	90.37	240.47
2. Garden/park waste	2.04	1.84	2.72	6.59
3. Paper and cardboard	1.52	1.68	2.27	5.47
4. Plastics—film	1.51	2.27	1.71	5.49
5. Plastics—dense	1.32	1.05	1.18	3.56
6. Metals	0.18	0.39	0.31	0.87
7. Glass	0.17	0.85	0.94	1.97
8. Textiles and shoes	0.63	1.22	0.77	2.61
9. Wood (processed)	0.31	1.11	0.44	1.86
10. Special wastes	0.24	0.20	0.19	0.63
11. Composite	0.16	0.92	0.28	1.36
12. Other	1.61	1.55	1.19	4.34

socioeconomic groups, waste generation rate and mixture of household waste have been computed in this study. The average waste-generation-rate per residence is found to be 1.91 kg daily and 0.46 kg per person daily. According to the World Bank (1999) (Hoornweg & Thomas, 2012) and (Sujauddin et al., 2008), the estimated value of waste was 0.15 kg and 0.25 kg per person daily, respectively (Hoornweg & Bhada-Tata, 2012), which differs from the result of this study. Considering the previous research, the higher socioeconomic group produces more solid waste than others (Masoud et al., 2011; Khan et al., 2016). However, this study has shown that the middle income-group has the greatest rate of waste creation (2.143 kg/HH/day). This study is similar to the other research findings in that the quantity of waste created per capita significantly increases as the number of the household expands (Masoud et al., 2011; Suthar & Singh, 2015). Several studies have shown similar results, such as the fact that kitchen garbage accounts for the majority of residential waste (87.37%), with the remaining 12.63% divided into 11 categories. The majority of residences do not discard recyclable garbage, such as metal, glass, plastics, and special waste, due to economic benefits, according to this survey (Masoud et al., 2011; Suthar & Singh, 2015; Ahsan et al., 2015; Choudhury et al., 2022)

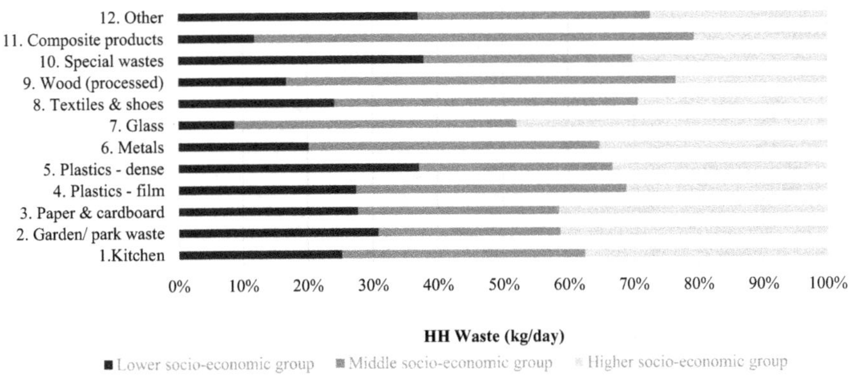

FIGURE 6.3 Percentage of waste consumption across the socioeconomic groups.

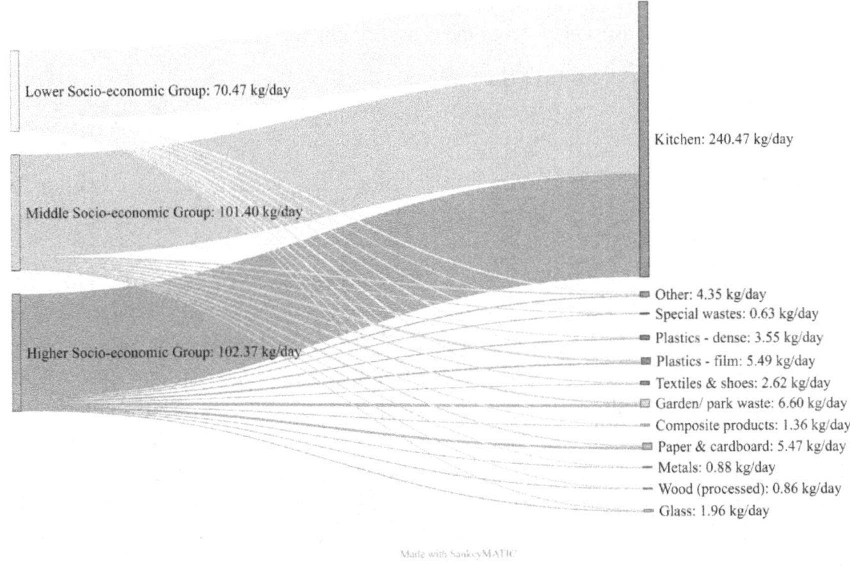

FIGURE 6.4 Daily waste consumption across all socioeconomic groups.

To achieve sustainable development goals (SDGs 7,11, and 12), we encourage investment in solid waste management, distribution of supply, and de-concentration of waste mechanisms using a bottom-top technique, involvement of resource recovery and eco-friendly restoration, and sectoral synchronization on the environment and its resources. Subsidies and taxation incentives would encourage waste-infrastructure development for SDG 12 (Sharma et al., 2021).

SWM offers provisions to help communities, which will enhance their livelihood opportunities and contribute directly to the SDGs (Wilson et al., 2015). Waste management practices must be developed related to geographical and

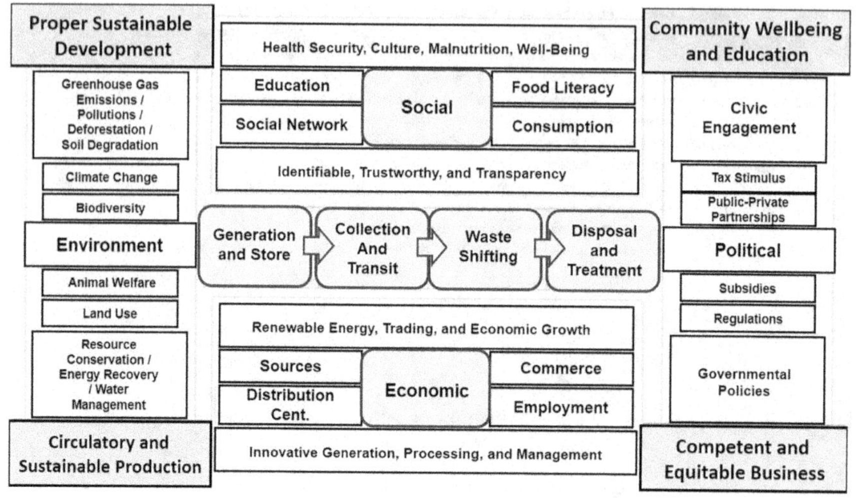

FIGURE 6.5 Different aspects of waste management policies and approaches.

Source: Authors, 2022 (adapted from Sharma et al., 2020).

cultural factors, taking into account community values, resource changes, waste-generation ratios, consumer engagement, and beliefs (Figure 6.5). To treat the rational, political, and socioeconomic aspects of waste management, policy guidelines recommend a multi-faceted solution (Sharma et al., 2020). The problems in implementing waste management policies include public participation, evaluation methods, political effort, and ambiguity about policy implications. The economic aspect of policy should stress concepts such as waste as an asset and environmental implications as well as other psychosocial interventions (Recuero Virto, 2018). Economic incentives such as tax benefits, subsidies to enhance packaging, labeling, setting waste management standards, developing waste collection mechanisms, and encouraging redistribution and making donations should all be included in these policies. Individuals are capable of understanding safety, hygiene, everyday maintenance, waste processing, and waste recycling because of the social aspect (Wan et al., 2019).

6.5 CONCLUSIONS

The current study emphasizes the importance of a municipal solid waste management system, which includes waste generating, collecting, transferring, disposal, and land filling or treatment. These solutions not only reduce environmental difficulties, but they also have the ability to improve the practical and economical parts of waste management. On the basis of the findings, it can be argued that environmental concerns and practices are important aspects of SWM that contribute to long term sustainability. In this context, the study finds that a large percentage of households have environmental concerns that might limit MSW generation as well as the amount of space necessary to store MSW. It has the potential to reduce the negative effects of MSW while also achieving the United Nations' Sustainable Development Goals.

The study also emphasized the priority of characterizing the potentials of various components of solid waste, such as food waste, plastic waste, construction waste, biomedical waste, and electronic waste, which introduce both challenges and issues for integrating the essential reactions that could assist the public and private fields in achieving both short-term and long-term UN-SDG goals. In Khulna City, municipal-waste management is a top-down and bottom-up strategy, with the KCC, Community Based Organizations (CBOs), and individuals all working together. KCC gathers and dumps municipal solid waste from municipal dustbins and other tertiary dumping stations, then distributes it without processing to the ultimate disposal site. This has major environmental consequences for both the surrounding environment and the surface water. Community-based garbage management is a relatively new approach in managing waste in Khulna City, but there are obviously more initiatives that need to be implemented to strengthen the city's waste management practices. Waste recycling is included as a system in municipal-waste management in the City Corporation Ordinance/Act. At the top level, the NGO operations in community-based solid waste management must be strongly aligned, regulated, and managed. It is an urgent need to establish a set of regulations for community-based waste treatment in Khulna City to achieve sustainable development goals.

ACKNOWLEDGMENTS

This study would not have been possible without the assistance and support of Md. Sajadul Alam Saimon and Asif Ahmed. The authors are grateful to them for the critical comments and helpful suggestions to improve the manuscript. We would like to thank all anonymous reviewers for their inspiring and constructive comments on the paper.

REFERENCES

Afroz, R., Hanaki, K., Tudin, R., Afroz, R., Hanaki, K., & Tudin, R. (2010). Factors affecting waste generation: A study in a waste management program in Dhaka City, Bangladesh. *Environmental Monitoring and Assessment*, *179*(1), 509–519. https://doi.org/10.1007/S10661-010-1753-4

Ahsan, A., Alamgir, M., El-Sergany, M. M., Shams, S., Rowshon, M. K., Daud, N. N. N., Bo, M. W., & Mai, L. (2014). Assessment of municipal solid waste management system in a developing country. *Chinese Journal of Engineering, 2014,* Article ID 561935. https://doi.org/10.1155/2014/561935

Ahsan, A., Alamgir, M., Imteaz, M., Shams, S., Rowshon, M. K., Aziz, M. G., & Idrus, S. (2015). Municipal solid waste generation, composition and management: Issues and challenges. A case study. *Environment Protection Engineering*, *41*(3), 43–59. https://doi.org/10.5277/EPEL50304

Alam, O., & Mosharraf, A. (2020). A preliminary life cycle assessment on healthcare waste management in Chittagong City, Bangladesh. *International Journal of Environmental Science and Technology*, *17*(3), 1753–1764. https://doi.org/10.1007/S13762-019-02585-Z/TABLES/5

Al-Khatib, I. A., Monou, M., Abu Zahra, A. S. F., Shaheen, H. Q., & Kassinos, D. (2010). Solid waste characterization, quantification and management practices in developing countries. A case study: Nablus district – Palestine. *Journal of Environmental Management, 91*(5), 1131–1138. https://doi.org/10.1016/J.JENVMAN.2010.01.003

Bari, Q. H., Mahbub Hassan, K., & Haque, R. (2012). Scenario of solid waste reuse in Khulna city of Bangladesh. *Waste Management, 32*(12), 2526–2534. https://doi.org/10.1016/J.WASMAN.2012.07.001

Bashar, S. (2021). *The Rohingya refugee crisis in Bangladesh: Environmental impacts, policies, and practices* [Master Thesis, Oregon State University]. https://ir.library.oregonstate.edu/concern/graduate_projects/m326m865f

Batool, S. A., & Ch, M. N. (2009). Municipal solid waste management in Lahore City District, Pakistan. *Waste Management, 29*(6), 1971–1981. https://doi.org/10.1016/J.WASMAN.2008.12.016

BBS. (2015). *Bangladesh bureau of statistics—Government of the People's Republic of Bangladesh-statistical-yearbook.* www.bbs.gov.bd/site/page/29855dc1-f2b4-4dc0-9073-f692361112da/Statistical-Yearbook

Bhuiyan, S. H. (2010). A crisis in governance: Urban solid waste management in Bangladesh. *Habitat International, 34*(1), 125–133. https://doi.org/10.1016/J.HABITATINT.2009.08.002

Choudhury, M., Sahoo, S., Samanta, P., Tiwari, A., Tiwari, A., Chadha, U., Bhardwaj, P., Nalluri, A., Eticha, T. K., & Chakravorty, A. (2022). COVID-19: An accelerator for global plastic consumption and its implications. *Journal of Environmental and Public Health, 2022.* Article ID 1066350, 17 pages. https://doi.org/10.1155/2022/1066350

Dangi, M. B., Pretz, C. R., Urynowicz, M. A., Gerow, K. G., & Reddy, J. M. (2011). Municipal solid waste generation in Kathmandu, Nepal. *Journal of Environmental Management, 92*(1), 240–249. https://doi.org/10.1016/J.JENVMAN.2010.09.005

Gomez, G., Meneses, M., Ballinas, L., & Castells, F. (2008). Characterization of urban solid waste in Chihuahua, Mexico. *Waste Management, 28*(12), 2465–2471. https://doi.org/10.1016/J.WASMAN.2007.10.023

Gu, B., Wang, H., Chen, Z., Jiang, S., Zhu, W., Liu, M., Chen, Y., Wu, Y., He, S., Cheng, R., Yang, J., & Bi, J. (2015). Characterization, quantification and management of household solid waste: A case study in China. *Resources, Conservation and Recycling, 98*, 67–75. https://doi.org/10.1016/J.RESCONREC.2015.03.001

Hande, S. (2019). The informal waste sector: A solution to the recycling problem in developing countries. *Journal of Field Actions Science Reports, 19*, 28–35. https://journals.openedition.org/factsreports/5143

Haque, M. N., Islam, M. R., & Bin Ansar, S. (2018). Assessing the water supply, sanitation and waste dumping condition of urban slum: A GIS based approach. *The Jahangirnagar Review, Part II: Social Sciences, XLII.*

Haque, M. S., & Islam, S. (2021). Effectiveness of waste plastic bottles as construction material in Rohingya displacement camps. *Cleaner Engineering and Technology, 3*, 100110. https://doi.org/10.1016/J.CLET.2021.100110

Harun, M., & Rashid, O. (2019). Sustainable municipal solid waste management in Dhaka city: Challenges and issues. *Journal of Bangladesh Institute of Planners, 12*, 97–107. www.researchgate.net/publication/342626180

Hoornweg, D., & Bhada-Tata, P. (2012). *What a waste: A global review of solid waste management.* https://openknowledge.worldbank.org/handle/10986/17388

Hoornweg, D., Bhada-Tata, P., & Kennedy, C. (2015). Peak waste: When is it likely to occur? *Journal of Industrial Ecology, 19*(1), 117–128. https://doi.org/10.1111/JIEC.12165

Hoornweg, D., & Thomas, L. (2012). What a waste : Solid waste management in Asia. *Urban Development Sector Unit (EASUR).* https://agris.fao.org/agris-search/search.do?recordID=US2012400855

Islam, F. A. S. (2016). Solid waste management system in Dhaka City of Bangladesh. *Journal of Modern Science and Technology*, *4*(1), 192–209.

Jadoon, A., Batool, S. A., & Chaudhry, M. N. (2013). Assessment of factors affecting household solid waste generation and its composition in Gulberg Town, Lahore, Pakistan. *Journal of Material Cycles and Waste Management, 16*(1), 73–81. https://doi.org/10.1007/S10163-013-0146-5

Jodder, P. kanti, Leya, R. S., Rana, M. S., & Sarkar, B. (2022). Generation and characteristics of household solid waste in Khulna city, Bangladesh. *Khulna University Studies*, *22*(2). https://doi.org/https://doi.org/10.53808/KUS.2022.19.01.2202-se

Kaza, S., Yao, L., Bhada-Tata, P., & Woerden, F. Van. (2018). *What a waste 2.0 A global snapshot of solid waste management to 2050.* https://openknowledge.worldbank.org/bitstream/handle/10986/30317/211329ov.pdf

Khan, D., Kumar, A., & Samadder, S. R. (2016). Impact of socioeconomic status on municipal solid waste generation rate. *Waste Management*, *49*, 15–25. https://doi.org/10.1016/J.WASMAN.2016.01.019

Manly, B. F. (2008). *Statistics for environmental science and management.* Chapman and Hall/CRC.

Masoud, S. M., Omrani, G. A., Karbassi, A., & Raof, F. F. (2011). The effects of socioeconomic parameters on household solid-waste generation and composition in developing countries (a case study: Ahvaz, Iran). *Environmental Monitoring and Assessment*, *184*(4), 1841–1846. https://doi.org/10.1007/S10661-011-2082-Y

Mondal, T., Choudhury, M., Kundu, D., Dutta, D., & Samanta, P. (2023). Landfill: An eclectic review on structure, reactions and remediation approach. *Waste Management*, *164*, 127–142. https://doi.org/10.1016/j.wasman.2023.03.034

Moniruzzaman, S. M., Bari, Q. H., & Fukuhara, T. (2011). Recycling practices of solid waste in Khulna City, Bangladesh. *Journal of Solid Waste Technology and Management*, *37*(1), 1–15. https://doi.org/10.5276/JSWTM.2011.1

Ogwueleka, T. C. (2013). Survey of household waste composition and quantities in Abuja, Nigeria. *Resources, Conservation and Recycling*, *77*, 52–60. https://doi.org/10.1016/J.RESCONREC.2013.05.011

Paul, S., Choudhury, M., Deb, U., Pegu, R., Das, S., & Bhattacharya, S. S. (2019). Assessing the ecological impacts of ageing on hazard potential of solid waste landfills: A green approach through vermitechnology. *Journal of Cleaner Production, 236*, 117643. https://doi.org/10.1016/j.jclepro.2019.117643

Rahman, M., Ruksana Sultana, K., & Ahasanul Hoque, M. (2008). Suitable sites for urban solid waste disposal using GIS approach in Khulna City, Bangladesh. *Proceeding of Pakistan Academy of Sciences*, *45*(1), 11–22.

Rahman, S. M. T., & Kabir, A. (2019). Factors influencing location choice and cluster pattern of manufacturing small and medium enterprises in cities: Evidence from Khulna City of Bangladesh. *Journal of Global Entrepreneurship Research*, *9*(1), 1–26. https://doi.org/10.1186/S40497-019-0187-X

Recuero Virto, L. (2018). A preliminary assessment of the indicators for Sustainable Development Goal (SDG) 14 "Conserve and sustainably use the oceans, seas and marine resources for sustainable development". *Marine Policy*, *98*, 47–57. https://doi.org/10.1016/J.MARPOL.2018.08.036

Rodić, L., & Wilson, D. C. (2017). Resolving governance issues to achieve priority sustainable development goals related to solid waste management in developing countries. *Sustainability*, *9*(3), 404. https://doi.org/10.3390/SU9030404

Saidan, M., Hamdan, S., & Linton, C. (2019). Solid waste management in Jordan: impacts and analysis. *Journal of Chemical Technology and Metallurgy*, *54*(2), 454–462. www.researchgate.net/publication/330925093

Sharma, H. B., Vanapalli, K. R., Cheela, V. S., Ranjan, V. P., Jaglan, A. K., Dubey, B., Goel, S., & Bhattacharya, J. (2020). Challenges, opportunities, and innovations for effective solid waste management during and post COVID-19 pandemic. *Resources, Conservation and Recycling*, *162*, 105052. https://doi.org/10.1016/J.RESCONREC.2020.105052

Sharma, H. B., Vanapalli, K. R., Samal, B., Cheela, V. R. S., Dubey, B. K., & Bhattacharya, J. (2021). Circular economy approach in solid waste management system to achieve UN-SDGs: Solutions for post-COVID recovery. *Science of The Total Environment*, *800*, 149605. https://doi.org/10.1016/J.SCITOTENV.2021.149605

Sikder, S. K., Asadzadeh, A., Kuusaana, E. D., Mallick, B., & Koetter, T. (2015). Stakeholders participation for urban climate resilience: A case of informal settlements regularization in Khulna city, Bangladesh. *Journal of Urban and Regional Analysis*, *7–1*(1), 5–20. www.jurareview.ro/resources/pdf/volume_15_stakeholders_participation_for_urban_climate_resilience:_a_case_of_informal_settlements_regularization_in_khulna_city_bangladesh_abstract.pdf

Sujauddin, M., Huda, S. M. S., & Hoque, A. T. M. R. (2008). Household solid waste characteristics and management in Chittagong, Bangladesh. *Waste Management*, *28*(9), 1688–1695. https://doi.org/10.1016/J.WASMAN.2007.06.013

Suthar, S., & Singh, P. (2015). Household solid waste generation and composition in different family size and socio-economic groups: A case study. *Sustainable Cities and Society*, *14*(1), 56–63. https://doi.org/10.1016/J.SCS.2014.07.004

UN Habitat. (2019). *Waste wise cities tool.* https://unhabitat.org/sites/default/files/2021/02/Waste wise cities tool—EN 3.pdf

Wan, C., Shen, G. Q., & Choi, S. (2019). Waste management strategies for sustainable development. *Encyclopedia of Sustainability in Higher Education*, 1–9. https://doi.org/10.1007/978-3-319-63951-2_194-1

Wilson, D. C. (2007). Development drivers for waste management. *Waste Management and Research*, *25*(3), 198–207. https://doi.org/10.1177/0734242X07079149

Wilson, D. C., Rodic, L., Modak, P., Soos, R., & Rogero, A. C. (2015). *Global waste management outlook.* www.uncclearn.org/wp-content/uploads/library/unep23092015.pdf

7 Managing Solid Wastes in Developing Countries

Insights From Three Indian Smart Cities

Shivkumar Vanjari and Anil Kumar Dikshit

7.1 INTRODUCTION

High population-growth coupled with rapid urbanization and economic development has led to several socio-economic and environmental challenges across the world. Urban areas are faced with a grave challenge in terms of provision of infrastructure for a rapidly growing population (Abdel-Shafy & Mansour, 2018). According to census 2011, India's population reached 1.21 billion out of which almost 31% reside in cities. As per estimates, till we reach 2050, approximately half of the human population will be residing in cities. This increasing urbanization has emerged as an issue not just due to environmental and beautification concern but also due to an immense amount of municipal solid wastes (MSW) which are generated every day. Recent figures as per the Central Pollution Control Board (CPCB) highlight that, during 2014–15 143,449 tonnes per day (TPD) of MSW was generated, which works out to be an average of 0.11 kg/capita/day. Out of this total waste, ~80% (117,644 TPD) was collected and only ~22% (32,871 TPD) was further processed. This clearly highlights that source segregation, waste collection, its transportation, treatment as well as scientific disposal was insufficient, resulting in environmental degradation and poor quality of life for citizens. Unsafe waste disposal generates poisonous gases and leachates as a result of microbial degradation, climatic conditions and characteristics of refuse as well as land filling operations (Singh, 2020). As per the 12th schedule of 74th Constitutional Amendment (1992), it is the Urban Local Bodies (ULBs) which are accountable for maintaining cleanliness in towns and cities. However, most of the ULBs are plagued with lack of infrastructure, poor institutional capability, scarce financial resources and absence of political will (Gupta & Gupta, 2015; Choudhury & Dutta, 2017).

Despite these challenges, it is commendable for municipal authorities to handle such massive quantities of waste in an effective way, dealing with the lowest level of health as well as environmental hazards (Joelsson & Lord, 2016; Fernández-Braña et al., 2019).

According to Joshi and Ahmed (2016), the lack of technical knowledge and awareness, scarce availability of financial resources and poor implementation of policies and legislations have resulted in the limited success in Municipal Solid Waste

DOI: 10.1201/9781003258377-7

Management (MSWM). Banerjee (2017) points out that the participation of private firms within SWM needs to be appreciated to lower the burden on government bodies and agencies. Chavan and Zambare (2013) highlight an important point in their research that the major inadequacy in India's SWM is mainly in the waste-processing methods and techniques. Krishna (2018) argues that, for improving SWM, there is a need to address institutional and financial issues. He recommends that the government rules should provide incentives to those systems where the producers attempt to undertake waste minimization and assume responsibility for reuse/recycling of used products. Another SWM approach which finds wide acceptance is that of decentralized SWM, which is believed to solve the issues associated with processing and treating the waste (Singh, 2015).

7.2 SOLID WASTE MANAGEMENT IN INDIAN CITIES

In the past decade, there has been a constant and rapid increase in the amount of waste generated in Indian cities. A recent estimate by Ministry of Housing and Urban Affairs (MoHUA) exposes that approximately 147,613 MT of solid waste is produced each day from 84,475 wards. As per the estimate, the state that produces highest amount of waste is Maharashtra amongst all Indian states, at 22,080 MT/day from its 7,322 wards, while the state of Sikkim produces the least amount, at 89 MT/day from 53 wards. In terms of overall quantity, Daman and Diu generate the lowest amount of waste within the country (MoHUA, 2020). As per the report of MoHUA, out of all the states, Chhattisgarh processed the maximum proportion of waste (90%), whereas Meghalaya processed the lowest proportion (4%) of its generated waste. Comparing amongst the UTs, it is very encouraging to note that Dadra & Nagar Haveli is able to process almost 100% of its waste; 95% of waste is processed by Chandigarh. However, Delhi is able to process only 55% of its generated waste, which is even lower than the average figure of 60% found across all the 84,475 wards.

7.2.1 Solid Waste Types and Characteristics of Waste

SW can be classified in different types based on the way they are generated, extracted and possibly treated. SW is comprised of solid and wet waste, which comes from municipalities, industries, mines and agricultural fields, etc. It includes organic matter, paper and plastic, glass and metals. Waste composition, however, changes, relying upon different factors; for example, climate and expectations for everyday comforts. Three main categories of SW include biodegradable or organic waste (kitchen waste), inert and non-biodegradable waste (generated due to construction and demolition) and recyclable waste (plastic, paper, bottles etc.). As per Ahluwalia and Patel (2018), the percentage of waste contribute is as follows: biodegradable waste (52%), inert and non-biodegradable waste (32%) and recyclable waste (17%).

7.2.2 Waste Segregation

One key factor of efficient management of solid waste segregation for municipal corporations in metro cities, waste segregation has been a cause for concern for many years. Developing nations are yet to advance their MSW techniques in order to meet the standards of countries like the US, Europe and Japan. The baseline methods that are followed in communities are segregation, composting and recycling (Paul et al., 2019). The waste transportation to disposal sites is done by municipality or via public-private partnership model (PPP). The main aim of waste management companies is to control, gather and process waste in such a way that the goal of sustainable development is fulfilled for the benefit of the whole community (Odonkor et al., 2020; Abdel-Shafy & Mansour, 2018). Wet waste serves the purpose of composting or biomethanation in a decentralized manner. Twenty out of 50 smaller municipalities in Tamil Nadu have achieved 100% segregation, while others have reached 80–90%. But still, the majority of Indian states are suffering the problem of mixing of segregated and unsegregated waste. According to the MoHUA's 2020 "Swachhata Sandesh Newsletter," 100% trash segregation is achieved at the household level as of January 2020 in 74.8% of wards. Chhattisgarh and Kerala have seen 100% waste segregation in all wards at the household level. Daman & Diu and Dadra & Nagar Haveli are the only Union Territories that have achieved 100% source segregation.

7.2.3 Waste Collection

According to Ministry of Environment, Forest and Climate Change (MoEFCC), only about 75–80% of solid waste is picked up, while only 22–28% of this waste actually gets processed and treated (MoEFCC, 2016). The remaining amount of waste gets either dumped in the ground or is clogging the drainage system as well as the sewerage systems. Indiscriminately dumped wastes also harbour disease-causing vectors. A recent study by ICRIER highlights that the national capital, Delhi, has the lowest percentage of waste collection, which stands at 39%, in contrast to the city of Ahmedabad, which has the highest waste collection, at 95% (Ahluwalia & Patel, 2018). With the amendment in the SWM rules in 2016, it is mandatory to practice door-to-door collection of separated waste, and the waste generators are obligated to pay a "user fee" to the collectors. But there are no details on how to decide the fee for waste collection—it is evaluated based on either the weight of waste or the type of waste. "Swachhata-Sandesh Newsletter" states that approximately 96% of the total wards in India are practicing door-to-door collection of waste as of January 2020. All Union Territories currently have a 100% rate of door-to-door collection. Mysuru has achieved tremendous success in this area as well as in-source segregation at the city level. Door-to-door collection is around 80% in Mumbai and Chennai. Warangal won the Clean Cities Championship in 2012; however Tirunelveli, Vengurla, and Uttarpara-Ktrung are still waiting for comprehensive rubbish collection door-to-door.

7.2.4 Waste Transportation

The movement of SW remains a challenge, as many cities' transit systems are underdeveloped. General vehicles used for waste transportation are as follows: handcarts or bin attached tricycles, tricycles with hydraulic tipping containers, small trucks used for commercial journeys with hydraulic tipping containers and, at a larger level, four-wheeled mini trucks are used that meet international standard of waste collection. The vehicle to be used is dependent of multiple factors such as the distance, the amount of waste, type of road, process technologies and so on. To make the transportation system efficient, companies have also installed technology such as GPS system, GIS as well as mobile communication systems to automate and regulate the waste collection from source to disposal site.

7.2.5 Solid Waste Treatment and Processing

Most common treatment techniques used in India are as follows: recycling, composting, incineration, biomethanation, best out of waste (wealth and energy), pyrolysis and refuse-derived fuel and so on. The use of a certain technology primarily depends on the waste type and quantity accessible, along with available of resource and funds, infrastructure capabilities, capital expenditure, land, calorific value and environmental issues of the neighbourhood. The most prevalent methods for processing huge amounts of trash for the formation of biogas, electricity, and compost include biomethanation and composting. Biogas contains 55–60% methane and can be used to generate electricity. In kitchen gardens, the compost is utilized as manure. Composting relies heavily on trash separation, which is a challenge in India because non-segregated waste is regularly deposited in open locations, contributing significantly to global warming. Hence, proper segregation is important to reduce waste transportation costs as well as harmful GHG emissions and leachate production. If waste is separated at the source, numerous recyclables can be used in a variety of manufacturing plants, resulting in commercial-use value (Mahanti, 2017).

7.2.6 Disposal of Solid Waste

In India, landfilling technology is rarely used for waste disposal. The collected SW is deposited on unprepared terrain that lacks foundations, liners, levelling, cover soil, leachate control and treatment facilities (Kumar et al., 2017). According to research, the majority of the country's landfills have now been depleted. According to Krishna (2018), the increasing demand to add newer landfill and waste-processing locations in the outskirts of municipal boundaries is expected to result in critical political consequences, although the 2016 SWM Rules are yet to specify the exact criteria for recognizing these facilities, as the current circumstances result in conflicts in land-use. For example, Jaitpur and Bawana were the proposed disposal sites in South East Delhi and North West Delhi, respectively. However, the ideas were met with opposition by locals and lower socioeconomic-status residents who were against the idea of garbage deposited in their neighborhood (Krishna, 2018).

7.2.7 Regulatory Framework for SWM

Large-scale concerns over inadequate MSWM methods were the subject of multiple public interest litigations (PILs) in the 1990s, which prompted the Supreme Court of India to form a committee dedicated to the country's MSWM situation. In 1999, the committee submitted its first report to the Supreme Court on SWM in India's Class I Cities. Following that, the Supreme Court ordered the MoEFCC to issue the MSW Management & Handling Rules, 2000, which mandated that all ULBs initiate a standardized waste management system, which would include project details and a timeline for installation of the processing units and proper garbage-disposal facilities. This was proposed to be accomplished by end of 2003, not only for metro cities but for all ULBs across the nation. The Ministry of Urban Development (MoUD) has produced an MSWM manual alongside the MSW Rules, 2000, to provide technical direction to all ULBs.

In 2005, funds were allocated under the 12th and 13th Finance Commission Grants to develop MSWM through programmes such as the Jawaharlal Nehru National Urban Renewal Mission (JnNURM) and the Urban Infrastructure Development Scheme for Small and Medium Towns (UIDSSMT). The National Urban Sanitation Policy was launched in 2008, and it covered urban sanitation and included SWM as a fundamental component. Despite positive pilots and accomplishments, the majority of ULBs still face challenges on multiple fronts such as in developing appropriate and sophisticated collecting and transportation systems, selecting technology and disposing of MSWM and maintaining long-term financial management of MSWM. Sixteen years later, since the notice of the MSW Rules, 2000, there is still non-compliance on MSWM. In 2014, the Ministry of Urban Development (MoUD) announced the Swachh Bharat Mission (SBM) to improve MSWM in Indian cities. SBM aims to promote cities as "engines of economic growth" by improving urban infrastructure quality, ensuring service standards and ensuring efficient governance. It also aims to address MSW management difficulties and provide assistance to cities in creating current and appropriate systems. The MSW (M&H) Rules, 2000, also received a change of name to "Solid Waste Management (SWM) Rules, 2016" by the MoEFCC. These rules include the following points:

- A list of authorities and their responsibilities in relation to management of solid waste;
- The state or union territory must establish a mandatory Municipal Solid Waste Management strategy and plan of action;
- Municipal Solid Waste Management Plans are required to be created by the municipality;
- Specific criteria for municipal solid waste management include segregation into wet, dry, and special trash as well as limits on what can be disposed of in landfills (only non-reactive, inert and pre-treated waste may be disposed);
- To make this service sustainable, the municipal government should impose service fees; people who trash in public spaces should be fined on the spot;
- Site selection and a necessary liner system are among the requirements for landfills;
- Environmental permits are required for the construction of municipal solid waste processing and disposal facilities, including landfills;
- Standards for composting of waste;
- Standards for treated leachate;

- Emission standards for incineration facilities; and
- Municipalities are required to report on MSW operations on an annual basis. The foregoing requirements of the Solid Waste Management Rules, 2016, must be carefully read by municipal authorities and other stakeholders, and concerted efforts must be made to develop solid waste management solutions and technologies in accordance.

7.3 INTEGRATED SOLID WASTE MANAGEMENT HIERARCHY

This concept is related with the 3R approach (reduce, reuse and recycle), which also targets at optimizing MSW management from all the waste-generating sectors. For concluding the processing or technology solutions for MSW, the planning process advises adapting the ISWM hierarchy (Figure 7.1-A). The most prevalent waste-avoidance measures are waste reduction at the source and product reuse, followed by recycling to recover material resources for the manufacture of new products. Waste disposal in open dumpsites is the least-desired choice. The 3R strategy is inextricably tied to integrated solid waste management. It is a helpful methodology for all ULBs to make sure that SW is managed in an environmentally sustainable manner and that resource recovery from waste is promoted.

7.4 TYPES OF SOLID WASTE TREATMENT

The various treatment strategies practiced for MSW and alternative, similar forms of waste are composting, aerobic digestion, biomethanation, thermal processes (incineration, pyrolysis), etc. These are the different processes that are performed at different stages in the waste treatment (Jagrutiben & Shelar, 2019).

7.4.1 Composting

Composting is one of the most traditional techniques. It is a biological process in which bacteria decompose and stabilize organic matter from solid waste, either with or without oxygen. One of the most common methods of composting is windrow composting. In this method, the organic MSW is stacked up in piles known as windrow, and sheds are installed for these windrows. Collected waste is applied with sanitrate and bioculum and formed into heap—windrow. The pile is placed loosely and turned frequently to allow aeration for better compost formation. It is important to control the C:N ratio, humidity and temperature for optimum growth of the microbial inoculants. Thus, frequent turnings are given to windrows, and this is screened into electro-mechanical trammels. Final end product is marketed which acts as a nutrient supplement for the agricultural crops.

7.4.2 Aerobic Digestion

Conversion of organics via bacteria and in the presence of air is referred to as aerobic digestion. It produces compost, which is often utilized as a fertilizer as a final product.

(A)

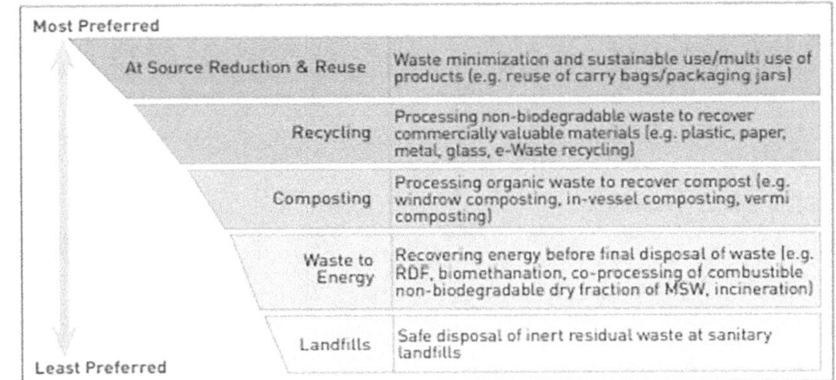

(B)

(C)

FIGURE 7.1 (A)—The hierarchy of integrated solid waste management. (B)—Transfer of solid waste from tipper vehicle to refuse compactor. (C)—Plastic waste to oil plant, and wood wastes to biofuel briquettes plant.

7.4.3 Biomethanation

In an oxygen-free environment, it is the process by which microbes transform organic materials into stable, inert residue. It generates methane-rich biogas for electricity and cooking as well as inert residue for manure.

7.4.4 Incineration

Incineration is a simple combustion process of drying and burning of waste. The end product of the process is carbon dioxide, water vapor, form and ash with a large amount of heat. It needs temperature between 980 to 2000 °C (Sharholy et al., 2008).

7.4.5 Pyrolysis

Pyrolysis is a waste-to-energy technique that involves distilling solid waste to recover its contents and energy.

7.4.6 Landfilling

Landfills are the sites where storage of inert waste is done, aggregated with techniques for control of gases, leachates and effluents into the open environment. Sanitary landfills are used for any SW that cannot be further processed. A typical sanitary landfill is a discrete area of land with a complex design consisting of pit liners, basically composite liners or geo-membrane and clay soil to protect groundwater from leachates. On the top of the liners, leachate collection and removal systems are present, which are meant for leachate removal for further treatment and disposal. Besides this, routine monitoring of the site takes place (soil and groundwater), along with compaction of the pit. Once the landfill reaches its capacity, scientific closure of the site is ensured for complete sanitization of the area and for environmental protection (Mondal et al., 2023).

7.5 CASE STUDY 1—THANE CITY

7.5.1 Location

Thane is a major metropolis in Maharashtra, India. It's also called as the "City of Lakes." Thane City lies within Thane *taluka*, one of Thane district's seven *talukas*, and serves as the district's headquarters. It is India's 15th most populous city, with a population of 1,841,488 people, spread over a land area of around 147 square kilometres (57 square miles), according to the 2011 census.

7.5.2 Demographics

As per 2011 census, the total population of Thane was estimated to be 11,060,148, of which male and female were 5,865,078 and 5,195,070, respectively. As per the census before 2011, the population was 8,131,849, of which males were 4,377,747 and the remaining 3,754,102 were females. The population of Thane district contributes to around 9.8% of Maharashtra's total population as per the 2011 census and is continuously growing.

7.5.3 Solid Waste Management in Thane City

Through various methods, the city produces roughly 950 metric tonnes of SW. Every year, approximately 7, 03,556 kg of biomedical waste is generated.

The city's major contributing factors of increase in solid waste generation include residential areas, malls, businesses, hotels, marketing complexes, film theatres, private hospitals and dispensaries, multiplexes as well as increased C&D operations, and are all contributing to an increase in SW generation inside the city. Thane Municipal Corporation (TMC) has split the city into nine administrative wards to carry out SWM for administrative purposes. In Ghantagadis, 212 regular staff and around 441 contract workers collect SW to carry out SWM. TMC has rented 109 large and 106 small *ghantagadis* from private companies for rubbish collection and transportation from house to house.

7.5.3.1 Cleaning Roads and Lifting Waste—Contractual Basis

Because cleaning the city's roads is such a large task and the Thane Municipal Corporation (TMC)'s manpower is insufficient, a mere 20 private contractors have been hired. For this project, these contractors have hired 1,645 people. In Ghantagadis, 1,204 people clean 236 kilometres of roads and 441 workers gather trash from various locations throughout the city.

7.5.3.2 Dump Yard

Presently, Thane City lacks a designated dumping yard. SW from Thane is currently being dumped on a 15-hectare plot of private property in Khardi. Survey Nos. 38/1, 38/2, 38/3, 38/4, 38/5, 38/6 and 38/7 are located at Khardi in Mumbra Ward. Another land from the C.P. Tank area is also used to store SW from regions other than Kalwa and Mumbra. The SW is temporarily stationed at the C.P. Tank area because Khardi is located away from the main city and the route of travel is not wide enough for big vehicles. The SW is then transported to the Khardi dumping zone by 10-wheel dumpers. Every day, roughly 700 to 750 tonnes of SW are transported from C.P. Tank to Khardi. The construction of a modern transfer terminal is underway to speed up the transit procedure.

7.5.3.3 Solid Waste Processing Center

TMC has set aside three pieces of land for SW disposal and processing, according to the Thane City Development Plan. MaujeBhayandarpada (8 ha in Sector 6), MaujeShil (14.3 ha in Sector 11) and MaujeDighar are the three pieces of land (Sector 11 total 18.89 hectares). TMC's request to build a SW processing centre on 18.89 hectares of land at MaujeDighar's Survey Nos. 15, 16, 17 and 18 was authorized in 2004. However, this project was not successfully appointed due to strong opposition from the local residents residing near the land. Since 2006, the effort of removing SW has been on hold. M/s. Thane Clean Environment Pvt. Limited has started a "Waste to Energy" project based on modern technologies in order to bring the situation back on track.

7.5.3.4 Use of Forest Land

A proposal has been presented to the Principal Secretary, Ministry of Revenue Forests, Mumbai, to make forest areas accessible for SWM. The goal is to utilize empty grounds on these lands for garbage and construction wastes, followed by tree planting. TMC will cover all costs and, once the trees have reached full maturity and the forest lands have been developed, the Forest Department will get them. As a result, on June 1, 2011, a final approval was awaited by the Forest Department that was, too, sent as a tripartite agreement. The Forest Department had requested a No-Objection Certificate from the Environment Department, which was later sent to the department. Once TMC has taken possession of the property, it is intended that it be used for solid waste disposal.

7.5.3.5 Treatment Plants

Due to the delay in the construction of the SW disposal project, TMC has begun the decentralized construction of bio/mechanical composting and bio-methanization projects with capacities ranging from 1 to 5 tonnes. Following the completion of the first phase's processing of 100 tonnes of SW, the second phase's processing of another 100 tonnes of SW will begin. Tenders for the same have been requested. The tender process for three decentralized projects based on plasma technology for processing 100-tonne capacity each has already begun.

Collection and disposal of horticulture waste uses a composting-cum-gardening strategy.

7.5.3.5.1 Wood Waste Collection

In Thane City, TMC initiated a wood waste to fuel briquettes operation. This project is primarily responsible for the city's horticultural waste and leaf litter. TMC's Garden Department has deployed separate vehicles to collect garden garbage throughout the city. This garbage is collected and transported to the treatment facility. Garden waste is collected, dried, de-moisturized and compressed into briquettes using compression machinery at the factory. The project's capacity is 10 TPD.

TABLE 7.1
Various Projects in Thane Based on Type of Solid Waste Generated

	Various Projects Solid Waste Management	
1)	green waste to fuel briquette	20 tonnes (implemented)
2)	kitchen waste to BioCNG	25 tonne (implemented)
3)	fertilizer from flower	1 tonne (implemented)
4)	bio-gas (CNG) from animal carcass	60 tonnes
5)	methane gas from garbage (Lodha complex)	5 tonnes (implemented)
6)	methane gas from garbage)	5 tonnes (implemented)
7)	plastic waste to oil	5 tonnes (started up)
8)	construction material from	100 tonnes (starting up)
9)	e-waste compilation	1 tonne (implemented)
10)	thermocol recycling	1 tonne (implemented)
11)	biomedical waste	1.5 tonnes (implemented)
12)	dry garbage collection centre	15 tonnes (implemented)
13)	bio-methaneshation capacity of 1 to 5 tonnes	100 tonnes (proposed)
14)	waste to electricity	awarded 600 tonnes per day

7.5.3.5.2 Plastic Waste to Oil

A total of 92 tonnes of plastic waste is generated within the area of TMC. 70 to 75% of that waste is picked up by waste pickers. The remaining plastic, which lies in the soil or which has very low reuse value, is left over in the waste. Such waste, when carried to dumping ground, becomes threat to the environment or, when drawn into drains, causes a flood-like situation. Therefore, TMC has proposed to produce oil from this plastic waste.

Plastic is basically made from petroleum substances. Therefore, obtaining oil by way of a pyrolysis method can be profitable. Hence, as per the government resolution in accordance with Plastic Waste and Management Handling Rules 2011, this project has been proposed to be undertaken as a pilot project (of 5 tonnes) in TMC area. In the pyrolysis method, plastic waste is melted in an oxygen-less plant, and oil is obtained by a distillation process. From plastic waste, 65–70% oil as a by-product and 10–12% carbon powder and hydrocarbon gas for fire are produced. The oil thus-produced is used in diesel generators and industries for greasing. Also, the hydrocarbon gas is used in the plant itself for fire, which reduces the daily use of electricity to the lowest. This plant is working at Kopri, in association with Samarth Bharat Vyaspeet (Figure 7.1-B).

7.5.3.5.3 Manure from Nirmalya

During the Ganesh festival, TMC has been actively performing the project of generating manure from Nirmalya (flowers as offerings) for the past three years. A sophisticated procedure was used to convert 120 tonnes of *nirmalya* harvested last year into manure. The shredding machine shreds this *nirmalya* into small bits. At the same time, bio-culture is added, hastening the degradation process and converting 3 tonnes of *nirmalya* into manure in 15 days. If this project's work is assigned to a social organization, it is envisaged that it will promote awareness in order to collect *nirmalya* at temples and from residential complexes in the temple's immediate vicinity, so assisting in the control of lake pollution.

7.5.3.5.4 Fuel Briquettes from Wood Waste

Thane is surrounded by greenery. The city's green space accounts for around 33% of its total area. In the city, there are 4 lakh trees. The city of Thane generates a lot of wood trash. Garden garbage from homes, twigs, branches, dry leaves, trees and other wood debris are all included in the wood waste category. This wood debris is currently disposed of in landfills. Scientifically, fuel briquettes can be made from such wood waste. In March 2016, Samarth Bharat Vyaspeeth began a project to transform wood waste into fuel briquettes on a PPP basis. In the project established at Kopri, they are already active in making manure from nirmalya accumulated from primarily temples, cemeteries and floral markets. The institute has initiated a project to make fuel briquettes from wood waste using a scientific process as an extension of this already-running project. This project has a capacity of 10 TPD and a 10-year duration. TMC has generously donated the institute with a block of land measuring around 20,000 square feet. TMC also provides sewage, toilet facilities, electricity, water connections and bill payment to its customers. This institute donates 1 tonne of fuel briquettes to the municipality every day for free, while selling the rest. The institute was responsible for both capital and operating costs (minus energy and water bills) (Figure 7.1-C).

7.5.3.5.5 Scientific Technique E-waste Disposal

Electrical and electronic garbage is referred to as e-waste. As the world becomes more urbanized, so does the amount of e-waste produced. Computers, printers, television sets, washing machines, mobile phones, air conditioners, typewriters, xerox machines, telephones and other electronic devices are included. Computers and mobile phones are typically utilized for three to five years before being discarded. These electrical and electronic items contain significant amounts of rare earth metals such as lead, copper, zinc, cadmium and other metals. As a result, the informal sector processes these items in enormous quantities in order to recover valuable metals. The items are often processed by melting them in acid, burning them and so forth. These processes not only pollute the environment to a large extent but also impair human health. The e-waste generated in the municipal corporation region is scientifically disposed of under this initiative by M/s Eco-recycling Private Limited on a PPP basis. TMC has placed 100 free-of-charge containers created by the proprietor firm over a 10-square-foot area in 100 locations for the next five years. TMC receives a profit share from the corporation when it disposes of collected e-waste and recovered items.

The e-waste collection centres have been designated at the following locations:

- TMC headquarter, Panchpakhadi, Thane;
- Naupada Ward Committee, Naupada, Thane;
- Uthalsar Ward Committee, Uthalsar, Thane;
- Rayaladevi/Wagle Ward Committee, Wagle Estate, Thane;
- Lokmanya Nagar/Vartak Nagar Ward Committee, Vartak Nagar, Thane;
- Majivada Ward Committee, Majivada, Thane;
- Kopri Ward Committee, Kopri, Thane;
- Kalwa Ward Committee, Kalwa, Thane;
- Kalwa Chhatrapati Shivaji Maharaj Hospital, Thane; and
- Mumbra Ward Committee, Mumbra, Thane.

7.5.3.5.6 Treatment of Construction and Demolition (C&D) Waste Generated

Debris and broken stones and bricks make up the majority of C&D waste. All of this garbage is dumped in rivers and bodies of water, along roadsides and in shrubs and thickets. Natural resources, aquatic life and biodiversity are all impacted. These wastes are frequently deposited on the creek's bank, in public locations and along roadsides in Thane. This has an impact on the city's environment and cleanliness. Currently, builders hire contractors to dispose of such garbage. These contractors separate the material that can be sold from the waste, such as broken bricks, RCC fractions after iron/steel rods have been removed and lime, among other things. In

Thane, almost 100 tonnes of such debris are generated every day. A project for the collection, transportation and reuse of C&D waste generated in Thane City has been suggested to be done on a public-private partnership (PPP) basis through a competitive contract process. The capacity of this project will be 300 TPD. TMC will provide a free 5-acres of land for this project, which will have a 20-year duration. By collecting, transporting and recycling C&D trash, useful by-products such as paver blocks, concrete blocks and RMC tiles will be produced. The selling of such products will generate revenue for the PPP partner.

7.5.3.5.7 Dry-Waste Collection Centre

Every day, 700 tonnes of urban SW are generated in TMC. With the growth of the city, there is an increase of building for residential as well as commercial purposes. In general, the urban SW has 40% wet waste and 60% dry waste, which includes paper, plastic, glass, metal scraps, wood, and other materials. According to the Urban Solid Trash Rule 1000, waste segregation of dry and wet waste is a must, and dry waste must be disposed of in accordance with proper standards to protect the environment. As a result, TMC has begun the process of separating wet and dry garbage. The collecting of dry garbage from housing societies has begun in three ward committees—namely, Vartak Nagar, Majiwada and Uthalsar—on an experimental basis. Similarly, TMC has made initiatives to raise public awareness and establish a PPP-based dry trash collection centre. A space of 10,000 to 15,000 square feet will be given as part of this project, along with a shed. The chosen organization will work to increase public awareness in the societies. The aim is to collect dry waste, sort it, and resell it to other organizations, with the help of rag-picker women. TMC's garbage transportation burden will be reduced, and waste pickers will benefit.

7.5.3.5.8 Recycling of Thermocol

TMC has taken steps to establish a thermocol-recycling scheme based on scientific principles. The project, which is being handled on a PPP basis, has an annual average capacity of 1 TPD. This helps in minimizing pollution produced by thermocol dumped on landfills. This project is in progress at TMC's C.P. Tank waste segregation facility.

7.5.3.5.9 Bio-Methanisation Plant at Chhatrapati Shivaji Hospital, Kalwa

TMC, in collaboration with Prerana Bahuddeshiya Sanstha, has set up a Bio-CNG generation project with a capacity of 125 TPD on the campus of Chhatrapati Shivaji Maharaj Hospital. Three of this organization's sub-institutes have been granted an international patent in this sector. TMC's SWM Department gathers wet waste from hotels, restaurants, banquet halls, hostels, messes and fruit and vegetable markets and delivers it to the Bio-CNG plant for processing. The fuel produced by this initiative is distributed to the Chhatrapati Shivaji Maharaj Hospital and the Thane Municipal Transport Service. Every day, 1,500 m3 biogas, 650 kg Bio-CNG and manure are produced. The project is planned for 20 years. A 5% property tax reduction has been announced for housing societies and establishments who adopt waste separation techniques to minimize waste, reuse and recycle garbage (3R).

7.5.3.5.10 Biomedical waste

Around 1,982 fee-charging medical institutions have been registered on behalf of Thane Municipal Corporation. A unique bar code has been supplied to each of these medical facilities. Similarly, cars transporting biological waste (BW) have GPS-based vehicle tracking systems installed. From January to December 2017, these facilities created around 7, 03,556 kg of biomedical waste.

7.5.3.6 Capital Expenditure on Solid Waste Management

TMC is working hard to achieve scientific sewage and waste management on all fronts. Sewage and SWM were budgeted for at 19.7% of total capital expenditure in the TMC Budget 2016–17. In comparison to the previous year's budget, a provision of 21% of total capital expenditures was set aside for sewage water and solid waste management in 2017–18.

7.5.3.7 Fine for Dumping Waste at Public Places

The rate of fines for depositing rubbish in public spaces and making the region dirty and unsanitary has been set by the TMC. The byelaws in question have been sent to the government for approval.

Public penalties for littering and waste dumping ranges from INR 200 to INR 10,000, the details of which can be accessed at the official website of Thane Municipal Corporation (thanecity.gov.in).

7.6 CASE STUDY 2—NASHIK CITY

7.6.1 Location

Nashik, the administrative capital of the Nashik Administrative Division (which includes the districts of Nashik, Ahmednagar, Nandurbar, Dhule and Jalgaon), is around 180 kilometres from Mumbai. Nashik Municipal Corporation (NMC) is located in the northwest corner of Maharashtra State, between 18°33' and 20°53' north latitude and 73°16' and 75°16' east longitude (Figure 7.1). Nashik City is located on both sides of the Godavari River, with Panchavati on the left bank and Tapovan on the right bank and the old Nashik on the right bank. The city's main functional areas are the Nashik core (the Nucleus), Satpur and Ambad industrial areas and the Nashik–Pune road area, which includes the Nashik-Road railway station and is supported by industrial activities at Eklahara Thermal Power Plant, Railway Traction Factory and other locations. Satpur is an industrial zone. The MIDC Satpur industrial estate is mainly responsible for industrial and commercial growth of Satpur area. The Civic Administration area of 267.48 square kilometres (26,747.75 hectares) is managed by NMC.

7.6.2 Demographics

Nashik has a population of 1,486,053 people in 2011, compared to 1,077,236 people in 2001, suggesting a growth rate of 37.95% between 2001 and 2011.

Nashik's population is expected to reach approximately 4 million by 2030 as a result of increased migration to metropolitan areas. NMC has a population density of 5,556 people per square kilometres, according to census data from 2011. Rapid motorization is taking place throughout the city, resulting in increased traffic and pollution.

7.6.3 Solid Waste Management in Nashik City

Almost 550 MT of waste is generated in NMC. Sweeping collects rubbish from the highways, while door-to-door collection by cars ("*ghantagadis*") with the help of 2,778 sweepers collects domestic waste. Two contractors have been awarded the collection and transportation of SW from the six divisions and 31 wards by the NMC. The collection and transportation contract comprises door-to-door SW collection via *ghantagadis* and transportation to an MSW treatment plant. SW is collected from families in various wards of the city using 260 vehicles, with NMC owning the vehicles. In an incinerating plant near Kannamwar Bridge, bio-hazardous waste generated by city hospitals is treated at 1000 °C (near core area). Municipal Corporation has installed *nirmalya kalash* at various locations along with rivers for collecting *nirmalya* from the city. Waste generated per capita per day in the NMC area is approximately 320 grams. Waste is collected via door-to-door collection method, and approximately 550 MT of waste is collected per day. It is then transported from the collection points to the landfill site near Pandav Leni Caves every day. Information about waste treatment plant installed in compost plant near Pandav Leni Caves by Municipal Corporation. Following table (Table 7.2) gives a brief of the present state of SWM in the city of Nashik.

7.6.3.1 Treatment Plants

7.6.3.1.1 Pre-Sorting Unit

It is an electro-mechanical segregation system with a capacity of 500 TPD for incoming non-segregated MSW. It has two lines with all mandatory specifications and materials. Compostable material is delivered to windrow composting after mechanical segregation, with higher calorific value material going to RDF plant and inert material going to an inert-processing facility for further processing.

7.6.3.1.2 Aerobic Composting Unit

Composting is done in this unit using the windrow composting method, and windrow buildings have been built. Currently, 3 to 5% of total MSW is transformed into compost. Within a 100-kilometer radius of Nashik, the compost has already earned a reputation among farmers. As the trend of waste segregation at source rises, the amount of compost generated will rise to 10 to 15% of total MSW. The functioning of the compost facility is based on a PPP model.

7.6.3.1.3 Inert Processing Unit

Mechanical sieves and an air density separator make up the inert processing unit, which has a capacity of 50 TPD. The main goal of this factory is to revive construction

TABLE 7.2
Solid Waste Management in Nashik City

Sl. No.	Particulars	2018–19
1	Population of NMC area (census 2011)	1,486,053
2	Total waste collection per day	550MT
3	Per capita quantity of waste generated	320 grams
4	Percentage of segregation	Wet 52% and dry 48%
5	Waste dumped at disposable site	550MT/day
6	Location of the dumping site	Sr. No. 278, Pathardi Shivar, Behind Pandav Caves, Pathardi Phata, Nashik—422010
7	Authorization for this site	Authorization no. BO/MSW/B-6253 dated 05/11/2011
8	Area of dumping site	48,000 square metres
9	Distance from city to dumping ground	12 km
10	Manpower required for management of waste	2,778
11	Number of vehicles at composting plant	27
12	Number of vehicles (private/NGO'S) for collecting solid waste	260
13	Present mode of treatment disposal	Aerobic composting unit, sanitary landfill, inert processing-unit, leachate-treatment plant, animal-carcass incinerator and refuse-derived-fuel plant.

materials from trash and recycle them for resale or for in-house construction projects. The primary goal is to reduce landfill impact on O&M costs while also conserving land.

7.6.3.1.4 Leachate Treatment Plant

A leachate treatment plant with a capacity of 0.4 MLD is created for the treatment of leachate from solid waste dumps, windrows, and sanitary-landfill sites. The pick-up and transport of leachate has been properly organized. Because leachate is primarily produced during the monsoon season, the same plant is also used to make biogas from organic waste during other times. The plant generates about 40 kW of power, which is used to run the pumps at the MSWM facility.

7.6.3.1.5 Refuse-Derived Fuel (RDF) Plant

From the step of receiving at the tipping level onwards, materials with significant calorific energy in MSW are handled individually. For the production of fuel pellets

TABLE 7.3
Showing Particulars of Various Waste-Treatment Plants of Compost Plants Near Pandhav Leni Caves

Sl. No.	Plant	Capacity	Present treatment
1	Electro-mechanical segregator	500 TPD	Average 500 MT
2	Composting plant	500 TPD	Average 12 to 15 TPD
3	Inert processing unit	50 TPD	Average 1.5 TPD (13 April to 14 March) fuel production 71,435 lit
4	Leachate treatment plant	0.4 MLD	Leachate treatment plant with capacity of 0.4 MLD leachate or 10 TPD organic wastes has been installed for treatment of leachate coming out from the windrows, the solid waste dumps and sanitary landfill site
5	RDF plant	150 TPD	RDF plant with capacity of 150 TPD is installed for generation of fuel pellets from high calorific value materials
6	Animal carcass incinerator	250 kg/h	Dead animal carcass incinerator with the capacity of 250 kg per hour is installed for the incineration of dead animals such as dogs, cattle, etc.
7	Landfill	2 hectares	

from high calorific-value materials, an RDF plant with a capacity of 150 TPD has been erected. The major ingredients of RDF, which is a significant source of alternate energy, are woody materials, paper products, textiles, jute and so on. The main focus of this plant is on the refinement of SW through a series of processes such as segregation, drying, size reduction, material recombination, blending and homogenization. It undergoes further refinement by separating sand, dust, metals, glass and others before grinding or shredding.

7.6.3.1.6 *Animal-Carcass Incinerator*

For the incineration of deal animals like dogs, cattle, etc., a dead animal's carcass incinerator with the capacity of 250 kg per hour has been installed (Figure 7.2-A).

7.6.3.1.7 *Waste to Energy (WtE) Plant*

The technology employed in a WtE plant funded by German Technical Corporation (GIZ Germany) is the bio-methanation process. The capacity of plant shown in Figure 7.5 is 30 TPD. Waste comprising of food and kitchen waste (around 15 tonnes) and septage of septic tanks (around 15 tonnes) is processed in the plant. The plant is capable of producing 3300 units electricity per day. Currently, roughly 500 kilowatt-

FIGURE 7.2 (A)—MSW treatment plant. (B)—Waste to energy plant.

hours (kWh) of power are produced per day. The factory has been operational for the past year (Figure 7.2-B).

7.6.3.1.8 Sanitary Landfill

The SW which cannot be further processed is sent to the sanitary landfill site. For this purpose, a sanitary landfill in an area of 2 hectares has been developed. All the necessary aspects of scientific landfilling were taken into account at the time of creation of sanitary landfill. Suitable ordering has been done for leachate collection and is further connected to the leachate treatment plant for its processing.

The open burning of waste is regulated by a strict punishment system. For this, a special team has been developed to detect open garbage or waste burning in the area. NMC has developed a grievance app called NMC e-connect to allow the public to participate in the process. A total of 62 cases have resulted in a fine of Rs. 3.10 L being collected from those who burn waste in the open. Every three months, public awareness activities are organized with the support of NGOs and educational institutions (Figure 7.3-A).

7.6.3.1.9 C & D Waste Storage, Segregation, Processing and Recycling

NMC has a facility for segregation of C&D waste of capacity 50 TPD shown in Figure 7.3 Collected C&D waste is segregated into stone, bricks, mortar, concrete, metal, sand, soil, steel, wood, etc. Figure 7.3 also shows road being constructed using recovered C&D waste (Figure 7.3-B).

(A)

(B)

FIGURE 7.3 (A)—Sanitary landfill. (B)—C&D waste segregation and processing unit of 50-TPD capacity. (C)—Chikalthana waste processing plant (capacity: 150 TPD; installed: 01 June 2019; make: Mayo Vessels & Machineries, Aurangabad). (D)—Kanchanwadi biogas plant (capacity: 30 TPD; construction, machineries and 10 years O/M; installed: 26th January 2021; expected output: 4KW of electricity generation).

(C)

(D)

FIGURE 7.3 (Continued)

7.7 CASE STUDY 3—AURANGABAD CITY

7.7.1 Location

Aurangabad is located on the banks of the Godavari River's tributary, the river Kham. The city is located at 19.8762 degrees north latitude and 75.3433 degrees east longitude (Table 7.1, Figure 7.4). It is located on the Mumbai-Nagpur Highway, about 335 kilometres from Mumbai and 475 kilometres from Nagpur, at an elevation of 512 metres above sea level.

7.7.2 Demographics

Aurangabad has a population of 565,910 females and 609,206 males according to the 2011 census. A total of 889,224 people were literate, accounting for 75.7% of the population, with male literacy at 79.3% and female literacy at 71.7%.

7.7.3 Solid Waste Management Plan in Aurangabad City

75% of city waste is processed at centralized processing plants at Chikalthana, Harsul-Sawangi and Padegaon, each having capacity of 150 TPD. The remaining 25% of city waste is treated at decentralized processing plants. A 30-TPD bio-methanization plant has been planned at Kanchanwadi. Collection of construction and demolition waste generated within Aurangabad Muncipal Corporation (AMC) limits is taken to a 100-TPD-capacity plant to utilize it on PPP basis for 10 years.

Aurangabad City generates around 450 to 500 metric tonnes of waste on a daily basis. AMC had decided to install the waste processing plant of 150 TPD capacity at Chikalthana, Padegaon and Harsul. In the future, waste generation will increase; the decentralized plants will be needed. The tenders for the civil construction and processing of waste for Chikalthana and Padegaon plants were approved earlier. The contract for civil construction was awarded to NK Construction, and the contract for waste processing was given to Mayo Vessels for both the sites. The Chikalthana plant was commissioned in June 2019, while the Padegaon plant has been commissioned in March 2020.

For transportation, a fleet of 337 vehicles has been pressed into service. 19 compactors (Figure 7.3-C) and 10 tippers, 200 auto tippers were purchased by the private agency, while 108 auto tippers and 12 skip bins and hook loaders have been provided by the AMC. This agency will be paid by the AMC as per the weight of the garbage it collects and transports.

7.7.3.1 Treatment Plants

As per the SWM rules 2016, decentralized processing is an attempt to maximize the biodegradable-waste processing and recovery of reusables by opening dispersed facilities near the waste generation source. AMC finalized on installing three decentralized units with a capacity to process 16 TPD of waste with one unit being started operations at Chikalthana and two others have been planned at Harsul Sawangi and Padegaon, around 7–8 km away from the city. Wet garbage is composted in a pit, while dry waste is disposed of in a material recovery facility, which involves segregation, shredding and bailing (Figure 7.3D). According to the AMC, due to a scarcity of land, decentralized waste processing plants could not be built within the city. The city's dry waste disposal is being handled with the support of the NGO CARPE. In addition, over 15 lakh tonnes of trash are dumped at the Naregaon depot. The waste cap is proposed in the SWM plan. An amount of Rs 25 crores has been set aside for the scientific capping of the Naregaon dump.

Separate vehicles have been arranged to pick up the wet waste coming out of the city, and this waste is being picked up from the city and disposed of in a scientific manner by pit composting at Padegaon. To prevent open garbage burning in the city, 27 ex-civilian servicemen's friend squads have been assigned to each of the nine zones. The team takes disciplinary action against people who burn waste in the city, litter openly or use plastic. The Kanchanwadi Biomethanation Project, with a capacity of 30 MT, has been completed. The work of producing gas and electric power by processing 30 MT of food waste per day has been started (Figure 7.3-B).

7.7.3.2 Digitalization

Use of the Letstrack application for efficient waste collection drive in the city was established. Letstrack application is a GPS-tracking AI-based system that works on providing real-time data to all commercial and non-commercial vehicles. Mapping door-to-door waste collection drives helps maximize the collection potential and monitor ward-wise collection drives. A mobile app is used to ease the tedious process of keeping track of multiple collection drives. Following images (Figure 7.4 and Figure 7.5) are an example of use of the application.

(A)

(B)

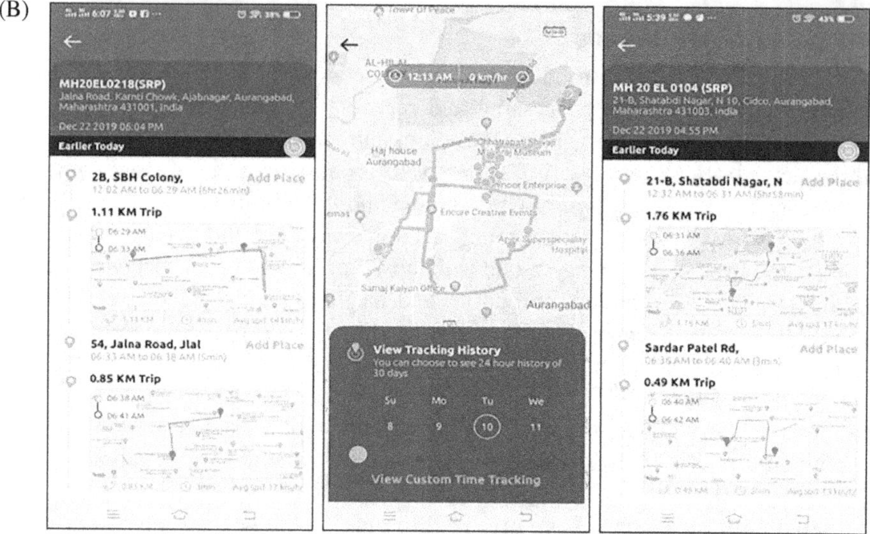

FIGURE 7.4 (A)—Typical screen of Letstrack software. (B)—Typical display output of software.

7.7.3.3 Construction and Demolition Waste

AMC contributes to about 80MT/day of C&D waste, the highest in the district for C&D waste. A copy of public notification by AMC for C&D waste management is given in Figure 15. Collection of C&D waste generated within AMC limits and Establishment & Operation of 100-TPD-capacity plant to utilize it is on PPP basis for 10 years. There are penalties for citizens for waste dumping on streets, spitting/urinating in public places and use of plastic (Figure 7.4 and Figure 7.5A–D). C&D waste is used by AMC for filling low-lying areas (Figure 7.4 and Figure 7.5A–D).

FIGURE 7.5 (A)—Waste collection door-to-door. (B)—Compost preparation. (C)—Fine against C&D waste on road. (D)—C&D waste being used by AMC for filling low-lying areas.

(D)

FIGURE 7.5 (Continued)

7.8 COMPARATIVE ANALYSIS BETWEEN THANE, NASHIK AND AURANGABAD CITIES

This section highlights the key comparative points for all the three cities discussed in this chapter; namely, Thane, Nashik and Aurangabad. It can be observed that Thane City has the most advanced SWM system in place, followed by Nashik and Aurangabad. However, Thane City has been facing challenges to meet the ever-increasing demand of waste collection due to lack of adequate manpower, vehicles and availability of dumping ground in the city. Presently, waste is transported outside Thane City to be treated. NMC has proposed to advance their treatment plants to process maximum percentage of the waste generated. Aurangabad City, new to SWM, has only begun treating SW in the past couple of years, and from collecting and treating zero MT/day to treating 480MT/day, has set a good example for other cities across the country. From proposed treatment plants, to provision of proper infrastructure and a digital website to track waste collection in real-time, Aurangabad City has led into collecting 100% of waste. There is a tremendous scope of improvement in the waste-processing systems for all the three cities, such as to increase decentralizing the SWM systems, segregating 100% of waste, implementation of advance technologies and policies that meet the standards of international cities, educational programs to increase awareness amongst the citizens, availability of appropriate infrastructure and funding to support these improvements.

7.9 RISING CHALLENGES IN SOLID WASTE MANAGEMENT FOR INDIA

India is third in the list of highest SW generators, after China and the US. This is a cause of serious concern, as most of the ULBs are not fully equipped to process the increasing amounts of waste per day. Some of the key challenges are discussed in the following paragraphs:

- The biggest challenge revolving around the waste management system in India is the conflicting data documented in various journals and reports, as there is no system of collecting data in a periodic format since the inception. This has led to extremely variable projections and estimations;

- As per the Solid Waste Management Rules 2016, it is highly recommended to segregate waste at the household level, dividing them into three categories of biodegradable, non-biodegradable and hazardous. Most ULBs have failed to carry and establish this system, along with a general lack of awareness of the importance of waste segregation within the public;
- According to research, garbage collection effectiveness in India is low due to informal and unstructured waste-collection practices. Waste collection efficiency has been perfected in those places and regions where private and non-governmental bodies are actively taking measures to collect waste;
- In most cities, the waste is being dumped at low-lying areas around the city due to lack of availability of land for landfill. This gap in the process of waste management has produced severe challenges for communities residing near these dumping grounds;

TABLE 7.4
Comparative Table on SWM in Thane, Nashik and Aurangabad

Particulars	Thane	Nashik	Aurangabad
Population of MC area (census 2011)	11,060,148	1,486,053	1,228,032
Total waste generated per day (MT/day)	1,039	600–650	500
Total waste collection per day (MT/day)	~ 1,000 (implemented and on-going)	550	480 + decentralized collection centres
Percentage of segregation	Wet 40%, dry 60%	Wet 52%, dry 48%	Wet 60%, dry 40%
Percentage of waste collection	80–90%	90%	~ 100%
Area of dumping site (sq.m)	150,000	48,000	NA
Distance from city to dumping ground (km)	13	12	6
Present mode of treatment disposal	Advanced—biomethanation, thermocol recycling, plastic to oil, fuel briquettes from waste, BioCNG, construction material from debridge, methane gas from garbage, e-waste treatment, animal carcass incineration, sanitary landfill	Moderate—aerobic composting unit, inert processing unit, leachate treatment plant, refuse-derived fuel plant, animal carcass incinerator, sanitary landfill	Beginner—Biomethanation, proposed treatment plants in progress

- Lack of detailed study for each city across the country has resulted in inadequate information on the actual status of waste generation and treatment. Due to unavailability and, several times, misleading information, policymakers find it challenging to provide appropriate solutions for a particular region;
- The infrastructure and resources available to local administrations are insufficient. As a result, they are unable to adopt cutting-edge garbage treatment and disposal methods;
- Trash to Energy (WtE) is a commonly utilized technique in India; however, it has a number of drawbacks, including unsorted waste and seasonal variations in waste composition. According to research, most WtE plants are inefficient due to operational and design flaws;
- Along with the concern of inadequate manpower and infrastructure, Covid-19 has also increased various additional challenges such as lack of availability of PPE/safety kits, reduced employees due to social distancing, etc., leading to additional pressure of corporations to process their waste;
- A lack of sophisticated waste processing facility has been a constant cause for concern, along with lack of proper planning and regular maintenance and evaluation of the performance of these processes.

7.10 CONCLUSIONS

The SWM system in Indian cities is still an issue, as urban local bodies (ULBs) are still unable to effectively manage solid waste. Due to their reliance for funding on the state government, local governments lack the funds to adopt new land or acquire the technologies needed for SWM. But it is encouraging to find that both waste collection and its processing are becoming better and will be even better in coming times. Garbage pickers face tougher situation due to lack of legal standing and protection. This results in slowing down the efficiency of waste collection and segregation. Institutional and financial issues must be prioritized in order for the situation to improve. While the 2016 SWM Rules provide solutions to a number of difficulties, compliance remains a problem. A comparative analysis of waste generation, waste collection, level of waste segregation, waste treatment technologies utilized and landfill management has been presented for the three cities of Thane, Nashik and Aurangabad. It can be observed that Thane City has the most advanced SWM system in place, followed by Nashik and Aurangabad. However, Thane city has been facing challenges to meet the ever-increasing demand of waste collection presently; waste is transported outside Thane City to be treated. NMC has proposed to advance their treatment plants to process maximum percentage of the waste generated. Aurangabad City, new to SWM, only began treating SW in the past couple of years, and from collecting and treating zero MT/day to treating 480MT/day, has set a good example for other cities across the country. There is a tremendous scope of improvement in the waste processing systems for all the three cities, such as to increase decentralizing the SWM systems, segregating 100% of waste, implementation of advance technologies and policies that meet the standards of international cities, educational programs to increase awareness amongst the citizens, availability of appropriate infrastructure and funding to support these improvements.

REFERENCES

Abdel-Shafy, H. I., & Mansour, M. S. M. (2018). Solid waste issue: Sources, composition, disposal, recycling, and valorization. *Egyptian Journal of Petroleum*, *27*, 1275–1290.

Ahluwalia, I. J., & Patel, U. (2018). *Solid waste management in India: An assessment of resource, recovery and environmental impact*. Indian Council for Research on International Economic Relations.

Banerjee, S. D. (2017). Scope of private participation in municipal solid waste management: The case of India. *Urban India, 117*.

Chavan, B. L., & Zambare, N. S. (2013). A case study on municipal solid waste management in solapura city of Maharashtra, India. *International Journal of Research in Civil Engineering*, *1*(2), 46.

Choudhury, M., & Dutta, J. (2017). A comparative study of municipal solid waste management status for three major towns of Upper Assam-India. *International Journal of Waste Resources*, *7*(291), 2.

Fernández-Braña, Á., Feijoo-Costa, G., & Dias-Ferreira, C. (2019). Looking beyond the banning of lightweight bags: Analysing the role of plastic (and fuel) impacts in waste collection at a Portuguese city. *Environmental Science and Pollution Research*, *26*, 35629–35647

Gupta, N., & Gupta, R. (2015). Solid waste management and sustainable cities in India: A case of Chandigarh. *Environment and Urbanization*, 573.

Jagrutiben, D., & Shelar, A. (2019). *Analytical study of municipal solid waste characteristics at Deonar dumping Yard in Mumbai region, Maharashtra, India. International Journal of Trend in Scientific Research and Development (IJTSRD)*, *3*(3), 1463–1466. ISSN: 2456-6470. www.ijtsrd.com/papers/ijtsrd23333.pdf

Joelsson, Y., & Lord. R. (2016). *Urban solid waste management in Mumbai current challenges and future solutions for urban development: Minor field study* [Bachelor thesis, KTH, School of Architecture and the Built Environment (ABE). Sustainable Development, Stockholm]. http://urn.kb.se/resolve?urn=urn:nbn:se:kth:diva-189137.

Joshi, R., Ahmed, S., & Ng, C. (2016). Status and challenges of municipal solid waste management in India: A review. *Cogent Environmental Science, 2*, 1139434.

Krishna, G. (2018). Why urban waste continuous to follow the path of least resistance. *Economic & Political Weekly*, *LII*(17).

Kumar, S. et al. (2017). Challenges and opportunities associated with waste management in India. *Royal Society Open Science*, *4*(3).

Mahanti, J. C. (2017). Clean India (Swachh Bharat) through Reuse and Recycling of wastes. *Journal of Pollution Effects & Control*, *5*(4), 1.

Ministry of Housing and Urban Affairs. (2020, January). *SwachhataSandesh newsletter*. India.

MoEFCC. (2016). *Report on recommendations of the committee on issues/challenges faced by municipalities related to implementation of solid waste management rules, 2016 and plastic waste management rules, 2016*. https://moef.gov.in/wp-content/uploads/2017/06/Final-Report-on-the-Issues-related-to-SWM-and-PWM-Rules-2016.pdf

Mondal, T., Choudhury, M., Kundu, D., Dutta, D., & Samanta, P. (2023). Landfill: An eclectic review on structure, reactions and remediation approach. *Waste Management*, *164*, 127–142. https://doi.org/10.1016/j.wasman.2023.03.034

Odonkor, S. T., Dei, E. N., Sallar, A. M. (2020). Knowledge, attitude, and adaptation to climate change in Ghana. *Scientific World Journal, 2020*, 3167317. doi: 10.1155/2020/3167317.

Paul, S., Choudhury, M., Deb, U., Pegu, R., Das, S., & Bhattacharya, S. S. (2019). Assessing the ecological impacts of ageing on hazard potential of solid waste landfills: A green approach through vermitechnology. *Journal of Cleaner Production*, *236*, 117643. https://doi.org/10.1016/j.jclepro.2019.117643

Sharholy, M., Ahmad, K., Mahmood, G., & Trivedi, R. (2008). Municipal solid waste management in Indian cities—A review. *Waste Management (New York, N.Y.), 28*, 459–467. https://doi.org/10.1016/j.wasman.2007.02.008.

Singh, S. (2015). *Decentralized solid waste management in India: A perspective on technological options in cities: 21st century India* (S. Singh, Ed.) (p. 289). BookWell.

Singh, S. (2020). *Solid waste management in urban India: Imperatives for improvement*. ORF Occasional Paper No. 283, Observer Research Foundation.

8 Impact and Revival of Sustainable Solid Waste Management for a Green Environment

Garima Kohli and Kumar Gourav

8.1 INTRODUCTION

Man's relationship with nature has historically been one of imbalance and overuse, resulting in environmental ruin. In the beginning, human have been incredibly in tune with their surroundings and early settlements through measurable impact on the environment, but it was relatively manageable (*Richard, 2014*). The waste produced by their routine activities did not pose any menace because of their small population size. With the technological advancements and innovation in agriculture, however, human beings began to find new ways to survive. This allowed them more permanent settlements, and hence, led to speedy growth in population. It was from here onwards that humans distanced themselves from nature and started exploiting it to satisfy all kinds of needs.

With the breakthrough in agriculture, settlements became more permanent, and cities began to take shape and led a step toward urbanization. This paved the way for the addition of many conveniences in the name of modernization, and man started extracting much from nature, hence creating an imbalance in man and nature's relationship. With such advancements, human beings started dominating nature by degrading and polluting the natural system. This, in turn, led to the accumulation of a huge volume of waste in rural as well as urban areas and has become a challenging issue for developing countries in light of the exponential growth of the human population globally.

Day-to-day activities produce solid waste that is not replaceable by any other processes that can't generate solid waste. These wastes are produced by the manufacturing of electronic gadgets like computers, mobiles, etc.; plastics, automobiles and so on have become extremely poisonous to the environment and, in turn, affect human health. Solid waste is also a major threat to the environment. Concerned area municipalities, local urban bodies and municipal councils are liable for managing the waste. They have to provide an operational and resourceful system to deal with solid waste that eases the general public. Nevertheless, they often face a lot of problems during the handling of the waste. There may be several reasons, but mention can be made of the availability of adequate financial resources, lack of organization and

DOI: 10.1201/9781003258377-8

complexity of the system, etc. (Choudhury & Dutta, 2017; Nathanson, 2002). Due to industrialization, there is an increase in the employment opportunities in towns; most of the population from villages have migrated, which has become a hindrance to the growth of developing countries such as the establishment of slum areas that increases the number of illegal tenants and informal lodging everywhere in the developing cities. This unexpected growth of population in the cities becomes a major threat for the municipal authorities to provide basic services, which leads to mismanagement and results in more waste production. With this shifting of people, solid waste management becomes a more tedious task due to the increase in population, changing lifestyles and increased consumption.

In most developing countries, there was inadequate infrastructure for collecting, transporting and dumping waste materials; lack of planning, scarce monetary funds, technological know-how and attitude of the society has worsened the situation. The practice of unrestrained clearance of waste materials on the roadsides and outer edges of cities or villages creates infectious health problems in the environment. Management of solid waste remains an ignored aspect of the development of urban areas. The re-utilization of solid waste is a feasible preference that can recuperate precious and monetary wealth that enhances energy consumption. To create useable goods, numerous methodologies are used for reprocessing wastes and thus offering various services for its recovery.

There are numerous methodologies for reprocessing wastes and offering various services for recovery and securing the remaining waste by using different tools and skills to create useable goods (Vazquez & Soto, 2017; Dev et al., 2019; Hereher et al., 2019; Berardi et al., 2020).

However, there are many adverse aspects related to solid waste, but certain issues increase the employment opportunities that diminish its negative impact. Thus, due to increased urbanization, solid waste management is a major issue in towns and cities. It is therefore desired that provision needs to be kept for an effective waste-management system before planning the new colonies so that the management of solid waste becomes a healthy, routine affair of the local urban bodies (ULBs). This will not only result in sustainable and strengthened community resilience but also pave the way toward acquiring the 'Green Environment Agenda,' which is mainly concerned with ecosystem protection and preservation of natural resources like soil, forests, biodiversity and water.

8.2 RELATED WORK

Management of solid waste has become a major challenge for developing countries that face the menace of municipal solid waste and its impact on the overall surroundings (Paul et al., 2019; Pattnaik & Reddy, 2010). Solid waste management is a daunting task, and there is definitely no doubt that it is the responsibility of the local bodies to manage it. Its management, however, needs to have organizational support as well as cooperation and coordination between the private and public sectors (Mcallister, 2015). Managing this increasing solid waste is very important and crucial, and the process involved must be environmentally effective, economically affordable and socially acceptable (Achanbeg, 2004). To accomplish an adequate performance of solid waste management, collaboration among the variety of participants tangled in managing waste, government policies for its management, disposal and its compilation of wastes carried out by companies and the society as a whole whose behavior must be enacted by the government strategies are required (Bartolacci et al., 2018).

Although environmental protection is extremely essential for the well-being of the community, solid waste management in the majority of cities is far from being satisfactory. Increased generation of MSW because of increasing population growth coupled with poor management practice create havoc (Idris et al., 2004). The problems associated with SWM in developing nations are very severe, when compared with other nations. India has been projected by World Bank to be the producer of the most waste globally. As per the survey, the waste generated in India is more than a tenth part of the waste produced by the whole world. According to a study conducted in 2016 in India, more than 277 million tons of waste was generated by Indians, which is approximately 80% of the 334 million tons produced every year. But interestingly, per person waste generated by Indians is just a small fraction (450 gm—620 gm on an average) (en.m.wikipedia.org) of that generated by other countries. Surprisingly, India produced 1.43 lakh MT of garbage per day (Times of India, Jadhav, 2018). According to Kaza et al. (2018), by 2050, waste production globally is expected to reach 27 billion tons per year. Presently, the Asian continent produces one-third of the whole waste; China produces 0–0.49 kg/capita/day, whereas India produces 0.50–0.9 kg/capita/day.

8.3 IMPACT OF SOLID WASTE

The rise in the quantity of waste generated in the present scenario has a massive influence on human health and the atmosphere as well. With the growth in population and rapid increase in the demand for food items and other necessities, there is a tremendous hike in the waste produced by every family per day (Joshi, 2016). Many of the waste materials are generally disposed of in the environment without any proper treatment. Such improper disposal can cause a serious impact on the health of human beings, on the one side, and the surrounding environment, on the other. Many of the raw materials disposed of in the general waste bin may take anywhere from hundreds to thousands of years to degrade.

Wastes coming through the households and community left unattended cause serious health infections, and then, converted to a heap of waste that attracts many insects, becomes the cause of many diseases. Populations, especially those residing near the dumped waste, waste workers and those who are associated with such work, are at risk of getting infected by deadly and contagious constituents generated through such wastes. Thus, biological domestic waste offers favorable circumstances for the development of microscopic pathogens that become hazardous to the environment. Moreover, direct contact can lead to serious diseases during the release of chemical waste into the environment and may become the cause of chemical poisoning. As per the study conducted by Ferronato and Torretta (2019) and Cheng et al. (2020), chemicals like sulfur oxides, toluene, benzene, dioxins, carbon monoxide, nitrogen oxides, ethyl benzenes, etc., released into the surroundings, have deteriorated the air quality index (AQI). Besides the previously mentioned, waste from agriculture and industries and burning the waste in the open air can have dangerous health problems. Burning in the open air is the major cause of air pollution and particulate substance transmissions (Li et al., 2016). The burning of trees is also a major source of polluting the atmosphere, and its leftover ashes also pollute the soil. Waste discarded nearby water resources also pollutes the groundwater as well as water bodies (Choudhury et al., 2022).

Presently, a huge amount of medical waste like bandages, discarded syringe needles, swabs, plasters and other types of infectious waste produced by hospitals, laboratories, health and research centers is mostly discarded with the general waste that is also the main reason for spreading infection. This infection includes blood infections, skin irritations, growth problems, respiratory problems, etc. Unmanaged and not-properly-designed waste-dump sites are the place where all rodents and various insects reside and cause multiple infections. To protect ourselves and our surroundings, these sites should be situated outside residential areas and must be properly fenced and enclosed so that there is no outflow into the nearby areas. To sum up, it can be said that unscientific treatment of solid waste impacts both human lives as well as affects the environment by spreading contagious diseases. Thus, unnecessary generation of solid waste should be organized, managed and controlled appropriately by taking precautionary actions on war-footing levels to further save human health and the environment.

8.4 CHALLENGES FACED BY SOLID WASTE MANAGEMENT

MSW management is undoubtedly an uphill task on account of various challenges and barriers. Some of the major challenges are as follows:

8.4.1 Urban Population Growth

Urban population growth is one of the serious problems associated with solid waste management in developing areas because of the fast growth rate and migration of people from outer reaches to cities to increase their standard of living. This increase in the growth rate of population in cities has consequently led to increased waste generation.

Further, it is also seen that the speedy development of urban areas leads to the formation of informal arrangement migration in towns that has resulted in the creation of slums with horrible housing and living conditions. Such an ever-growing population puts pressure on the infrastructure, and this overburdens the provision of urban services. Urban local bodies, hence, remain under a burden to fulfill the necessities of the public. This leads to devastating consequences on the economic, social and environmental well-being of the surroundings.

8.4.2 Insufficient Infrastructure

The first internal barrier in hampering solid waste management is the ineffective municipal solid waste system itself. There is insufficient infrastructure for the operation of waste, and in the absence of separate bins in public places, people throw plastic bags full of waste beside the streets. Also, the general public does not know how and where to dispose of their waste because there were no proper places to discard the waste. This is the reason for spreading pollution, home to various insects that spread diseases and odors.

To set up an operation unit of waste management, infrastructure facilities in the form of land and labor are needed. Government and local municipalities alone cannot provide the required capital, and hence, must develop the participation of the private enterprises. The waste needs to be segregated, collected and transported to recycling centers. So, it is very urgent and important to develop an effective and user-friendly mechanism between the waste producer and the waste-processing facility for effective SWM.

8.4.3 Lack of Adequate Finance

Finance is the lifeblood of any organization, and so, is an integral part of solid waste management. Lack of financial resources causes common problems like waste handling, waste collection and its disposal, which leads to differences in revenue and expenditure.

Generally, in developing countries, the framework of SWM encompasses around 50% of the population residing in urban areas, and the budget for SWM accounts for 20 to 50% of the municipal budget (Chandak, 2010). Such budget allocation results in the sufficiency of services, which, in turn, is reflected in the determination of the urban environment, and hence, poses a threat to human health as well as the environment. This, coupled with an unwillingness to pay for the retrieval services by causing loss to the revenue collection of local bodies, further deteriorates the quality of services. The impact of worsening services is felt as able to be seen with the rise in waste dumps in the locality, which is harmful to both environment and humans.

8.4.4 Environmental Problems

Reprehensible solid waste dumping can wreak chaos on the environment, chiefly because of the release of harmful greenhouse gases. Weitz et al. (2002) expressed that the waste management sector is responsible for approximately 4% of the total greenhouse gases emissions, toxic waste sites being the main resource of methane contributing around 90% to the entire GHGs discharged from the waste sector in the US. Landfills globally have been stated to be the third largest anthropogenic emission source and account for 13% of the global methane emission (Methane to Markets Partnership, Website, 2004). This is one such case of the SWM system considerably influencing the troubles connected to changes in the climate.

8.4.5 Socio-Cultural Barriers

Segregation of waste by households and establishments has become an obstacle to successful waste management because of their negative attitudes and behavior. Some of the population from rural areas has the worst hygiene and sanitation ethics. By a large, we have the habit of littering garbage in the open. Most of the drains in major cities are clogged with garbage, leading to severe epidemics of chikungunya, malaria, etc. The situation has been further worsened by the massive use of items made from plastics. Lack of awareness about waste management has led to such indiscriminate garbage dumping. Governments and courts, though, have imposed heavy penalties for polluting the environment, but sadly, it has no impact on the change in the behavior of individuals.

8.4.6 Climatic Changes

Releasing carbon dioxide and other harmful gases into the atmosphere increases their absorption, which, in turn, leads to progressive warming. This is the physical problem underlying climate change or global warming. The solution to this is that humanity must lessen its radiation of these gases.

8.4.7 No Primary Waste Collection System

The collection of waste from the public is the primary activity in the waste-management system. It is obligatory to make sure that waste gathered at resources is collected regularly and is not thrown on the drains, roadsides, water bodies, etc. The waste creation spot should be at the prime location for waste collection. Door-to-door waste collected from the public usually does not coordinate with the ability of storage deports and haulage of garbage. The streets are normally designed as a wastebasket, barring a few posh localities of cities and towns, and are shown that the major compilation of waste is completed. Thus, there is mismanagement and a lack of cooperation between both of the parties involved in the system, which, in turn, pollutes the environment and the overall health of individuals.

8.4.8 Awareness about Segregation

Segregation is mainly the serious stage in the entire procedure of municipal solid-waste management wherein the reusables and recyclables are separated from waste going into landfills. In India, there is no proper provision for segregation of wastage from the pickup source, as it becomes a problematical task to separate the wastes at the dumping ground, which otherwise leads to various health hazards. Moreover, because of unplanned development of cities and towns, no proper points for waste disposal are earmarked, and consequently, the residents waste offensively. Hence, bins should be placed in every community, and the government should employ more ULB to get out of this problem.

8.4.9 Disposal of Waste

In developing nations, disposal of waste is a serious problem, as it is stored over the surface of the land in an inappropriate way that creates many dangerous issues for the health of inhabitants. The rise in population and lack of financial resources available for the local municipalities are the major threats behind inefficient waste disposal. For most of the population, disposal of waste becomes an issue because there was no proper disposal of waste from their surroundings, even though they are ready to pay for the services of which they want to avail themselves. The current condition of disposal of wastes is predicted to decline more because, till now, the dumping sites were near the waste gathering area, and this enabled local bodies to discard the waste at a low price. Thus, waste disposal sites are becoming increasingly difficult to find new sites.

8.4.10 Irregular Waste-Collection Street-Sweeping

Due to irregular waste collection, residents resort to dumping waste at nearby vacant plots and in open spaces. Many people even dump garbage in nearby water bodies, including drainage channels, and clogging them. Moreover, there is no fixed day and time for waste collection because the waste collection randomly comes off and on at their own will. This irregular waste collection system ultimately leads to littering on streets and roads, giving birth to many diseases. Conventional techniques are still used by the municipalities and by cleaning staff, which reduces the efficacy of services.

8.4.11 Failure of Waste-to-Energy Projects

The idea of waste-to-energy (WTE) is presently in its infancy in India. Many initiatives were taken but were not successful, owing to severe opposition from the public in locating these plants in the neighborhood due to pollution concerns. According to Lahiraja et al. (2018), there are several countries where WTE plants are well suitable for the development of the nation, but still in India, there is a lack of such resources. In India, 22 states have submitted 53 WTE project proposals to the Ministry of Urban Development (MoUD) under the Swachh Bharat Mission (SBM), which are presently under different stages of approval (www.bestcurrentaffairs.com). Only eight functional WTE thermal plants with a total capacity of 94.1 MW currently are operational in India. By 2022, India is aiming for the installation of 175GW of renewable energy capacity (60 GW from wind power, 100 GW from solar power, 5 GW from small hydropower) (Kumar et al., 2019).

8.5 RECOMMENDATIONS

For maintaining quality of life and to ensure better standards of health and sanitation, SWM is one of the most essential services. With the rapid growth in the industrial sector and thus rise in the population, there has been a mass migration of the population from villages to towns. This further has resulted in tremendous pressure on government and local bodies to meet the ever-increasing waste generation and its management. This has posed many problems and challenges for authorities, and solutions need to be worked out. With a detailed-eye view of the challenges/barriers (as discussed) that cause hindrances in the proper disposal of waste, the following recommendations have been made to bring improvements in the existing Indian municipal solid waste management practices. These recommendations are discussed under the following heads:

8.5.1 4R Concept

Das et al. (2019) and Ruslinda et al. (2019) have defined the 4Rs theory as a part of their study that comprises reducing, reusing, recycling, and recovery, which plays a significant role in managing waste. The reduce aspect of 4R recommends buying environmentally friendly materials instead of toxic materials that are not good for the health of individuals. Reuse is defined as the recycling of waste material for supplementary use instead of discarding it outside unreasonably. Recycling means emphasizing reprocessing waste plastics, waste metals or e-wastage into valuable goods rather than throwing it away. Recover means waste materials are affluent sources of the substrate to get better resourceful materials (nutrient and energy) from it; therefore, using them contributes to sustainability.

These standards supply a consistent waste severance method at the starting place—i.e., a house-to-house solid waste compilation service (Das et al., 2019). Applying the 4R concept in our day-to-day activities will help efficiently run MSW.

8.5.2 Community Participation

It is for certain that the quality and quantity of waste produced at a particular place, city, town, village, and more specifically, at the community level, varies after their social-economic class, customs, population and business activities. The SWM method has to be established on the composition and the amount of waste created therein. Community participation is a sociological process where residents were concerned about a residential place or nearby locality to contribute and further enhance the way of living. In the context of SWM, community participation comprises a varying degree of involvement, ranging from contributions in the form of cash and labor, involvement in administration and decision-making, adaption, etc. To keep the SWM system running smoothly, community involvement is essential.

8.5.3 Management of Municipal Solid Waste

8.5.3.1 Waste Storage and Primary Disposal Practices

Storage of waste (being the first step of SWM) at the source and its collection from the point where it is placed by persons and organizations that produce it are the first and second necessary steps of the solid waste management system. For storage of waste, waste bins must be provided in all the municipal wards that will reduce the discarding of waste everywhere. Thus, it is necessary to segregate the waste by using different colors of waste bins. After on-site storage, the next step is a primary collection from where the waste is transported to transfer stations to local disposal sites.

8.5.3.2 Waste-Collection Process

Collection of waste is an important and forecast point in the solid waste management process. Due to the inefficient waste collection system, numerous issues are acknowledged for the surroundings. It is, therefore, necessary to collect solid waste, and its related inspection on regular basis should be observed, which should be done in all the areas. This practice will keep the surroundings clean and minimize the risk of spreading various diseases.

8.5.3.3 Waste Minimization

This practice highlights reducing the quantity and the impact of harmful wastes. At this level, recycling of waste as well as its minimal consumption were taken into consideration. By adopting this technique, the wasted material can be reused and therefore its recycling, its incineration for energy and disposal of waste to dumping areas were not required. This further lowers the cost involved in managing the waste. This can be efficiently achieved through proper arrangement, development and use of resources in such a way that reduces the quantity of garbage produced so that very few natural resources are consumed besides less creation of waste.

8.5.3.4 Waste-Sorting

To have proper waste sorting, the three-bins technique ought to be adopted and used by local bodies. This technique provides awareness among the public for waste sorting by using three colored waste bins for discarding waste in respective bins in an efficient manner. Sorting of waste is beneficial for the ecosystem to use and discard the waste one by one for sending to the recycling units for further processing, which will, in turn, reduce time, labor and cost.

8.5.3.5 Compositing

Compositing is the decay of natural resources by bacteria in a humid, hot, aerobic and anaerobic atmosphere. For degradable waste such as food garbage, vegetables garbage, etc., backyard composting technology can be adopted. It is also recommended that the government favor and encourage the usage of advanced technology for the transformation of municipal waste to compact, and further, do its marketing for sale in the marketplace. Inhabitants can also be educated to use this compost as a soil conditioner to grow vegetables and crops. This will lead to sustainability, but it demands patience and cooperation.

8.5.4 Solid Waste Management—Best Practices

8.5.4.1 Enforcing Law

For sustainable waste management policy, the authorities must initiate strict punishments for people who don't follow the rules and regulations by throwing the garbage on the streets, roads and water bodies, etc. It should be the prime responsibility of the local bodies to have proper and effective enforcement of the New Solid Waste Management Rules (SWM) 2016 in letter and spirit. The local bodies should also decide, at their levels, the number of fines and penalties be imposed on the defaulters. Besides these, the government should also direct the urban local bodies to have a separate and independent wing/department of Municipal Solid Waste Management with full-time offices and staff for better results. The local bodies should ensure the full implementation of the new rules at the grass-root level.

8.5.4.2 Education and Awareness

The local bodies should organize regularly the education and awareness performance in all municipal wards for giving proper training to the general public for proper handling of municipal solid waste. It is important that the waste should be properly bifurcated at the source and should be sent to different units for its recycling. As such, the local bodies should give top priority to the segregation of waste at the source. So much so, these bodies should not accept the waste if not segregated. And in addition, they must impose a fine on defaulters. There should be proper publicity through print and electronic media. In addition, the local bodies should also organize awareness meetings about the New Rules (2016) for the residents of different areas/ localities. Conducting awareness programs can eliminate the lack of knowledge concerning SWM among the people. The message 'Waste generates are responsible for waste treatment and disposal' should be communicated to the general public through efficient awareness programs, so that the wrong belief that it is the major duty of the municipalities should be altered.

8.5.4.3 User's Fee

To pay the user's fee, mandatory as per New Rules (2016) for the collection services provided by the municipalities, should be implemented strictly. It will strengthen the financial condition of these bodies that, by and large, are functioning with a weedy financial system. They are usually dependent on central and state government finances for MSWM. Moreover, the major portion of the funds the local bodies receive is generally spent just on the collection and transport of the waste to the dumping sites. So, it is recommended that the local bodies should become financially self-sufficient by taking user's fees.

8.5.4.4 Foster Waste Management Culture in Children

It is almost a universal truth that what people study and learn in early childhood is remembered and gets embossed till death. We should teach and inculcate in our children right from their childhood to manage waste. Just as we teach them to wash their hands before and after meals and to brush their teeth twice a day, similarly, they should also be made aware to dispose of the waste in bins. Imparting training in respective schools is important for children, which will encourage the management of waste.

8.5.4.5 Research and Development

To ensure a productive, economical and environment-friendly solid waste management system, the SWM staff in local bodies must be imparted with proper training. Keeping in view the seriousness of the waste generation problem, it is recommended that the waste management institutes be set up at the state level so that staff can be well trained variably from time to time to implement the latest know-how regarding waste management technology. In addition to the previously mentioned, the Government of India should issue a notification through the Ministry of Education that research should be carried out by the universities and research institutes in the country on different aspects of municipal solid waste management, at least for 10 coming years in the near future to upgrade our waste management system and save our country from drowning under its garbage.

8.6 CONCLUSION

Sustainable waste management can be accomplished through monetary benefits, economically and scientifically viable techniques as well as public-private association. It should, however, take into account the long-run goals to serve local needs effectively by well-known approaches of the 4Rs—i.e. Reduce, Reuse, Recycle and Recover. To make it effective, stakeholder involvement, accompanied by appropriate financial human-resource management, is required. Waste, thus, can prove to be an asset rather than a menace and can be altered and used as an important resource; waste, hence, is undoubtedly a 'hidden treasure' and, if managed properly, will no more be a blot on the landscape. From the foregoing account of challenges faced in the context of solid waste management, specific solutions to overcome the challenges have been proposed. If taken care of in the true spirit, this can improve and make the waste management system sustainable.

REFERENCES

Achankeng, E., (2004). *Sustainability in municipal solid waste management in Bamenda and Yaounde, Cameroon* [PhD thesis, University of Adelaide, Adelaide].

Bartolacci, F., Paolini, A., Quaranta, A. G., & Soverchia, M. (2018). Assessing factors that influence waste management financial sustainability. *Waste Management*, *79*, 571–579. https://doi.org/10.1016/j.wasman.2018.07.050

Berardi, P., Almeida, M. F., Lopes, M. d. L., & Maia Dias, J. (2020). Analysis of Portugal's refuse derived fuel strategy, with particular focus on the northern region. *Journal of Cleaner Production*, *277*, 123262. https://doi.org/10.1016/j.jclepro.2020.123262

Chandak, S. P. (2010). Trends in solid waste management–issues, challenges, and opportunities. In *The International consultative meeting on expanding waste management services in developing countries* (pp.18–19). Tokyo, Japan.

Cheng, K., Hao, W., Wang, Y., Yi, P., Zhang, J., & Ji, W. (2020). Understanding the emission pattern and source contribution of hazardous air pollutants from open burning of municipal solid waste in China. *Environmental Pollution*, 263, 114417. https://doi.org/10.1016/j.envpol.2020.114417

Choudhury, M., & Dutta, J. (2017). A comparative study of municipal solid waste management status for three major towns of Upper Assam-India. *International Journal of Waste Resources*, *7*(291), 2.

Choudhury, M., Jyethi, D. S., Dutta, J., Purkayastha, S. P., Deb, D., Das, R., . . . Bhattacharyya, K. G. (2022). Investigation of groundwater and soil quality near to a municipal waste disposal site in Silchar, Assam, India. *International Journal of Energy and Water Resources, 6*(1), 37–47. https://link.springer.com/article/10.1007/s42108-021-00117-5

Das, S., Lee, S. H., Kumar, P., Kim, K. H., Lee, S. S., & Bhattacharya, S. S. (2019). Solid waste management: Scope and the challenge of sustainability. *Journal of Cleaner Production*, *228*, 658–678. https://doi.org/10.1016/j.jclepro.2019.04.323

Dev, S., Saha, S., Kurade, M. B., Salama, E. S., El-Dalatony, M. M., & Ha, G. S. (2019). Perspective on anaerobic digestion for biomethanation in cold environments. *Renewable* and *Sustainable Energy, 103*, 85–95. https://doi.org/10.1016/j.rser. 2018.12.03

Ferronato, N., & Torretta, V. (2019). Waste mismanagement in developing countries: A review of global issues. *International Journal of Environmental Research and Public Health*, *16*, 1060. https://doi.org/10.3390/ijerph16061060

Gunning, P. M. (2004). The Methane to Markets Partnership: An international framework to advance the recovery and use of methane as a clean energy source. *Environmental Sciences*, 2(2–3), 361–366. https://doi.org/10.1080/15693430500402390

Idris, A., Inanc, B., & Hassan, M. N. (2004). Overview of waste disposal & landfills/dumps in Asian countries. *Journal of Material Cycles and Waste Management*, *6*, 104–110. https://doi.org/10.1007/s10163-004-0117-Y

Hereher, M. E., Al-Awadhi, T., & Mansour, S. A. (2019). Assessment of the optimized sanitary landfill sites in Muscat, Oman. *Egyptian Journal of Remote Sensing and Space Science*, *23*, 355–362. https://doi.org/10.1016/j.ejrs.2019.08.001

https://en.wikipedia.org/wiki/Waste_management_in_India

https://en.wikipedia.org/wiki/Waste_management

Jadhav, R. (2018). 75% of municipal garbage in India dumped without processing. The Times of India post. https://timesofindia.indiatimes.com/india/75-of-municipal-garbage-in-india-dumped-without-processing/articleshow/65190477.cms

Joshi, R., & Ahmed, S. (2016) Status and challenges of municipal solid waste management in India: A review. *Cogent Environmental Science*, *2*, Article ID: 1139434. http://dx.doi.org/10.1080/23311843.20161139434

Kaza, S., Yao, L., Bhada-Tata, P., & Van Woerden, F. (2018). What a waste 2.0: A global snapshot of solid waste management to 2050. *The World Bank*. https://doi.org/10.1596/978-1-4648-1329-0

Kumar, J. F. C., Mary, A. B., Jenova, R., & Majid, M. A. (2019). Sustainable waste management through waste to energy technologies in India-opportunities and environmental impacts. *International Journal of Renewable Energy Research*, *9*(1), 309–342.

Lahiraja, A. C., Panna, P. K., Kroezen, S., & Nallachernvu, V. (2018). *Waste to energy in India: A study on the store peace of adoption in Delhi* (pp. 1–42). Delft University of Technology.

Li, L., Lei, Y., Pan, D., Yu, C., & Si, C. (2016). Economic evaluation of the air pollution effect on public health in China's 74 cities. *SpringerPlus, 5*, 1–16.

Mcallister, J. (2015). *Factors influencing solid-waste management in the developing world. All graduate plan B and other reports 528*. https://digitalcommons.usu.edu/gradreports/528

Nathanson, J. A., (2019). *Solid waste management*. Retrieved 2019, from https://www.britannica.com/technology/solid-waste-management.

Pattnaik, S., & Reddy, M. V. (2010). Resources, conservation & recycling assessment of municipal solid waste in Puducherry (Pondicherry), India. *Resources, Conservation & Recycling*, *54*(80), 512–520.

Paul, S., Choudhury, M., Deb, U., Pegu, R., Das, S., & Bhattacharya, S. S. (2019). Assessing the ecological impacts of ageing on hazard potential of solid waste landfills: A green approach through vermitechnology. *Journal of Cleaner Production, 236*, 117643. https://doi.org/10.1016/j.jclepro.2019.117643

Richard. (2014). *Our role and relationship with nature*. Retrieved December 2, 2021, from https://you.stonybrook.edu/environment/sample-page/

Ruslinda, Y., Raharjo, S., Dewilda, Y., & Hidayatullahand Aziz, R. (2019, July 30). Minimization of household hazardous solid waste (HHSW) with 4R concepts (reduce, reuse, recycle and recovery) in Padang City, Indonesia. *IOP Conference Series Material Science and Engineering*, *602*, 012055. https://doi.org/10.1088/1757-899x/602/1/012055

Solid waste management rules (2016). Retrieved December 12, 2021, from https://vikaspedia.in/energy/environment/waste-management/solid-waste-management-rules.

Vazquez, M. A., & Soto, M. (2017). The efficiency of home composting programmes and compost quality. *Waste Management, 64*, 39–50. https://doi.org/10.1016/j.wasman.2017.03.022

Waste to energy plants operational in India. Retrieved November 12, 2021, from www.bestcurrentaffairs.com/waste-energyplants-operational-india/

Weitz, K. A., Thorneloe, S. A., Nishtala, S. R., Yarkosky, S., & Zannes, M. (2002). The impact of municipal solid waste management on GHG emissions in the United States. *Journal of the Air and Waste Management Association, 52*(9),1000–1011.

9 Recent Trends in Sanitary Landfilling and Future Prospects

Muhammad Sajid

9.1 INTRODUCTION

The increasing demand for energy and materials due to escalating population growth and continuous improvement in living standards leads to higher rates of waste generation (Sajid et al., 2021). This continuously increasing waste generation needs serious scientific efforts for proper disposing; otherwise, it will be a significant environmental concern in near future. Waste generally collected from the municipalities in the form of garbage and refuse is a mixture of various organic and inorganic species (M. S. U. Rehman et al., 2016). It has been estimated that current global waste generation is approximately 2×10^9 metric tons (MT) with an average rate of 0.74 kg per capita per day (Atlas, 2018), out of which 67% is collected, whereas 33% remain uncollected, mainly in the least developing countries, especially in South Asia. It is estimated that the waste volume will escalate to 3.4×10^9 MT by the year 2050 under the prevailing scenario, with a rate of 0.11 to 4.54 kg from low-income to high-income regions (Bank, 2020). Currently, conventional methods such as burning and dumping are still widely used waste management practices; however, the adoption of scientific waste management practices is increasing with awareness and legislation by the authorities. Currently, the major waste management technique is still landfilling, and 70% of waste collected is disposed of through landfills, while other major techniques are recycling and energy production, with 19% and 11% share, respectively (Bank, 2020). According to the waste atlas scenario, nearly 50% of the world population (7.6 billion people) are not covered with waste collection and management facilities, which may increase to 5.6 billion by 2050 (Atlas, 2018). The implication of waste management is mostly governed by the community standard and national gross income of a country. Waste management has always been a paradigm between developed and developing societies. Mostly, developed countries employ advanced scientific techniques such as waste to energy (WTE) to cope with the problem of waste disposal (Nanda & Berruti, 2021b). These techniques have the advantages of energy recovery and sustainability, along with efficient waste disposal (Sajid, Bary, et al., 2022). However, WTE demands high capital investment, which the least-developed or developing countries hesitate to invest in. Uncontrolled population growth, high population density, poverty, lack of commitment, and inefficient planning and execution are major barriers to facilitating the community with a sustainable, efficient waste management system in developing countries. The United Nations

DOI: 10.1201/9781003258377-9

(UN) has emphasized the importance of municipal waste management through sustainable development goals as sustainable cities and communities (Goal 11) and responsible consumption and production (Goal 12) in order to develop a campaign for effective waste management in developing countries (Nations, 2015). Municipal solid waste (MSW) has the potential to produce energy and platform chemicals through various processes. Various WTE techniques have been developed and are currently being employed in developed countries to convert waste into useful heat and energy. Similarly, some thermal and thermochemical processes have been developed to convert organic waste into useful energy and materials (Mulk et al., 2022; Sajid, Ahmad, Shafqat, Pasha et al., 2020). Thermochemical techniques such as incineration, gasification, liquefaction, and pyrolysis are frequently used processes that facilitate the production of syngas, bio-oil, and biochar from waste materials (Dastyar et al., 2019; Sajid, Ahmad, Shafqat, Mulk et al., 2020). These techniques not only provide energy but also shrink the waste volume up to 96%. Additionally, the organic-rich MSW can be converted to biofuels (biomethane, biogas) and organic fertilizers employing biological processes such as composting, anaerobic, and aerobic digestion (Nanda & Berruti, 2021a; Pasha et al., 2021). However, currently, only 25% of MSW has been treated through these techniques. Therefore, an effective approach is still needed to implement a proper MSW management model ecologically and economically.

The landfill technique, a traditional MSW management model, has been used for a long time to dump the waste collected as a whole in the least-developed and developing countries and for the dumping of non-recyclable parts in developed countries (Mondal et al., 2023; Nanda & Berruti, 2021a). Sometimes, landfills also serve as a temporary storage bin to facilitate transfer processing such as sorting, recycling, and treatment before final disposal. Similarly, the non-recyclable waste from waste-to-energy processing facilities needs additional disposal arrangements, and landfilling is the most preferred one. Sometimes, the mixing of these leftovers with concrete as a building material has also been practiced. With the evolution of sustainability and ecological concepts, the landfill sites are manufactured and managed through engineering principles and termed sanitary landfilling. The processing of waste in landfills is by engineering control, which reduces the soil and underground water contamination potential with improved capacity (Satyanarayana & Chandra, 2014; Paul et al., 2019). This chapter will focus on sanitary landfill operations, design considerations, opportunities, challenges, and prospects considering the future environment-sensitive scenario.

9.2 COMPARISON OF SANITARY LANDFILLING AND OPEN DUMPING

Sanitary landfilling is an upgraded engineering approach based on the dumping of waste materials. The major issues of waste management of open dumping have been solved by the concept of sanitary landfilling. Here, the major difference between sanitary landfilling and open dumping will be compiled:

i. Well-planned dumping of waste with continuous monitoring is sanitary landfilling, whereas open dumping is just haphazard piles of garbage without any design or monitoring.

ii. Sanitary landfilling is planned and location-specified activity with built-in purpose location, whereas open dumping is sometimes piled on a place that is inappropriate and not meant to be.
iii. Sanitary landfilling is a systematized waste management facility according to the environmental management system, whereas open dumping is a continuous environmental hazard.
iv. In a sanitary landfill, the biogas produced during the biological decomposition of organic waste is properly collected and used for fruitful purposes, whereas, in open dumping, produced biogas pollutes the local environment that may cause health and safety issues.
v. Leachate formed in sanitary landfilling is collected and disposed of after neutralization, following the local and international standards, whereas no arrangement for leachate management is possible in open dumping of wastes.
vi. Open dumping can be a breeding home for parasites, harboring syndromes, and inviting animals along with a source of unpleasant odors, whereas all of these have been solved significantly in sanitary landfilling operations.
vii. Sanitary landfilling is an organized activity that needs approval from authorities, whereas roadsides or unused land is mostly used for illegal, open dumping.

9.3 METHODS OF SANITARY LANDFILLING

The process by which the materials that can neither be recycled nor further used are dumped in the soil is called landfilling, and the place where the material is physically placed is called a landfill site (Andreottola et al., 2018). The MSW materials and remaining materials of the recovery facilities after the recovery of convertible products and energy is usually dumped on land. Due to the negative public opinion and harmful effects of refuse materials, sanitary landfilling was developed because every waste management practice ends up with materials desired for landfills. Initially, sanitary landfilling was just the coverage of waste materials at the end of each day. But, with the evolution and implication of environmental regulations, the engineering design is upgraded to reduce the environmental and health impact to a minimum level. This process includes monitoring of the arriving waste streams, engagement and the compaction of materials received, and implementation of landfill environmental observing and control protocols, considering the local and international legislation (Ministry of the Environment, 2012).

Waste materials are compacted after collection and sent to landfill sites in closed containers. Compaction arrangements are generally provided within or near the residential area. Materials are compacted in order to reduce the waste volume for economical transportation. Sorting is performed before compaction and materials suitable for recycling and reuse are stored separately. Materials that are categorized unfit for landfilling are also separated and processed according to engineering procedures. Generally, waste received is categorized as (i) hazardous waste, (ii) designated waste or special waste, and (iii) MSW. At the site, pulverizing, high compaction, and sorting are performed, if required. Mostly, waste materials are baled with high compaction to further reduce the waste volume. It will increase the waste handling capacity of the landfilling site. Relatively loose material is placed at bottom of the cell or lift, compacted again, if not baled, to a layer of an appropriate thickness (generally 0.5 meters). When the layer thickness is about 2–3 meters or at the end of the working day, the covering is placed either with clay or with some synthetic material.

With time and temperature, various physicochemical changes take place (Ahamad et al., 2020; Diejomaoh Abafe et al., 2022). Waste decomposition produces biogas and leachate through various chemical and biological reactions as shown in Figure 9.1-A. Leachate composition varies according to the reaction

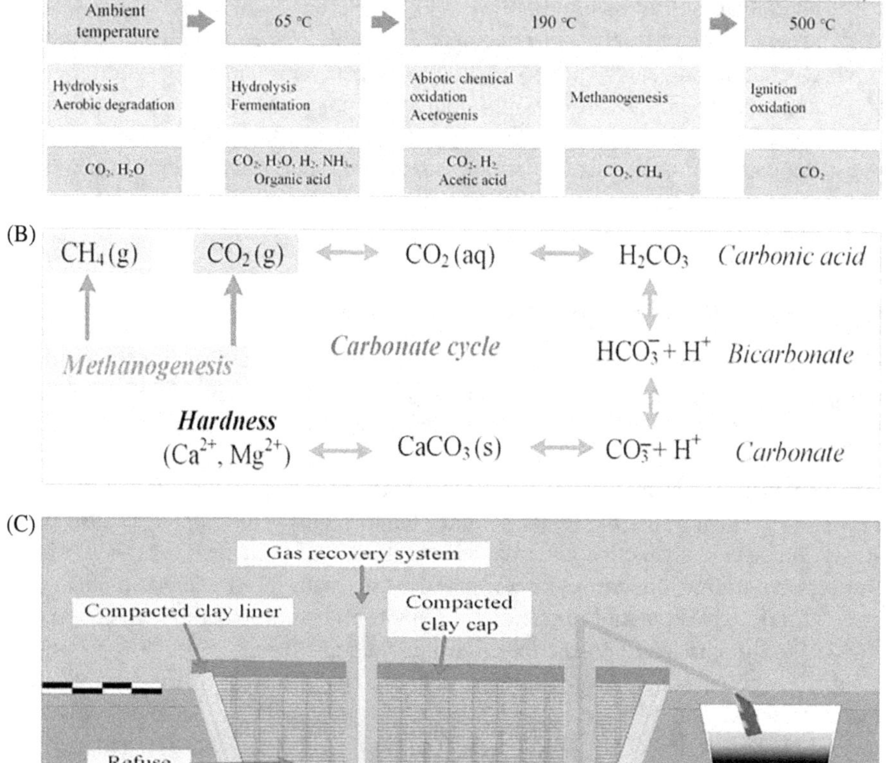

FIGURE 9.1 (A) The sequence of physiochemical reactions in a landfill (Reinhart & Stegmann, 2018) (Carey, 2000). (B) Formation of carbonates during waste decomposition in a landfill (Reinhart & Stegmann, 2018). (C) A schematic illustration of a sanitary landfilling site.

TABLE 9.1
Composition of Leachate (EPA, 2003)

	Acetic phase		Methanogenic phase	
Parameters	**Range**	**Average**	**Range**	**Average**
pH	4.5–7.5	6.1	7.5–9	8.0
COD ($\times 10^3$ mg L^{-1})	6–60	22	0.5–4.5	3.0
BOD ($\times 10^3$ mg L^{-1})	4–40	13	0.02–0.55	0.18
Mg	50–1150	470	40–350	180
Ca (mg L^{-1})	10–2500	1200	20–600	60
Mn (mg L^{-1})	0.3–65	25	0.03–45	0.7
Fe (mg L^{-1})	20–2100	780	3.0–280	15.0
Zn (mg L^{-1})	0.1–120	5.0	0.03–4.0	0.6
SO_4 (mg L^{-1})	70–1750	500	10–420	80

temperature, waste composition, and stage of decomposition (A. Rehman et al., 2022). Generally, the waste digestion and leachate formation process is categorized into the *acetic phase* and *methanogenic phase*. The produced leachate has a distinct composition, as shown in Table 9.1. In the *acetic phase*, degradable waste materials are decomposed into organic matter, CO_2, H_2O, and heat. Larger organic moieties are digested to produce the simplest molecules such as H_2, NH_3, H_2O, CO_2, and acids via aerobic/anaerobic digestion. Finally, the decomposition of organic acids produces methane and other products. Additionally, the methane may also be produced from carbonates present in the waste materials through the carbonate cycle, as shown in Figure 9.1-B. The soluble products of these biochemical reactions, along with other soluble species, are the constituents of leachate.

The landfilling methodologies which are commonly applied for waste management are expanded upon in the following sections.

9.3.1 Trench Method

This methodology is mostly applied in an area where the water bed is deep and an adequate amount of covering materials are available. It is also known as the excavated cell method. The trenches are excavated and waste is placed there, covered by the soil collected from digging. These quarries are lined by employing membranes or clay having the least permeability (Sener, 2004). Sometimes, the combination of membrane and clay is also used, depending on the typical requirements due to waste compositions and depth of the water bed focusing on the principal objective of decreased mobility of both landfill gases and leachate.

These landfill sites are generally constructed below the water bed. The site is so designed to eliminate the chances of water entering and water contamination from produced leachate and biogas.

9.3.2 Area Method

This method is applied when the terrain is unfit for the excavation of trenches. Mostly when the water bed is high, then the area method is preferred over the trench methodology (Beaven & Woodman, 2018). The landfill site is prepared by lining to control the leachate flow, and covering is provided through some externally transported materials. Clay borrowed from adjacent lands or transported from a remote location can also be used. Alternatively, geomembranes are a selective option. If the synthetic membranes are the choice, these can be removed before the next lift and reused after that.

9.3.3 Canyon Method

This method is also known as the depression method. In this process, quarries, dry pits, and canyons are used to dump the waste materials after compaction. Compaction and placement processes are designed according to the site geology, hydrology, geometry, and availability of lining materials (Bell, 1973). Covering is designed to retain the leachate and facilitate an efficient biogas collection system.

Waste filling in a lift is made from a rare end to an open end in order to avoid the water accretion behind the landfill. Similar to the area method, canyons are filled in several lifts. If the available floor is reasonably uniform, the filling can be made by employing the trench approach, primarily. Lining materials' availability and the covering are critical aspects. Covering clay can be managed from walls or by digging before filling if the site geology is suitable.

9.3.4 Bioreactor Method

This method is designed to increase the rate of decomposition of organic species present in the collected waste (Andreottola et al., 2018). In order to enhance the rate of decomposition, the collected leachate is periodically recycled into the dumped materials. In contrast to ordinary design, the biogas collection system is quickly installed because biogas production will start rapidly. Moisture contents are increased, either with the mixing of sewage sludge or with some other source to increase the decomposition rate. Other parameters are also adjusted to make landfills a practical bioreactor. These designs are made with a proposal to decrease the length of taking care of closed landfill sites. With quick decomposition, buried waste will be stabilized swiftly, sinking the monitoring period. Covering and lining design is still improving to make this design more practicable and appropriate, with minimum cost.

9.3.5 Special Methods

Sometimes, a landfill site is specially designed for a specific purpose. For example, if there is major demolition work in any specific area, a specially designed landfill site is prepared to accommodate that concrete waste (Kreith, 2019). Mostly, these are purposefully built sites, and no other material is dumped herewith. Similarly, a landfill site is designed if there is a high volume of typical waste from foundries, paper mills, mines, power plants, and mines. Each of these landfills has distinctive design considerations; therefore, they are specially designed. These designs may have or don't have all the features of common landfill designs, depending on the specific nature of the waste material. A typical landfill design for the management of ash from a power plant doesn't need any biogas collection system because there are no ash decomposition reactions.

9.4 TYPICAL TERMINOLOGIES OF SANITARY LANDFILLING

The following are common terminologies that are regularly used in landfilling process of waste management (EPA, 2003).

9.4.1 Cell

The volume of waste materials managed in any landfill site daily or during any appropriate operating period is called a cell. It includes the material received and cover material surrounding it during the set period. Material is spread lightly after receiving at the landfill site and covered after compaction during a specified time frame—usually daily. Normally, a 6–12-inch layer of soil or any other substitute, which is being used as a working face, is placed at the end of each working day. This facing layer is placed in order to control the invasion of pests, limit windblown wreckage, shield unpleasant waste, and inhibit the infusion of rain and snowmelt into compacted waste.

9.4.2 Lift

The complete layer of cells in any landfill site is called a lift. There can be several lifts in any landfill site, and the final lift is usually composed of a cover layer.

9.4.3 Bench or Terrace

If the height of the landfill exceeds 15.24 meters to 22.86 meters, then it is called a bench or terrace. The slope of the landfill site is managed with benches. Slope sustainability is necessary in order to provide water drainage and gas recovery channels. The last layer is called the cover layer, which is usually composed of various layers of geomembranes and soil. These layers are provided over the entire surface of the landfill site.

9.4.4 Leachate

The fluid collected at the bottom of the landfill site is called leachate. Leachate formation is due to (i) precipitation percolation, (ii) uncontrolled runoff, (iii) inward water spillage, (iv) permeating groundwater, and (v) water contents of waste collected and dumped. Its compositions vary according to the products of chemical and biological reactions in specific wastes. It may contain heavy and rare earth metals due to the decomposition of organometallic compounds present in the waste under treatment.

9.4.5 Biogas

The gas evolved from the landfill site is usually a mixture of different gases. Generally, methane (CH_4) and carbon dioxide (CO_2) are major constituents of biogas collected from the sanitary landfill site. These gases are the results of the anaerobic digestion of the organic portion of waste collected. Other constituents' gases may be atmospheric oxygen (O) and nitrogen (N), along with minor contents of ammonia (NH_3) and carbon monoxide (CO). Gas rich in methane is considered good for energy and chemical production. This gas is enriched in methane before forwarding it to consumer units.

9.4.6 Liners

The materials, either synthetic or natural, are placed at the bottom and sides of the landfill site in order to protect the seepage of fluids from the waste collected are called liners. Compacted clay and reinforced membranes are usually used as liners having the potential to stop the leakage of gases and leachate produced during waste decay. It helps to retain the leachate and collection of biogas.

9.5 DESIGNING OF SANITARY LANDFILL SITE

Designing a landfill site is an interactive process that incorporates conceptual design proposals, environment assessment remarks, results of environmental monitoring systems, and risk assessment outcomes. The vital objective of sustainability must be evolved through the designed site. This objective can be achieved by designing an intrinsically connected development and operating process. The proposed design is optimized, keeping in view (Sivapullaiah & Naveen, 2016):

i. To diminish water permeation into dumped waste from covering or walls.
ii. To accumulate gas and leachate and their safe flow to a processing facility.
iii. To eliminate the release of leachate into the water bed and/or subsurface.

The concept of sanitary landfilling is to protect the environment and life from hazardous effluents, whether biogas or leachate, caused by the decay of MSW. This objective is achieved through engineering design. The lining is made to retain the leachate, which is then collected and disposed of scientifically, under no harm conditions. The arrangements for the collection of biogas are also provided for effective recovery and utilization of biogas produced (Kjeldsen & Scheutz, 2018). Mostly, sanitary landfills are designed with a lifespan of 20–30 years, followed by 30 years of monitoring after closure to ensure environmental protection.

A double layer of lining material is provided to trap the leachate. Different synthetic materials such as plastic membranes, textiles, mats, and grids having bentonite as a base material can be used safely (Tn-doer-r et al., 2004). The collected waste is dumped in cells, and these cells are covered after working. Cell size is designed to manage the daily volume of waste collected at a specific location (Carey, 2000). Typically, the nominal cell size is 4.6 meters, but it may vary between 2.4 to 9.1

meters, depending on the waste volume and covering materials' availability. The slope is provided with a ratio of 3:1, horizontal to vertical, in order to provide sufficient compaction and easy capping. It will also help to establish vegetation on landfill slopes. The working face usually has a width of < 1 meter. The first lift is maintained at ≤ 1.5 meters. The first layer is managed carefully, keeping in view of lining protection. Daily thickness is usually maintained between 0.15 to 0.3 meters through compaction, and the direction of compaction equipment should be from bottom to top. Intermediary cover may be used if any lifting surface is likely to be exposed over one month. Materials with higher resistivity than routine covering materials are applied as an intermediary cover with a typical thickness of 0.33 meters (Oweis & Khera, 1990). When the landfill capacity is exhausted, the final covering is provided with layers of soil and synthetic materials. Finally, the layer of vegetative cover is provided to border the erosion and percolation. A schematic illustration is provided in Figure 9.1-C.

Local authorities and communities as well as state governments issued guidelines to establish sanitary landfill sites. Standards for ventilation, distance from residential areas, provision of local facilities, and establishment of the buffer zone are managed in accordance with the proposed legal design-criteria. According to guidelines issued by the Ontario environmental protection authority for the designing of the landfill site (Ministry of the Environment, 2012), a buffer area from 30–100 meters is currently recommended for safe handling and monitoring of landfilling sites. Contamination of groundwater from chemicals and leachate must be protected, providing a 1.5-mm layer of high-density polyethylene, followed by a 0.75-meter layer of clay. Additionally, the provision of a 3-meter attenuation layer below the initial lift is also essential. The covering must be with 0.6 meters of covering materials, followed by 0.15 meters of topsoil.

9.6 SITE SELECTION PROCESS FOR SANITARY LANDFILLING

Site selection for sanitary landfilling is a critical decision that needs careful techno-economical analysis. Minimum environmental impact, economics, and capacity are critical parameters that are typically addressed during the decision-making process following the local regulatory guidelines (Asim et al., 2022; Ahamad et al., 2020). The best location that can make a reliable balance between these factors is the principal choice.

A substantial process of factorization and evaluation is performed numerically in order to analyze the suitability of available sites. Various techniques such as Geographic Information Systems (GIS) and Multicriteria Decision Analysis (MCDA) can be applied for the selection of suitable landfill sites (Aksoy & San, 2017; Barakat et al., 2017). Evaluation strategies are structured for efficient analysis of available information. Reliable results are considered applicable if they are reproducible and according to the local social and legal framework. Lane et al. (1983) elucidated that any technique that has the following characteristics can be used for the selection of landfill site:

i. If the technique has the potential to evaluate all possible locations impartially in a systematic way.
ii. If the technique can concisely establish numerical criteria for minimum and absolute suitability. The standard values may vary from community to

community according to local constraints of social and legal infrastructures. The decision standard should clearly define the location selection as selective or non-selective, rather than vague information.

iii. If the technique can be applied and established on generally obtainable statistics.

iv. If the technique can be applied for computerized statistical analysis.

v. If the technique can explicate the results of the statistical analysis clearly, in an understandable format, for officials as well as for the general community.

The mapping technique, long used for site selection in different projects, similarly can be used effectively for the selection of landfill sites (Lane & McDonald, 1983). Maps prepared with geological data and floodplains information are common tools to analyze site location and development of the final landfill plane. Initially, tracing papers and acetates were used to build maps and overlays manually with local information. But now, with the advancement in technology, these time-consuming processes have been replaced with GIS and MCDA (Aksoy & San, 2017; Şener et al., 2005). GIS eased the local analysis because it has the potential to manage and analyze large data volumes from different resources efficiently and accurately. These processing tools are not only helping in data analyses in a short time but also assist to increase objectivity and flexibility. Furthermore, the information and conclusions can be presented in an objective way, supported by scientific data analysis. Hence, they are more reliable and selective.

9.6.1 Selection Criteria

Various standards have been established by different environmental protection agencies and regulation departments for the selection of landfill sites. A summary of generally important standards is provided in Table 9.2.

TABLE 9.2
Summary of Suitability Standard for the Selection of Landfill Sites

	Suitability			
Parameter	**High**	**Moderate**	**Low**	**Ref**
Depth to groundwater (meters)	> 60	15–60	< 15	(Sener, 2004)
Groundwater quality (TDS in mg/l)	> 10000	1000–10000	< 1000	(Bagchi, 1994)
Buffer zone (meters)	> 100	30–100	< 30	(Ministry of the Environment, 2012)
Rock type	Unfractured, crystalline shale and clay	Limestone	Sandstone	(Oweis & Khera, 1990)

Dolui and Sarkar (2021) have defined some additional standards according to local regulations for the analysis of landfill sites. Some new factors, such as distance from a residential and agricultural area, education institutes, as well as historical and religious places, have been added. Different weightage methods were applied for factorial analysis, and results are presented in highly suitable to unsuitable classes (Figure 9.2).

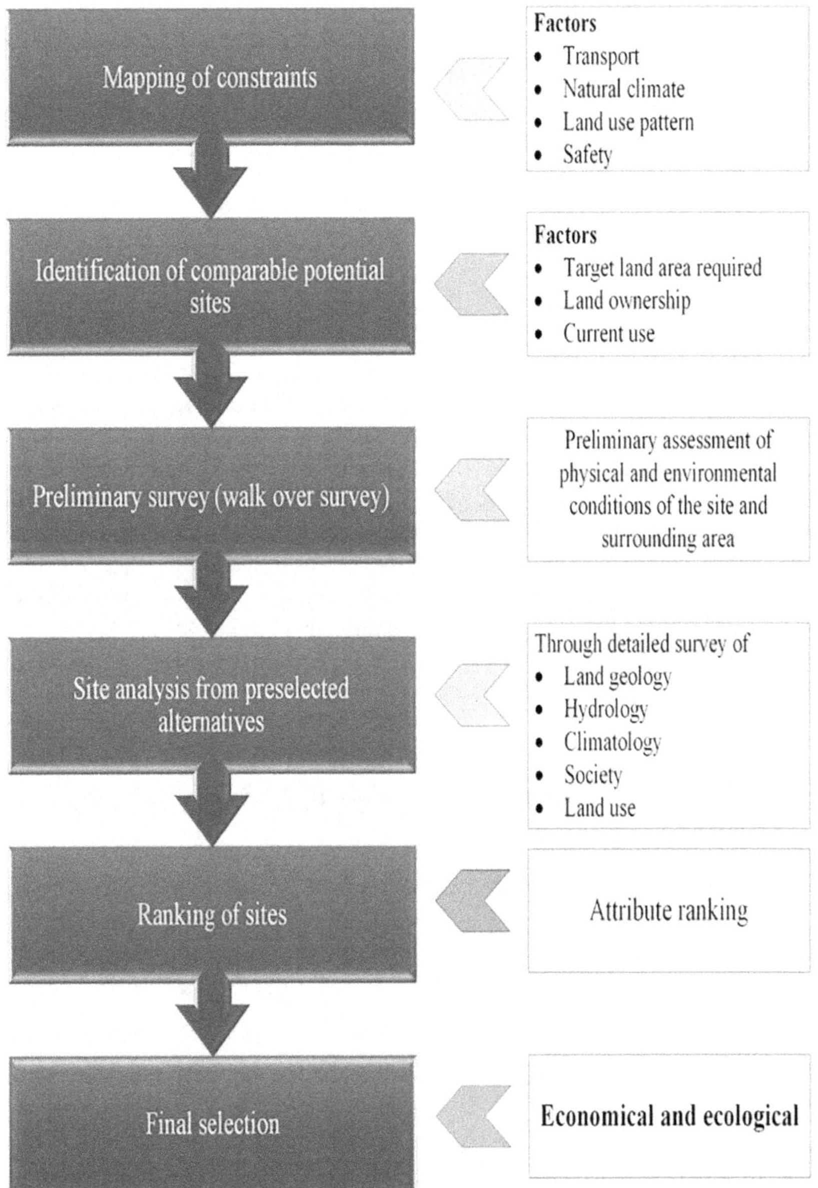

FIGURE 9.2 Selection procedure for the sanitary landfilling site (Carey, 2000).

TABLE 9.3
Buffer Standards for the Selection of Sanitary Landfill Sites in Morocco (Barakat et al., 2017)

Parameter	Value
Distance from isolated dwellings residential communities (meter)	500
Distance from planned communities (meters)	2,000
Distance from roads (meters)	300
Slope distance to fault (degree)	5
Geological distance to faults (meters)	500
Distance from water (river/streams/lakes/wells) reservoirs (meters)	500
Distance from agricultural and forest landscape (meters)	300

Barakat et al. (2017) used the GIS-based analysis tool to evaluate the suitability of landfill sites for Be´ni Mellal-Khouribga, Morocco. Different factors were categorized as environmental and economic factors, with a 0.75 and 0.25 weight average proportion. Elevation, slope, and distance from roads were categorized as economic factors, whereas all others were grouped as environmental factors. The evaluated suitability criteria are provided herein in Table 9.2, and the provision of compulsory buffer zones is provided in Table 9.3. The distance from the road was considered a critical aspect with major weightage in the economic sector, whereas land used, distance from any residential area, and surface water were given nearly equal weightage in the environmental sector. Similar factorization techniques and critical-factors division was also made in many other studies (Ali & Ahmad, 2020; Satyanarayana & Chandra, 2014). Hence, it is pertinent to mention that environmental factors are more significant than economic factors in the selection of landfill sites.

9.7 MAJOR CONCERNS AND FUTURE PROSPECTS OF SANITARY LANDFILLING

Although the management of solid waste through sanitary landfills is considered relatively safe, as compared to open dumping, and economical, as compared to scientific waste treatment proposals such as waste to energy technologies, still, there are significant concerns that can affect the future of sanitary landfill options, if not properly addressed.

9.7.1 Usability

The land needed to develop a sanitary landfill site is usually large enough to handle MSW for 10–30 years; continuous population growth needs a place to live as well as land to cultivate food; therefore, it will not be feasible to spare such a large piece of land permanently for unproductive use of sanitary landfilling. In the future, a compact design with some reusable options can be a viable solution. The provision of some microorganisms that can catalyze the decomposition with higher potential and high kinetics may help to recover and reuse land within a short span.

9.7.2 Economics

Underdeveloped and developing countries are mostly investing in product options and consider waste management as a non-productive expenditure. They usually don't consider the harmful effect on human life and medical expenses hereafter as a liability. Therefore, they are reluctant to assign a specific required budget for municipal waste management. Municipalities suffer due to the unavailability of financial and human resources. The unavailability of waste collection containers and insufficient bins is also because of financial constraints, which results in improper waste collection and disposal systems. As a result, waste piles can be witnessed on roadsides, around unused lands, and sometimes in between residential communities. Proper budgeting can solve this issue, with a focus on local municipalities.

9.7.3 Legislation

Many countries, especially the least developed countries, lack proper waste-management legislation. Therefore, communities use nearby vacant spaces as dumping sites without developing any sanitary-landfill sites. Additionally, due to the lack of waste management and collection laws, collected wastes have no classification, which results in an additional problem in sanitary landfilling. Lack of waste taxation results in uncontrolled waste generation from commercial units and reduced funding to sustain the waste management facilities. Moreover, the privation of regulations on take-back responsibilities or extended manufacturer accountabilities results in an increased number of materials in waste streams that can be recycled.

This issue can be solved with the development of enforcing laws for waste collection and management systems. It can be fixed for residential colonies as well as for industrial units to develop a waste collection system, forwarding this collected waste to the centralized sanitary landfilling site with proper segregation and a baling mechanism. Furthermore, the buffer zones for the sanitary landfilling site must be specified according to local land geology and hydrology. The approval guidelines must cover the parameter for operation and after seal monitoring.

9.7.4 Engineering and Technical Knowledge

The deficiencies in engineering and technical knowledge are a key obstacle to the development and implementation of sanitary landfilling and other waste management facilities scientifically. Therefore, people are practicing illegal and unspecified dumping of waste instead of sustainable waste management. A little technical knowledge can urge them to manage waste properly for their own health and safety. Additionally, the provision of sustainable waste management courses and guidelines in education institutions can change the scenario toward efficient waste management. In the future, it can be practiced, especially in the least-developed countries, to incorporate waste management as an essential social responsibility.

9.7.5 Underground Water

The major concern with sanitary landfills is the potential for groundwater pollution. If the leachate collected at the base leaks, it will pass toward underground water, resulting in the mixing of chemicals and pathogens into groundwater (Sajid, Ahmad, Shafqat, Pasha et al., 2020). It will cause a serious health issue for a long time because its removal from underground will be a big issue.

Its solution is specific guidelines for lining as well as for monitoring. Some countries have specified monitoring guidelines to record the underground water parameters after 24 hours. These parameters must be tested from different distances, as specified in the guidelines, from the sanitary landfill site. Similar guidelines with continuous updating can help to manage the operation of sanitary landfilling in the least developing countries.

9.7.6 Biogas

It has been observed that more than 70% of collected waste material is organic. A tremendous amount of biogas is generated through the anaerobic decomposition of this organic waste, which is a highly combustible gas. The produced biogas in the depth of the landfill may percolate horizontally and even trigger an explosion if accumulated heavily. Leakage to the surface may eradicate fauna and flora by harming the roots. The provision of a proper biogas-collection and the -transferring system will not only protect against fire and explosion danger but will also provide a continuous source of fuel for the local community. Different municipalities have already installed biogas wells on old and newly constructed landfilling sites in order to collect and distribute biogas properly. These will collect biogas, purify it, and then shift it to the distribution system on developing a specified pressure. Hence, the harmful effect and chances of explosion can be eliminated.

9.7.7 Decomposition

Various polymeric materials, which are being used as packing materials, are resistant and take a very long time to decompose. These materials are mostly fossil-derived, long-chain structures that can't be decomposed by natural microbes. These materials require special attention and need to be handled separately, and a recycling option must have been opted. Alternatively, these should be managed separately, according to their decomposition rate. Decomposition of waste results in settling and compaction, which creates space. This shallow depression may cause deep holes, which will collect water or provide an uneven surface. This problem will be serious if the landfilling site has been converted into a playground, park, or golf course. This problem can be addressed through continuous monitoring and after closure services. The developed holes and depression must be filled properly in order to avoid any mishaps/accidents.

9.8 KEY FEATURES OF SANITARY LANDFILLING

As sanitary landfilling is managed scientifically by employing engineering principles conforming to environmental protocols, it has plenteous advantages over conventional dumping of waste materials (Beede & Bloom, 1995). It has the advantages of eliminating fire probabilities and health hazards due to rodents and insect breeding, along with decreased odors. As compared to waste-to-energy technologies, sanitary landfilling has the advantages of low cost and operational simplicity. It doesn't require any separation of refuse, and trash from waste-to-energy technologies can be accommodated easily with collected wastes. Additionally, it can be performed with low energy-inputs and without demanding any specific skilled labor. The comparative pros and cons of sanitary landfilling are described as follows (Sener, 2004);

9.8.1 Advantages

i. It doesn't require any specific facilities.
ii. Post-closure land can be used to develop forests, parks, parking, etc.
iii. It can be developed near residential areas, as opposed to waste-to-energy technologies.
iv. Collected biogas can be used as domestic fuel.
v. No specific labor or technology is required.
vi. It can generate local employment, along with local cleaning.
vii. A disposal strategy can be developed for waste disposal for 30 years.
viii. It is flexible and can handle population growth effectively.
ix. Machinery can also be used for other community purposes.

9.8.2 Disadvantages

x. Wastage of land.
xi. Chances of garbage spread due to wind or any other natural disaster.
xii. It can generate unpleasant odors; hence, it is opposed by the general public.
xiii. Needs movement of heavy machinery for waste collection and dumping.
xiv. It has the potential to contaminate groundwater and waterways.
xv. Needs continuous observation and monitoring, even after closure.

ACKNOWLEDGMENT

I would like to thank the Faculty of Materials and Chemical Engineering, Yibin University, China for its support to complete this work.

REFERENCES

Ahamad, M. I., Zafar, Z., Arsalan, M., Rehman, A., Sajid, M., Zulqarnain, R. M., Mehmood, M. S., Abdal, S., & Aslam, M. (2020). Effects of temperature and pressure on reservoir fluids and seismic properties of reservoir rocks. *International Journal of Pharmaceutical Sciences Review and Research*, *63*(2), 36–43. https://globalresearchonline.net/journal-contents/v63-2/06.pdf

Aksoy, E., & San, B. T. (2017). Geographical Information Systems (GIS) and Multi-Criteria Decision Analysis (MCDA) integration for sustainable landfill site selection considering dynamic data source. *Bulletin of Engineering Geology and the Environment*, *78*(2), 779–791. https://doi.org/10.1007/S10064-017-1135-Z

Ali, S. A., & Ahmad, A. (2020). Suitability analysis for municipal landfill site selection using fuzzy analytic hierarchy process and geospatial technique. *Environmental Earth Sciences, 79*(10). https://doi.org/10.1007/S12665-020-08970-Z

Andreottola, G., Cossu, R., & Ritzkowski, M. (2018). Landfill Gas Generation Modeling. In Solid Waste Landfilling. Elsevier Inc. https://doi.org/10.1016/b978-0-12-407721-8.00020-6

Asim, M., Zhang, S., Wang, Y., Maryam, B., Sajid, M., Shi, C., Pan, L., Zhang, X., & Zou, J. J. (2022). Self-supporting NiCoP for hydrogen generation via hydrolysis of ammonia borane. *Fuel, 318*, 123544. https://doi.org/10.1016/J.FUEL.2022.123544

Atlas, W. (2018). *What a waste: An updated look into the future of solid waste management.* www.worldbank.org/en/news/immersive-story/2018/09/20/what-a-waste-an-updated-look-into-the-future-of-solid-waste-management

Bagchi, A. (1994). *Design, Construction and Monitoring of Landfills*, John Wiley & Sons, New York.

Bank, T. W. (2020). *Trends in solid waste management.* https://datatopics.worldbank.org/what-a-waste/trends_in_solid_waste_management.html

Barakat, A., Hilali, A., Baghdadi, M. El, & Touhami, F. (2017). Landfill site selection with GIS-based multi-criteria evaluation technique. A case study in Béni Mellal-Khouribga Region, Morocco. *Environmental Earth Sciences, 76*(12). https://doi.org/10.1007/s12665-017-6757-8

Beaven, R. P., & Woodman, N. D. (2018). Landfill hydraulics. In *Solid waste landfilling*. Elsevier Inc. https://doi.org/10.1016/b978-0-12-407721-8.00014-0

Beede, D. N., & Bloom, D. E. (1995). The economics of municipal organic waste. *The World Bank Research Observer, 10*(2), 113–150. http://documents.worldbank.org/curated/en/116201468331246520/pdf/770780JRN0WBRO0Box0377291B00PUBLIC0.pdf

Bell, J. M. (1973). *Sanitary landfill method of solid waste disposal.* School of Civil Engineering, Purdue University.

Carey, P., Carty, G., Donlon, B., Howley, D., Nealon, T. (2000). *Landfills Manual, Landfills Site Design.* Environmental Protection Agency, Wexford, Ireland. https://www.epa.ie/pubs/advice/waste/waste/aguidetolandfillsitedesign.html#. Vxn4xzHUV5Y

Dastyar, W., Raheem, A., He, J., & Zhao, M. (2019). Biofuel production using thermochemical conversion of heavy metal-contaminated biomass (HMCB) harvested from phytoextraction process. *Chemical Engineering Journal, 358*, 759–785. https://doi.org/10.1016/j.cej.2018.08.111

Diejomaoh Abafe, O. T., Azim, M. M., & Sajid, M. (2022). Ionic liquid-mediated CO2 conversion. In *Ionic liquid-based technologies for environmental sustainability* (pp. 225–233). Elsevier. https://doi.org/10.1016/B978-0-12-824545-3.00014-3

Dolui, S., & Sarkar, S. (2021). Identifying potential landfill sites using multicriteria evaluation modeling and GIS techniques for Kharagpur city of West Bengal, India. *Environmental Challenges, 5*, 100243. https://doi.org/10.1016/J.ENVC.2021.100243

Environment Protection Agency. (2003). *Guidelines for the Siting, Design and Management of Solid Waste Disposal Sites in the Northern Territory.* http://www.nt.gov.au/nreta/environment/waste/codes/pdf/landfillguidelines.pdf

Kjeldsen, P., & Scheutz, C. (2018). Landfill gas management by methane oxidation. In *Solid waste landfilling* (pp. 477–497). Elsevier Inc. https://doi.org/10.1016/b978-0-12-407721-8.00024-3

Kreith, F. (2019). *Handbook of solid waste management* (2nd ed.). McGraw-Hill. https://doi.org/10.2307/j.ctvjz80w7.13

Lane, W. N., & McDonald, R. R. (1983). Land suitability analysis: Landfill siting. *Journal of Urban Planning and Development, 109*(1), 50–61. https://doi.org/10.1061/(ASCE)0733-9488(1983)109:1(50)

LeBoeuf, E., Thackston, E., Schroeder, P., & Palermo, M. (2004, December). *Liner design guidance for confined disposal facility leachate control* (pp. 1–25). US Defense Technical Information Center. https://archive.org/details/DTIC_ADA430421

Ministry of the Environment, C., & P. (2012). *Landfill standards: A guideline on the regulatory and approval requirements for new or expanding landfilling sites | ontario.ca.* www.ontario.ca/page/landfill-standards-guideline-regulatory-and-approval-requirements-newexpanding-land

Mondal, T., Choudhury, M., Kundu, D., Dutta, D., & Samanta, P. (2023). Landfill: An eclectic review on structure, reactions and remediation approach. *Waste Management, 164*, 127–142. https://doi.org/10.1016/j.wasman.2023.03.034

Mulk, S., Sajid, M., Wang, L., Liu, F., & Pan, G. (2022). Catalytic conversion of sucrose to 5-hydroxymethylfurfural in green aqueous and organic medium. *Journal of Environmental Chemical Engineering*, *10*(2), 106613. https://doi.org/10.1016/J.JECE.2021.106613

Nanda, S., & Berruti, F. (2021a). Municipal solid waste management and landfilling technologies: A review. *Environmental Chemistry Letters*, *19*(2), 1433–1456. https://doi.org/10.1007/s10311-020-01100-y

Nanda, S., & Berruti, F. (2021b). Thermochemical conversion of plastic waste to fuels: A review. *Environmental Chemistry Letters*, *19*(1), 123–148. https://doi.org/10.1007/s10311-020-01094-7

Nations, U. (2015). *Sustainable development goals: 17 goals to transform our world.* United Nations. https://www.un. org/sustainabledevelopment/energy/.

Oweis, I. S., & Khera, R. P. (1990). *Geotechnology of waste management.* United Kingdom: Butterworths. http://inis.iaea.org/search/search.aspx?orig_q=RN:22023798

Pasha, M. K., Dai, L., Liu, D., Guo, M., & Du, W. (2021). An overview to process design, simulation and sustainability evaluation of biodiesel production. *Biotechnology for Biofuels*, *14*(1), 129. https://doi.org/10.1186/s13068-021-01977-z

Paul, S., Choudhury, M., Deb, U., Pegu, R., Das, S., & Bhattacharya, S. S. (2019). Assessing the ecological impacts of ageing on hazard potential of solid waste landfills: A green approach through vermitechnology. *Journal of Cleaner Production, 236*, 117643. https://doi.org/10.1016/j.jclepro.2019.117643

Rehman, A., Song, J., Haq, F., Mahmood, S., Ahamad, M. I., Basharat, M., Sajid, M., & Mehmood, M. S. (2022). Multi-hazard susceptibility assessment using the analytical hierarchy process and frequency ratio techniques in the northwest Himalayas, Pakistan. *Remote Sensing*, *14*(3), 1–31. https://doi.org/10.3390/rs14030554

Rehman, M. S. U., Kim, I., Rashid, N., Umer, M. A., Sajid, M., & Han, J.-I. (2016). Adsorption of brilliant green dye on biochar prepared from lignocellulosic bioethanol plant waste. *Clean—Soil, Air, Water*, *44*(1), 55–62. https://doi.org/10.1002/clen.201300954

Reinhart, D., & Stegmann, R. (2018). Physical/chemical reactions in landfills. In *Solid waste landfilling*. Elsevier Inc. https://doi.org/10.1016/b978-0-12-407721-8.00007-3

Sajid, M., Ahmad, M. I., Shafqat, S. S., Pasha, M. K., & Asim, M. (2020). De-colorization of azo dyes: C.B-10X Blue and C.B.M-10X blue by ozonation for waste water treatment. *University of Wah Journal of Science and Technology*, *4*(1), 1–7. http://uwjst.org.pk/index.php/uwjst/article/view/49

Sajid, M., Ahmad, M. I., Shafqat, S. S., Mulk, S., & Pasha, M. K. (2020). Study of phosphorous oxide (P2O5) and Iron oxide (Fe2O3) in rock phosphate of Hazara basin of Pakistan. *International Journal of Agrochemistry*, *6*(1), 46–51. https://doi.org/10.37628/ijac.v6i1.937

Sajid, M., Bai, Y., Liu, D., & Zhao, X. (2020). Organic acid catalyzed production of platform chemical 5-hydroxymethylfurfural from fructose: Process comparison and evaluation based on kinetic modeling. *Arabian Journal of Chemistry*, *13*(10), 7430–7444. https://doi.org/10.1016/j.arabjc.2020.08.019

Sajid, M., Bari, S., Saif Ur Rehman, M., Ashfaq, M., Guoliang, Y., Mustafa, G., Saif, M., Rehman, U., Ashfaq, M., Guoliang, Y., & Mustafa, G. (2022). Adsorption characteristics of paracetamol removal onto activated carbon prepared from Cannabis sativum Hemp. *Alexandria Engineering Journal*, *61*(9), 7203–7212. https://doi.org/10.1016/j.aej.2021.12.060

Sajid, M., Bary, G., Asim, M., Ahmad, R., Ahamad, M. I., Alotaibi, H., Rehman, A., Khan, I., & Guoliang, Y. (2022). Synoptic view on P ore beneficiation techniques. *Alexandria Engineering Journal*, *61*(4), 3069–3092. https://doi.org/10.1016/j.aej.2021.08.039

Sajid, M., Dilshad, M. R., Rehman, M. S., Liu, D., & Zhao, X. (2021). Catalytic conversion of xylose to furfural by p-Toluenesulfonic acid (pTSA) and chlorides: Process optimization and kinetic modeling. *Molecules*, *26*(8), 2208. https://doi.org/10.3390/molecules26082208

Satyanarayana, D. N. V., & Chandra, K. R. (2014). Municipal Solid waste management by sanitary landfill. *International Journal of Engineering Sciences & Research Technology*, *3*(6), 811–818.

Sener, B. (2004). *Landfill site selection by using geographic information system* [Middle East Technical University]. http://onlinelibrary.wiley.com/doi/10.1002/cbdv.200490137/abstract

Şener, B., Süzen, M. L., & Doyuran, V. (2005). Landfill site selection by using geographic information systems. *Environmental Geology*, *49*(3), 376–388. https://doi.org/10.1007/S00254-005-0075-2

Sivapullaiah, P. V, & Naveen, B. P. (2016). Municipal solid waste landfills construction and management-a few concerns. *International Journal of Waste Resources*, *6*(2), 1–8. https://doi.org/10.4172/2252-5211.1000214

10 Potential for Applicability of Decentralized Approaches in Sewage Treatment Systems Globally

Aditi Roy

10.1 INTRODUCTION

Water, as a universal solvent, has the property of assimilating various conservative, persistent and toxic impurities and transporting them to various destinations. Wastewater generally explains the by-product of domestic, industrial and agricultural run-offs containing organic and inorganic matter in suspended, colloidal and dissolved forms (Arceivala et al., 2006). Figure 10.1-A represents a comprehensive layout describing types and sources of wastewater discharged into the environment. Sewage forms the subset of wastewater, which incorporates all types of waste generated from domestic dwellings, including pollutants like fecal matter, urine, soaps, detergents, food particles and other things that are disposed of in a drain (Gogoi & Choudhury, 2022; Kamble et al., 2017). Any kind of untreated sewage may contain water, along with nutrients, organic matters, pathogens, oils and greases, runoffs, along with heavy metals. In order to improve and purify, sewage generally travels to sewage treatment plants through sewer systems (Rajpal et al., 2022). This decontaminated water is reused and released, either in surface water or in agricultural lands for various purposes. However, the absence of proper sewage treatment facilities and their maintenance compromise the treatment efforts. This results in limiting treatment efficiency, and thus, generating huge loads of untreated wastes. Mostly in developed countries, the treatment capacity is nearly 70%, which drops down drastically to 8% in developing nations (Gondhalekar et al., 2019). The direct discharge of sewage, either treated or untreated, poses a threat to human health and quality of life and environmental conditions. Many studies have reported municipal effluents as a major source for the discharge of regulated compounds like polychlorinated biphenyls (PCBs), polycyclic aromatic hydrocarbons (PAHs) or heavy metals into the environment. Several emerging pollutants pertain to the environment due to hydrophobicity and their bio-accumulative property (Díaz-Garduño et al., 2017). Sewage, when drained in large quantities into nearby surface waters, reduces the dilution

DOI: 10.1201/9781003258377-10

of the constituents of water. This further makes it stagnant and accelerates various water-borne diseases.

10.2 SCENARIO OF SEWAGE GENERATION AND TREATMENT

The data vital for estimating total sewage generated and treated is omnipresent through various published and web-based online resources. The approximate volume of wastewater generated, treated and recycled at global, regional and country levels is made available through the Food and Agriculture Organization of the United Nations, European Commission supported Eurostat, United States Environmental Protection Agency and various published reports. These data can be variedly used by policy makers and practitioners to form various action plans regarding wastewater treatment and its productive use in distinct ecosystems. However, this available information is sometimes limited or outdated in various cases.

According to a study by Qadir et al. (2020), approximately 380 billion cubic meter (380 trillion L) of wastewater are generated annually in a global scale. Out of this, only 55% of the domestic wastewater is treated, with the rest of industrial and agricultural wastewater getting directly released into nearby locations (WHO, UN-Habitat). Figure 10.1-B shows the proportion of domestic and industrial wastewater flow that is safely treated globally. However, these figures need to be taken with caution. There are only 55 countries that are reported to have complete data regarding wastewater generation, treatment and reuse. Even the data provided is not recent and dates back to 2017. The lower-income countries lack sufficient data, and thus, are not reported in AQUASTAT. Second, the majority of middle-income countries have treatment plants that are below average. Third, while most of the countries report the effluent from tertiary treatment as treated wastewater, some include primary and secondary treated wastewater too. This makes the dataset vague and strenuous.

10.3 RATIONALE OF CENTRALIZED SEWAGE TREATMENT SYSTEMS (CSTS)

Currently, there is a need for sustainable water-resource management because of elevating pollution load and inefficiency in treatment systems. The overall concept of wastewater treatment through a centralized system was initially developed in highly urbanized and industrial sectors of Central Europe and the USA around the mid-nineteenth century (Figure 10.2-A). The massive outbreak of catastrophic diseases due to open disposal of household wastes led to the technical solution of forming a public sewer system for collecting all wastewater and transporting it for treatment outside the locality. Thus, the first inclusive sewer network was built in Hamburg around 1842, followed by other cities as well. Later, many advanced technologies were developed that were considered to be suitable for large cities on a large scale (Burian et al., 2000). As in the developed countries, this conventional method was highly preferred and considered to be the standard tool of environmental protection and control, and so, was extended to developing countries as well. Therefore, the centralized management system was applied to various large cities and secondary towns in the industrialized as well as in middle- or lower-income countries (Bakir, 2001).

(A)
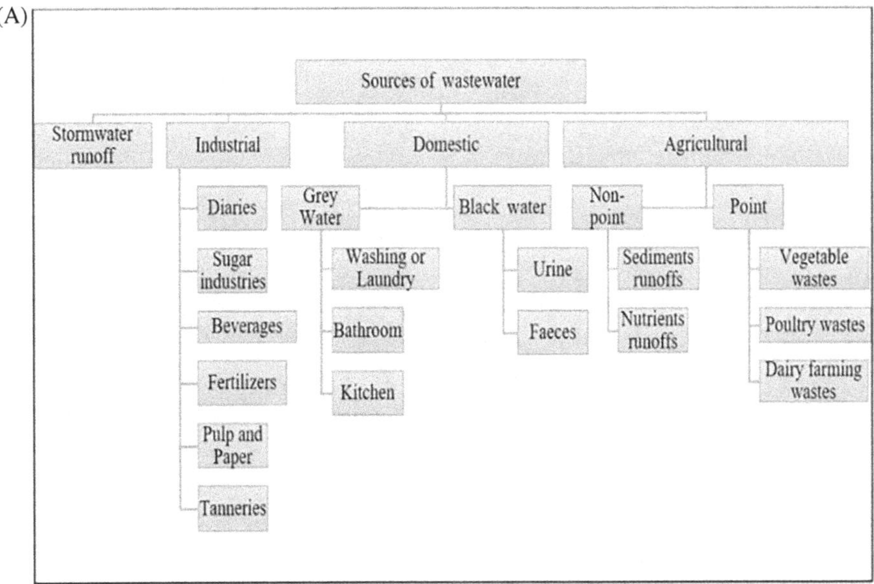

(B)
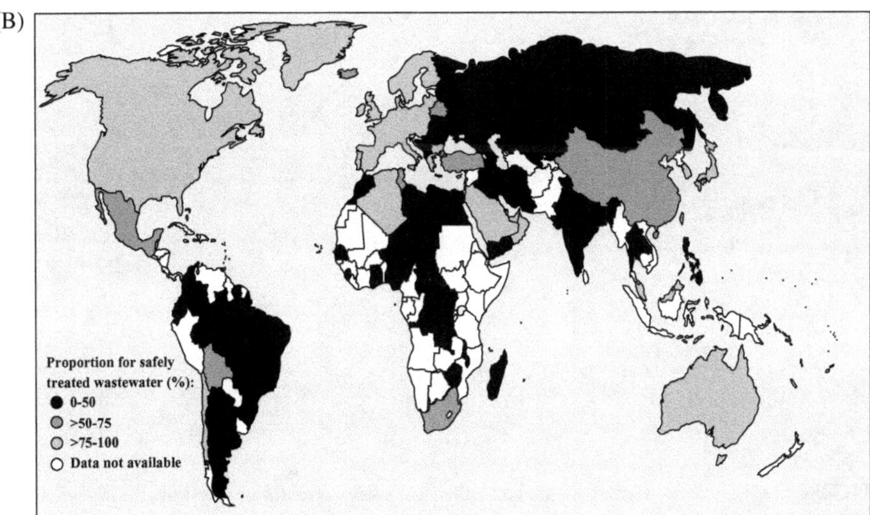

FIGURE 10.1 (A)—Types and sources of waste water. (B)—Proportion of domestic and industrial wastewater that is safety treated (UN-water, 2020).

10.3.1 Insights of Centralized Sewage Treatment Systems

The centralized system of wastewater or sewage treatment handles wastewater generated from various sectors of society all over a particular place. The major objective is to treat and reduce the levels of contaminants in wastewater as per the effluent standards (Figure 10.4). The treatment is done at a single place where all the wastewater of that area is collected and treated. The centralized sewage treatment systems

(A)

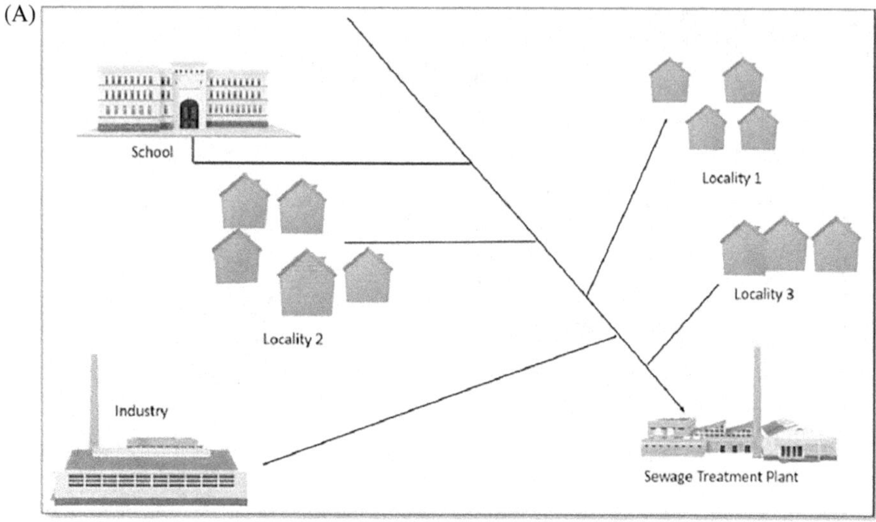

(B)

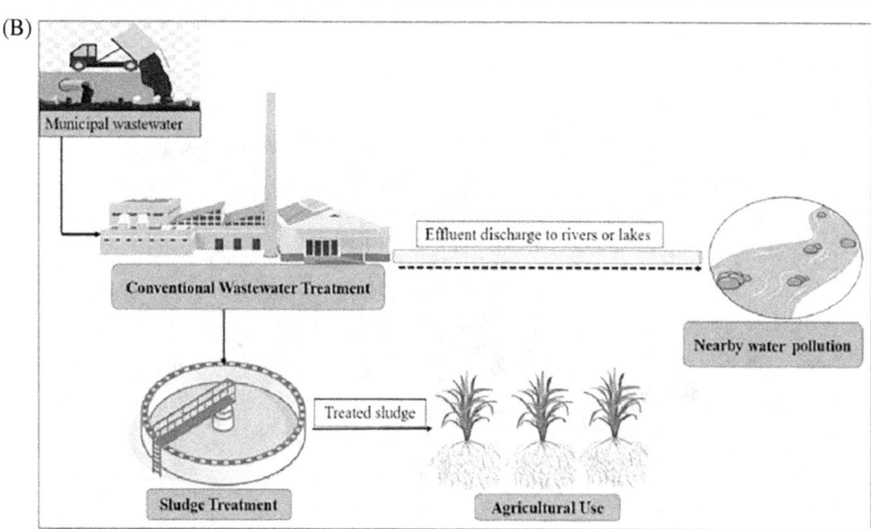

FIGURE 10.2 (A)—Concept of centralized sewage treatment systems. (B)—Fate of wastewater through a conventional, centralized treatment facility.

mainly focus on the municipal treatment plants applied to a much larger scale. These centralized systems are publicly owned and are used for treating the sewage for the entire community. The wastewater is collected through a series of sewer pipes and pumping sets, which is then transported to a central treatment-plant. The sewer pipes can be combined or separate, depending on the advancement of area. Centralized systems are more planned, designed and operated by governmental agencies. Since these centralized plants are solely responsible for treating the total waste generated in any area, this increases the load on their efficiency. Due to this, most of

(A)

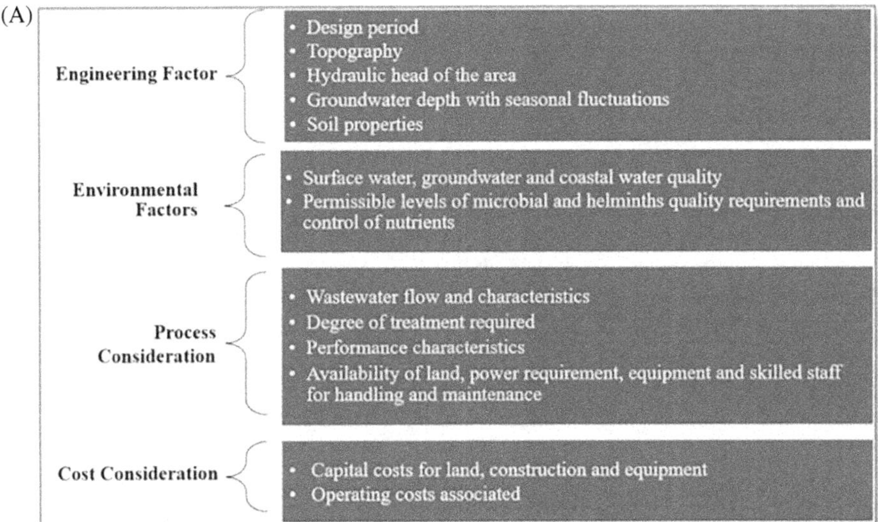

(B)

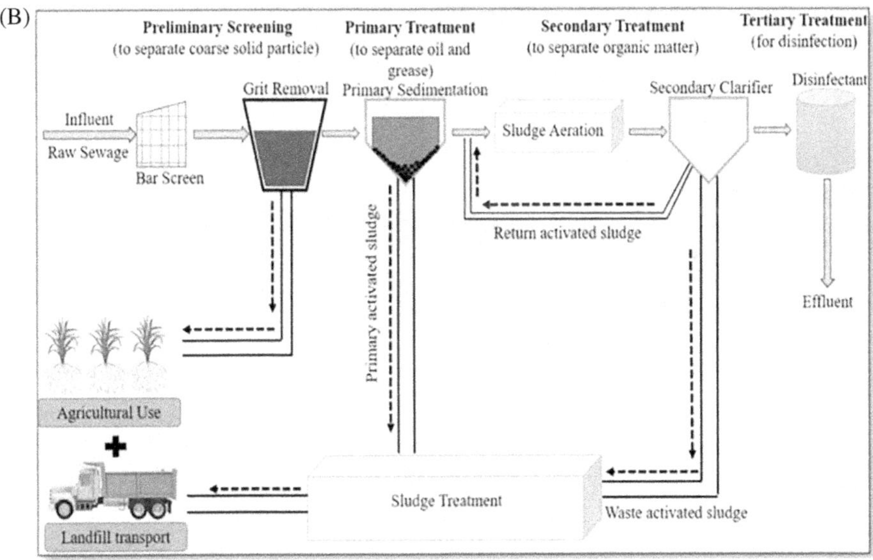

FIGURE 10.3 (A)—Factors responsible for designing sewage treatment plant. (B)—Stages of operation in a sewage treatment plant.

the centralized systems could not meet the effluent standard before its discharge. This leads to direct discharge of the untreated wastewater into the nearby facilities (Figure 10.2-B). The gradual increase in population has overburdened the treatment system globally. In most of the cases, treatment capacity remains underutilized due to inadequacy in operation and maintenance. The large slum areas of developing countries have poor connectivity with the sewer lines, and hence, pose greater challenges for proper treatment and disposal.

(A)

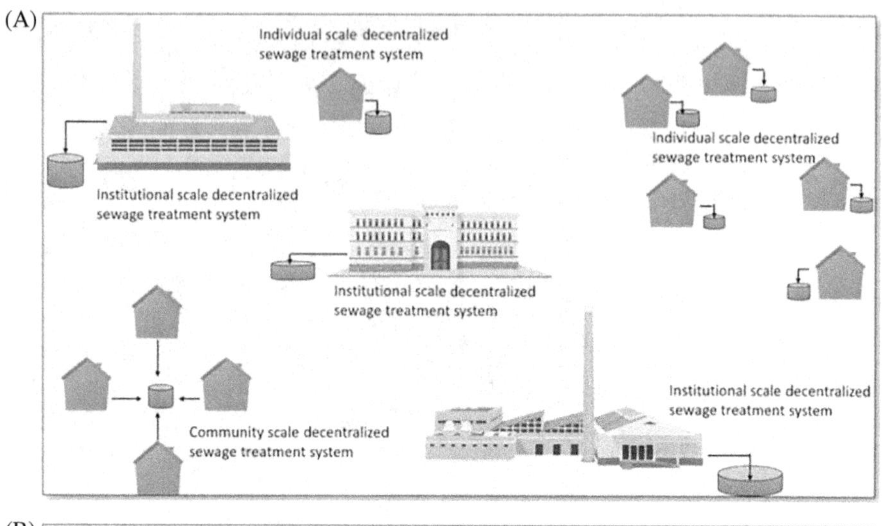

(B)

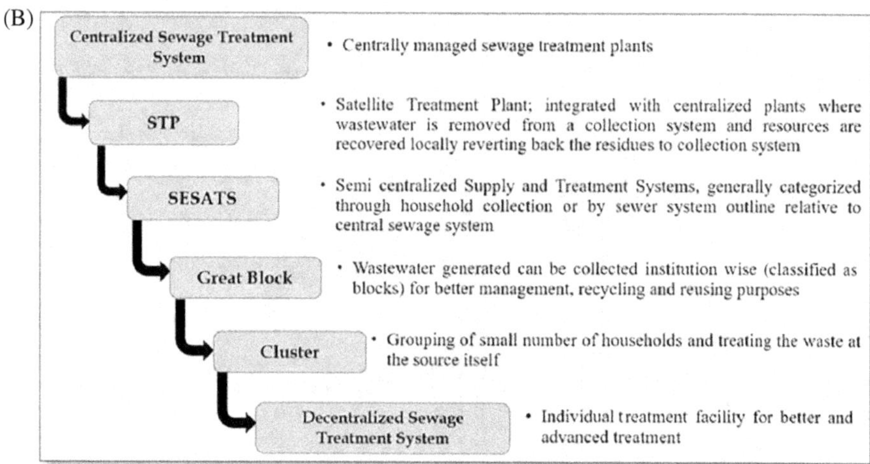

FIGURE 10.4 (A)—Concept of decentralized sewage treatment systems. (B)—Transition from centralized to decentralized sewage treatment facility.

10.3.2 Design of Treatment Plants

The purpose of sewage treatment plants is to provide a low-cost process for eliminating pollutants through physical, chemical and biological means. Initially, design procedures posed challenges to reducing cost by assessing the steady-state concepts for unit processes and design guidelines. Once entered into the treatment plants, the sewage undergoes different stages before being suitable for disposal. Centralized wastewater treatment plants are designed based on demography, design period, water consumption pattern, degree of treatment required, the sewer materials, their economy, the topography, limiting depth of sewer, minimum velocity for self-cleaning, availability of land for treatment, degree of treatment required for recycling, nature of outfall, quality and quantity of effluent and receiving water (Figure 10.3-A and B).

The collection, treatment and disposal of sewage and other related wastes form the three foremost components of any sewage treatment management system. Out of this, the centralized system of treatment spends around 80% of total investment for collection only. This contains a common and centralized sewer system for collecting all the sewage from households and other commercial and industrial places. This is considered to be an off-site management system, as the location of the treatment plant and the disposal of the treated residue is often far off from the point of origin (Wilderer & Schreff, 2000). However, many non-point sources of waste generation could not be handled properly in this case.

10.4 CONCEPT OF DECENTRALIZED SEWAGE TREATMENT SYSTEMS (DSTS)

The shortcoming of centralized sewage treatment systems is progressively surfacing in recent times. The lack of technical efficiency and proper management of the treatment plants aids in the downfall of these centralized systems. This creates a new prospect of on-site sewage treatment facilities. The major costs associated with the functioning of the off-site treatment facilities are collection from different point and non-point sources along with their connection and transportation to the treatment plants through sewer pipelines and pumping sets. Also, the overburdening of centralized treatment plants has failed to meet the standards of effluent, thus contributing to pollution more. On contrary, decentralized sewage treatment systems can help in reducing the associated operational costs, thus increasing the affordability of sewage treatment systems. Even small and remote settlements in low-income countries can be highly benefitted by these on-site treatment systems. The concept of centralized, off-site treatment is ideal for developed countries, but for the developing nations, just the reverse is required. However, the effectiveness of the decentralized approach depends on the setting up of a management program to ensure regular inspection for the proper functioning and maintenance of the system (Figure 10.4-A).

10.4.1 Details of Decentralized Sewage Treatment Systems

The decentralized sewage treatment system is basically an on-site treatment facility that allows the purification and reuse of wastewater for various purposes. The collection network is small, receiving a smaller quantity of sewage that can be easily managed with simple and natural treatment methods (Rohilla & Dwivedi, 2013). Thus, the system will allow for more independent, self-maintained and sustained facilities, ensuring the restoration of the wastewater resources. The decentralized sewage treatment systems can be of different scales like for every individual household, for a cluster of homes, in a neighborhood, in public facilities, in any commercial area or industrial parks and even for small portions of large communities (Figure 10.4-B) (Bakir, 2001).

Any decentralized system for sewage treatment is set up using a septic tank at every waste-generation point-source. Here, the solids are eliminated and liquids are further conveyed to a central treatment facility through sealed and small-diameter plastic piping. This reduces manholes and other potential storm-water and ground-water infiltration. In order to maintain a constant flow of solid wastes, it is, however, required that the discharge point should be below the inputs. The construction design of the decentralized systems is kept simpler, less disruptive and leak free, which makes it easier to install and maintain specifically in developing countries. Currently, the EPA has been running a decentralized wastewater program that focuses on technical guidance and advocates the concept of utility-based system management. Relatively small energy- and nutrient-recovery projects have been promoted by ecological engineering- and environmental-remediation groups that have contended with curtailing resource wastage for wastewater treatment (Banerjee & Choudhury, 2022).

10.4.2 Process Under Decentralized Systems

A study for decentralized treatment-systems has reported a few processes, including natural systems (i.e., constructed wetlands (CWs)) and improved aerobic or anaerobic biological systems (both suspended or attached growth) (Singh et al., 2015). The application of these processes solely depends on general climatic conditions, wastewater strength (dilution) and variability, land requirements and availability and local recycle/reuse requirements and may strongly influence the feasibility and operational outcome of an installation. The main influencing factors in a decentralized technology choice are: specific treatment efficiency in that specific condition, low O&M requirements, operational reliability and future, gradual expansion possibilities and favorable economics. In Table 10.1 are some the decentralized processes discussed with advantages in brief:

10.5 PARADIGM SHIFT FROM CENTRALIZED TOWARD DECENTRALIZED

The main advantage for a centralized system of sewage treatment is the centrally managed municipal governance, which has restrained the reclamation and reuse of wastewaters. The majority of developing nations lack financial funding for construction and technical expertise for operation and management of these off-site treatment plants. The emerging constraints and complications regarding centralized treatment is increasing progressively. Off-site treatment has led to elevating the pollution load due to non-functioning of the facilities. The continuous and expanding inefficiency of centralized systems has led to direct disposal of semi- and untreated effluent to surface-water bodies or nearby vicinity. Unscientific dumping increases the threat on different ecological and environmental factors. Potential eutrophication phenomena may occur in the receiving water body due to the large volumes of untreated sewage discharged. Thus, the present scenario demands a total paradigm shift from centralized to decentralized approaches, where the sewage could be treated in the source itself. This opens up an alternate approach from a disposal-based, linear system to a recovery-based, closed system (Figure 10.4-B).

TABLE 10.1
Description and Advantages of Decentralized Sewage Treatment Systems

Process	Description	Advantages
Intermittent/ recirculating sand filters	Modified version of single pass open sand filters. Sewage passes through septic tank for primary treatment and then to recirculating tank. The sand filter filtrates are mixed with sewage here and then pumped back to the sand filter bed.	• No chemicals are used for treatment. • High effluent quality with 95% removal of Biological Oxygen Demand (BOD) and Total Suspended Solids (TSS). • Treatment capacity can be modified. • Easy to operate and maintain with less skilled manpower. • Significant conversion of nitrogen to nitrates.
Facultative lagoons	Type of waste stabilizing ponds. Sewage is catabolized in constructed ponds used for biological treatment.	• Effective in removing settled solids, disease causing pathogens, ammonia. • High nutritive effluent which can be used for agricultural purposes. • Cost-effective and easy to operate.
Anaerobic lagoons	Deep earthen basins used to anaerobically digest soluble organic substrate and sludge. These are further followed by aerobic or facultative lagoons for achieving proper treatment.	• Eliminates disease causing pathogens. • Efficient for treating strong organic wastes. • Production of methane. • High nutritive effluent for agricultural utility.
Aerobic lagoons	Fully aerated condition which allows sunlight to penetrate and promote algal growth for sewage treatment.	• Easy to operate and maintain. • Do not require skilled labor. • Effluent can used further for irrigation purposes.
Constructed wetlands	Natural engineered treatment where microbes are used which consumes the wastewater as their energy sources. For complex systems, plants are used for filtration and other bio-chemical reactions.	• Low organic and suspended solids content of the effluent. • Reduces land area needed for wastewater treatment. • Reduces odor. • Provide wildlife habitat. • Inexpensive to operate. • Able to handle various wastes type.
Sequencing batch reactor	Here, sewage is added to a single batch reactor for treatment. The working principle is similar to that of activated sludge process.	• Feasible in treating all types of wastewaters. • Suitable for enhanced treatment. • Flexible operation and easy to control • Minimum ecological footprint. • Equalization, primary treatment and biological treatment is done using a single reactor.

(*Continued*)

TABLE 10.1 *(Continued)*
Description and Advantages of Decentralized Sewage Treatment Systems

Process	Description	Advantages
Membrane bio-reactor	Hybrid of conventional biological sewage treatment process. Incorporates advanced level of organic and suspended particle removal through membrane filtration.	• Maximum microbial removal from MBR membranes. • High rate of ammonia and nitrate removal. • Ample reduction in the suspended solid particle concentration in effluent. • Easily expandable and also incorporated in any type of conventional treatment plants. • Less sludge production. • Long loading rate capability.
Soil biotechnology	Modern green technology assimilates process of sedimentation, infiltration and biodegradation for the bioconversion of the sewage. The bacterial interaction helps in the processing of organic and oxidizable inorganic through the natural oxygen supply.	• Maximum microbial removal from MBR membranes. • 90% removal of COD and BOD from influent sewage. • Treatment process can run on batch or on continuous mode. • No sludge generation. • No skilled labor required. • Low annual operation and maintenance cost and is eco-friendly. • Treatment process can run on batch or on continuous mode.

10.5.1 Challenges Faced in Centralized Systems

Centralized sewage treatment plants generally showcase greater dependency on technical expertise and skilled man-power. There are a comparatively higher construction and operational costs associated with setting up of a treatment plant. Because of heavy inflow rate of sewage, the treatment efficiency is reduced and less maintained, which becomes a major factor for untreated effluent. Thus, there is a huge load of sludge generated that gets dumped in surrounding agricultural fields. Also, huge cost is associated with establishing grids of pipelines and pumping chambers that connect the generation points to the treatment point. Sometimes, due to physical hindrances like plastics, the sewer lines get choked, thus causing leakages and blockages. Although there is collection system present, there is a periodic maintenance of every 50–60 years. Since bigger in size, a huge volume of portable water is required for cleaning the area, which indicates wastage of natural resources for waste handling and treatment.

TABLE 10.2
Comparative Analysis between Centralized and Decentralized Systems

Parameters	Centralized systems	Decentralized systems
Collection system	Large diameter sewer pipelines installed over long distance	Generally, acts on gravity flow and small pipelines
Area required	Installed in larger area in outskirts having enormous units	Comparatively smaller in size and can be installed at individual level
Total Capacity	Able to treat larger volume of sewage	Built for less capacity
Electrical requirement	Largely dependent on electricity for all purposes	Less electrical usage due to smaller units
Cost associated	Huge operational and maintenance required	Low operation cost associated
Dilution grade	More dilution	More concentrated
Ecological risk	Risk on larger scale	Risk distributed
Manpower	Skilled labor required	Unskilled labor can also operate
Ease of operation	Complex system; cannot be easily modified	Can be easily modified as per requirement of treatment
Potential to reuse	Effluent is directly discharged to surface water bodies	Reused for irrigational purposes
Sludge generation	Huge toxic sludge is generated which gets dumped in nearby fields	Limited sludge generation which can be used in agriculture
Working efficiency	Weaken with time	Gets enhanced with time
Technology involved	High-cost technology applied	Low-cost eco-friendly technology is used
Odor	Excessive use of chemicals results in foul smell	No operating chemicals thus no odor

10.5.2 Advancement of Decentralized Over Centralized Approaches Toward Sewage Treatment

Various governmental agencies are responsible for the collection, operation and maintenance of sewage treatment plants on a larger scale. However due to inefficiency in the treatment process, the centralized systems lag behind. Association of huge operational and maintenance costs, along with skilled labor, makes the functioning more difficult. This is the major factor due to which most of the treatment plants remain under-utilized. Remotely located areas are neither connected with the sewer lines nor with treatment plants. This allows the improper dumping and discharge of wastes, which is a greater challenge for proper treatment. Alternatively, decentralized systems provide the benefit of fulfilling the constrains of centrally owned plants by reducing potential residual effluent contamination. This on-site treatment facility provides a better solution that can easily comply with the environmental standards. Low-cost, environment-friendly technologies used in the decentralized approach make it suitable for the longer run without hampering the ecosystem. Contradictory to the off-site treatment, the treated sewage could be used for irrigation, gardening and other purposes. Thus, the system will allow for more independent, self-maintained and sustained facilities, ensuring the restoration of the wastewater resources.

10.6 ON-SITE APPLICABILITY OF DECENTRALIZED SEWAGE TREATMENT SYSTEMS

The target of the decentralized system is to treat wastes generated from the non-point sources of peri-urban and rural sectors. These small-scale treatment facilities may assure greater environmental sustainability by reusing recycled wastewater for various purposes. It allows treatment of wastes at the source itself, thus reducing the impact of emerging pollutants. Waste segregation at the source and ease of treatment with potential reuse, at the same time, elevates treatment efficiency and minimizes potential cumulative negative impact. Four factors govern for setting up of a decentralized treatment plant: cost, flexibility of land use, maintenance and environmental protection. All these factors can be easily handled with less involvement, as compared to the large-scale, off-site treatments. The lower ecological footprint makes the decentralized system a secured alternative for treating sewage and other wastes. There are various case studies showcasing the applicability of decentralized wastewater all over the world. A study by Parkinson and Tayler (2003) reported the factors for establishing and managing decentralized wastewater plants, especially in peri-urban areas of developing countries. Another study by Singh et al. (2015) examined performances of these plants with respect to Indian scenarios. Many reports have been demonstrated to integrate and implement both centralized and decentralized approaches for sewage treatment, depicting a transition from an existing approach to an alternate one. Authors have also reviewed various treatment processes for implementing an eco-friendly system like membrane filters for eliminating contaminants. Apart from developed countries, the pertinency of the approach needs to be validated in the context of developing nations, as these are the potential areas of larger sewage generation.

10.7 CONCLUSION

The inefficiency of installed sewage treatment plants requires an urgent shift to more eco-friendly techniques. Nowadays, the dichotomy of centralized and decentralized systems is the center of attraction for wastewater treatment sciences. Although having a vast area and large capacity for treatment, the centralized plants are not cost effective and are inefficient for the treatment process. Also, the requirement for skilled and trained labor, along with advanced and heavy machineries, makes the CSTS a bigger challenge in the upcoming years. This disadvantage can be easily overcome by using decentralized techniques, which, although being less in capacity, possess higher and better working efficiency for treatment. However, there is very little investigation for the application of decentralized systems, and only ideas regarding its establishment are speculated about. It is the need of the hour for shifting from a conventional, disposal-based, linear system to a better, recovery-based, closed system alternative. The decentralized approaches for treating sewage allow flexibility in management and possess greater benefits in terms of technical, economic, environmental and social aspects. Being maintenance free, it provides a feasible on-site treatment solution that is more reliable, cost-effective and also complies with the environmental standards, having better treatment efficiency of municipal wastewater.

REFERENCES

Arceivala, S. J., & Asolekar, S. R. (2006). *Wastewater treatment for pollution control and reuse*. Tata McGraw-Hill Education.

Bakir, H. A. (2001). Sustainable wastewater management for small communities in the Middle East and North Africa. *Journal of Environmental Management, 61*, 319–328.

Banerjee, A., & Choudhury, M. (2022). Green technology for wastewater treatment and remediation. In C. Baskar, S. Ramakrishna, & A. Daniela La Rosa (Eds.), *Encyclopedia of green materials*. Springer. https://doi.org/10.1007/978-981-16-4921-9_30-1

Burian, S. J., Nix, S. J., Pitt, R. E., & Durrans, S. R. (2000). Urban wastewater management in the United States: Past, present, and future. *Journal of Urban Technology, 7*(3), 33–62.

Díaz-Garduño, B., Pintado-Herrera, M. G., Biel-Maeso, M., Rueda-Márquez, J. J., Lara-Martín, P. A., Perales, J. A., . . . Martín-Díaz, M. L. (2017). Environmental risk assessment of effluents as a whole emerging contaminant: Efficiency of alternative tertiary treatments for wastewater depuration. *Water Research, 119*, 136–149.

Gogoi, N., & Choudhury, M. (2022). Green materials synthesis for wastewater treatment. In C. Baskar, S. Ramakrishna, & A. Daniela La Rosa (Eds.), *Encyclopedia of green materials*. Springer. https://doi.org/10.1007/978-981-16-4921-9_31-1

Gondhalekar, D., Al-Azzawi, M., & Drewes, J. E. (2019). Urban Water reclamation with resource recovery as a cornerstone of urban climate change resilience. In *Handbook of climate change resilience* (pp. 1–22). Springer International Publishing.

Kamble, S. J., Chakravarthy, Y., Singh, A., Chubilleau, C., Starkl, M., & Bawa, I. (2017). A soil biotechnology system for wastewater treatment: Technical, hygiene, environmental LCA and economic aspects. *Environmental Science and Pollution Research, 24*(15), 13315–13334.

Parkinson, J., & Tayler, K. (2003). Decentralized wastewater management in peri-urban areas in low-income countries. *Environment and Urbanization, 15*(1), 75–90.

Qadir, M., Drechsel, P., Jiménez Cisneros, B., Kim, Y., Pramanik, A., Mehta, P., & Olaniyan, O. (2020, February). Global and regional potential of wastewater as a water, nutrient and energy source. In *Natural resources forum* (Vol. 44, No. 1, pp. 40–51). Blackwell Publishing Ltd.

Rajpal, A., Ali, M., Choudhury, M., Almohana, A. I., Alali, A. F., Munshi, F. M. A., Khursheed, A., & Kazmi, A. A. (2022). Abattoir wastewater treatment plants in India: Understanding and performance evaluation. *Frontiers in Environmental Science, 10*, 881623. https://doi.org/10.3389/fenvs.2022.881623

Rohilla, S., & Dwivedi, D. (2013). *Re-invent, recycle, re-use: Toolkit on decentralized wastewater management*. CSE.

Singh, N. K., Kazmi, A. A., & Starkl, M. (2015). A review on full-scale decentralized wastewater treatment systems: Techno-economical approach. *Water Science and Technology, 71*(4), 468–478.

UN-water. (2020). *Indicator 6.3.1 'proportion of domestic and industrial wastewater flow safely treated'*. UN. https://www.unwater.org/our-work/integrated-monitoring-initiative-sdg-6/indicator-631-proportion-domestic-and-industrial

Wilderer, P. A. (2001). Decentralized versus centralized wastewater management. In P. Lens, G. Zeeman, & G. Lettinga (Eds.), *Decentralized sanitation and reuse* (pp. 39–54). IWA Publishing.

Wilderer, P. A., Schreff, D. (2000). Decentralized and centralized wastewater management: A challenge for technology developers. *Water Science and Technology*, *41*(1), 1–8.

11 A Comprehensive Characterization of Wastewater and Sewage Sludge in SBR-Based STPs of India

Ghazal Srivastava, Ankur Rajpal, Anwar Khursheed, and Absar Ahmad Kazmi

11.1 INTRODUCTION

Wastewater contains an ample amount of organic matter, mainly identified as COD. The COD fractions have a significant impact on the biological nutrient-removal capacity of wastewater-treatment plants. Understanding the contribution of organic contaminants liable to biodegradation and non-liable to biodegradation is very essential in the designing for biological removal of nitrogen and phosphorus, as it affects the dynamics of the activated sludge process, as the oxygen demand, maintaining a constant sludge age or system's SRT and kinetic parameters of the biological reactor functions. Recognizing COD as the key parameter for deciding the amount of organic carbon present in wastewater and the distribution of COD into fractions explaining diverse degrees of their biodegradation is, at present, an important addition to the characteristics of wastewater (Drewnowski et al., 2019). The different fractions of COD in municipal wastewater have distinguished functions; readily biodegradable COD (rbCOD) determines the metabolism of the microbiome, slowly or particulate biodegradable COD consumes in adsorption, hydrolysis, and cellular respiration; non-biodegradable soluble COD is found in biologically clarified effluent, and non-biodegradable suspended COD takes part in the sludge production (Srivastava et al., 2023; Płuciennik-Koropczuk & Myszograj, 2019). Selector-attached SBR plants have their benefits in biological nutrient-removal (N and P) and bulking control.

The rbCOD/TKN ratio determines the denitrification rates/efficiency of the SBR plants (Srivastava et al., 2023). Similarly, the rbCOD/TP proportion determines the diversity of biological phosphorus removal species. Higher rbCOD to P ratios of >40 mg-rbCOD/mg-P show a distinguished relation with GAO-led culture and lower values (<10–20 of mg-rbCOD/mg-P) were observed relayed to PAO diversity (Broughton et al., 2008). Generally, simultaneous nutrient removal of N and P occurs during the subsequent feast and famine stages: in the feast phase (in the

DOI: 10.1201/9781003258377-11

selector zones anoxic phase), PHBs are produced within the bacterial cell by sequestering rbCOD, and poly-phosphates are released as soluble phosphorus, while, in the famine phase in the aeration tank, PHB, as storage material, helps the particular bacteria to survive in the midst of other microorganisms competing for food and phosphorus uptake occurs as polyphosphates and enhanced biological phosphorus removal (EBPR) takes place (Mino et al., 1987; Third et al., 2003). The derivative of sewage treatment is known as biological sludge, which is usually a partially solid waste, which needs additional treatment before being appropriate for dumping or land applications (Alalayah et al., 2017). The phosphorus and potassium in the sewage sludge have an elevated compost-value. It has been observed that sludge characterization is quite an essential analysis for determining the further sludge management strategies applicable for the type of dewatered sludge produced. The different sections of the study include:

- Typical wastewater characterization based on COD, TN, and TP fractions.
- Relationship between these influent parameters with biological nutrient removal, microbiota, and intracellular storage products formation.

Specific sludge production calculations and sludge characterization for the objective of determining the dewatered sludge potential in producing organic compost and other sludge management practices.

11.2 PERFORMANCE OF SELECTOR-BASED AND NON-SELECTOR-BASED STPs HAVING SIMILAR WASTEWATER CHARACTERISTICS

The overall fifteen SBR plants have been analyzed in this study, in which twelve plants are cyclic-technology-based (C-Tech) SBR plants, two plants are Intermittent Cycle Extended Aeration System (ICEAS) technology-based plants (Xylem's Sanitaire) from Mussoorie and one with Aquatech SBR process in Gurgaon. Table 11.1 gives the information about these SBR plants in detail.

11.2.1 Performance Evaluation of SBR Plant

The plants have performed quite well, except for Mussoorie SBR plants, as they are devoid of selectors to control the filamentous growth. Also, the SBR plants at Dwarka and Gurgaon were unable to work better for nitrogen removal. The COD and BOD removal efficiencies were >90% in all the plants, except the 7.4 MLD SBR plant at Dwarka. This plant was getting saline water from the Arabian Sea passing to Dwarka, Gujarat, which can be the cause of the lower treatment performance of the plant. Excellent simultaneous TN and TP removal were only achieved by the 120-MLD capacity SBR plant (Varanasi, Uttar Pradesh) and 87.5-MLD capacity SBR plant (Koparkhairane, Mumbai), respectively. The 25-MLD capacity SBR plant near Abohar, Punjab was getting a high concentration of COD, BOD, and TN due to dairy wastewater inclusions in domestic sewage in the plant, exceeding the design capacity,

TABLE 11.1
SBR Plants and Their Characteristics

S. No.	SBR capacity (MLD)	Plant's location	Configuration	Features	Sludge management techniques
1	3	Roorkee, Uttarakhand	Pre-anoxic selector attached	Raw sewage	Centrifuge and belt press
2	27	Haridwar, Uttarakhand	Pre-anoxic selector attached	Primary sedimentation tank+ SBR	Sludge holding tank, centrifugal dewatering
3	25	Abohar, Punjab	Pre-anoxic selector attached	Dairy wastewater ingress	Centrifugal dewatering system
4	70	Rajkot, Gujarat	Pre-anoxic selector attached	Mixing of Storm water in raw sewage	Centrifugal dewatering system
5	56	Rajkot, Gujarat	Pre-anoxic selector attached	Mixing of Storm water in raw sewage	Centrifugal dewatering system
6	60	Ahmedabad, Gujarat	Pre-anoxic selector attached	Raw sewage	Centrifugal dewatering system
7	7	Karnal, Haryana	Pre-anoxic selector attached	Mixing of cow-dung from drains in raw sewage	Centrifugal dewatering system
8	10	Karnal, Haryana	Pre-anoxic selector attached	Raw sewage	Centrifugal dewatering system
9	7.4	Dwarka, Gujarat	Pre-anoxic selector attached	TDS in wastewater>2100 mg/L	Centrifugal dewatering system
10	120	Goithaha, Varanasi	Pre-anoxic selector attached	Raw sewage (N and P removal both)	Centrifugal dewatering system
11	87.5	Koparkhairane, Mumbai	Pre-anoxic selector attached	Raw sewage (N and P removal both)	Centrifugal dewatering system
12	80	Airoli, Mumbai	Pre-anoxic selector attached	Raw sewage	Centrifugal dewatering system
13	50	Gurgaon, Haryana	No selector	Industrial ingress	Centrifugal dewatering system
14	3.2	Mussoorie, Uttarakhand	No selector	Raw sewage	–
15	1.2	Mussoorie, Uttarakhand	No selector	Raw sewage	–

and ultimately, deteriorating the treatment efficiency of the plant. Table 11.2 and Figure 11.1 (A–F) show the performance evaluation of all fifteen SBR plants.

11.2.2 Analysis of COD Fractions

The rbCOD content in total COD affects the treatment performance of the plant in terms of biological removal of nitrogen and phosphorus. The rbCOD fraction in COD was calculated by the flocculation-filtration method (Wentzel et al., 2000; Srivastava & Kazmi, 2022). Among all the fifteen treatment plants, the SBR plant at Koparkhairane, Mumbai has the highest percentage of rbCOD—i.e., 32%—which

TABLE 11.2
The Observed Effluent Values and Discharge Effluent NGT Standards for COD, BOD, TN, TP, and TSS Parameters in Sewage Treatment Plants

	Satisfied NGT standards or not				
SBR plants capacity	**Effluent COD (30 mg/L)**	**Effluent BOD (10 mg/L)**	**Effluent TN (10 mg/L)**	**Effluent TSS (20 mg/L)**	**Effluent TP (1 mg/L) (For discharge into Ponds, Lakes)**
70 MLD	Yes	Yes	No (11 mg/L)	Yes	No
60 MLD	Yes	Yes	No (12 mg/L)	Yes	No
56 MLD	Yes	Yes	Yes	Yes	No
27 MLD	Yes	Yes	Yes	Yes	No
7.4 MLD	No (56 mg/L)	No (21 mg/L)	No (19 mg/L)	No (23 mg/L)	No* (1.4 mg/L)
3 MLD	Yes	Yes	Yes	Yes	No
1.2 MLD	Yes	Yes	No (38.4 mg/L)	Yes	No
10 MLD	Yes	Yes	No (13.5 mg/L)	Yes	No
3.2 MLD	Yes	Yes	No (19.9 mg/L)	Yes	No
7 MLD	Yes	Yes	Yes	Yes	No
80 MLD	Yes	Yes	No (10.6mg/L)	Yes	No
120 MLD	Yes	Yes	Yes	Yes	No* (1.05 mg/L)
87.5 MLD	No (35 mg/L)	Yes	No (10.5 mg/L)	Yes	No* (1.5 mg/L)
50 MLD	No (39 mg/L)	Yes	No (28.9 mg/L)	Yes	Yes
25 MLD	No (180 mg/L)	No (33 mg/L)	No (17 mg/L)	No (67 mg/L)	No

* depicts that the effluent value of TP obtained and the standard value according to NGT guidelines are very close, and the typically observed effluent concentration is written in brackets.

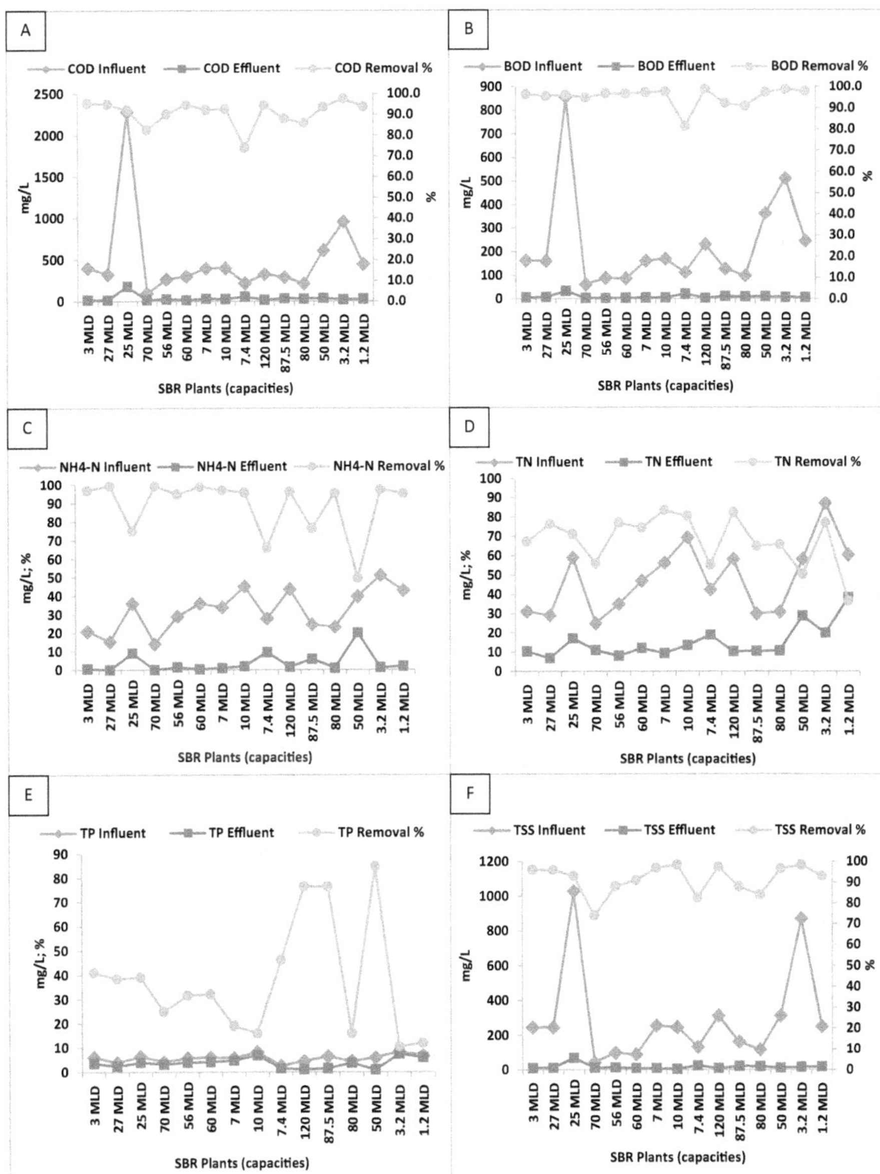

FIGURE 11.1 Performance evaluation of SBR plants. Panels (A–F) show COD, BOD, NH_4-N, TN, TP and TSS in the influent (mg/L), treated effluent (mg/L), and their removal efficiencies (%).

helped the plant in achieving excellent SND (89.3%) as well as EBPR (70.8%), simultaneously. The rbCOD content directly controls the nitrate-reduction rate as well as the BPR mechanism (Metcalf & Eddy, Inc. 2003). In the plant, 65% TN removal (>75% ammonia removal) and 77% TP removal (>70% orthophosphate removal) were observed. In the SBR plant at Haridwar, lower rbCOD% was observed (8.14), and it contributed to good SND (81%), but no EBPR. In Varanasi and Gurgaon plants, the rbCOD fraction in total COD was >16%, SND% were 94% and 100%, and similarly, EBPR% were 51% and 60%, respectively, after managing all the essential parameters permitting SND, along with BPR. These parameters have been discussed in detail. The Gurgaon SBR plant was working as a no-selector Aquatech SBR plant. Here, due to the absence of a selector, high-ammonia concentration was present in the effluent. Similarly, slowly biodegradable substrates represent great potential for being available as slow electron donors during denitrification and BPR. In the majority of the plants, the sbCOD portion in COD was >50%, except in 25 MLD SBR at Abohar (Punjab), 56 MLD SBR at Rajkot (Gujarat), and 60 MLD SBR at Ahmedabad (Gujarat) as shown in Figure 11.1. Bhattafall and Landhor plants in Mussoorie were also without selector ICEAS based SBR plants. Biological phosphorus removal was nearly absent in them; however, nitrogen removal was consistent.

11.2.3 Detailed Analysis of TN and TP Fractions in the Wastewater of STPs

The wastewater characterization in SBR plants was performed in terms of TN and TP fractions. The TN was classified in the form of organic nitrogen, ammonia-N (combined as TKN), and nitrites and nitrates. In the influent, the contribution of TKN varied from 97±2% of TN, and ammonia-N contributed to 70±9% of total TKN in the SBR plants. Also, soluble phosphorus in terms of orthophosphates (PO_4-P) shares 54±14% of total phosphorus. It determines the way of the removal of nutrients from wastewater treatment plants. If the concentration of ammonia-N is higher, biological wastewater treatment with aerobic-anoxic sequencing can be successful. Anammox process also proves excellent for the elimination of ammonia-N and TN. Total phosphorus can be removed by enhanced biological phosphorus removal or chemically using alum or any coagulant. In these SBR plants, only biological phosphorus removal has been focused on.

11.3 POSSIBLE MICROBIAL-COMMUNITY PREVALENCE AND INTRACELLULAR STORAGE-PRODUCTS-BASED PERSPECTIVES WITH WASTEWATER TYPE

To ensure the optimal operation in SBR plants, technology enhancement, optimization, and troubleshooting, there is a necessity to comprehend the configuration and role of the microbial communities and the factors governing their diversity and movement. The knowledge of the functioning of the microbial population that performs the common tasks (to remove carbon, phosphorus, and perform nitrification and denitrification) should be complete as well as engaging. According to several kinds of literature available, six kinds of functional groups can be observed in these kinds of treatment plants: polyphosphate accumulating organisms (PAOs), glycogen accumulating organisms (GAOs), nitrifiers, denitrifiers, protein hydrolyzers, fermenters, and filamentous bacteria. These kinds of studies need to associate quantitative microbiological analysis, like quantitative polymerase chain reactions,

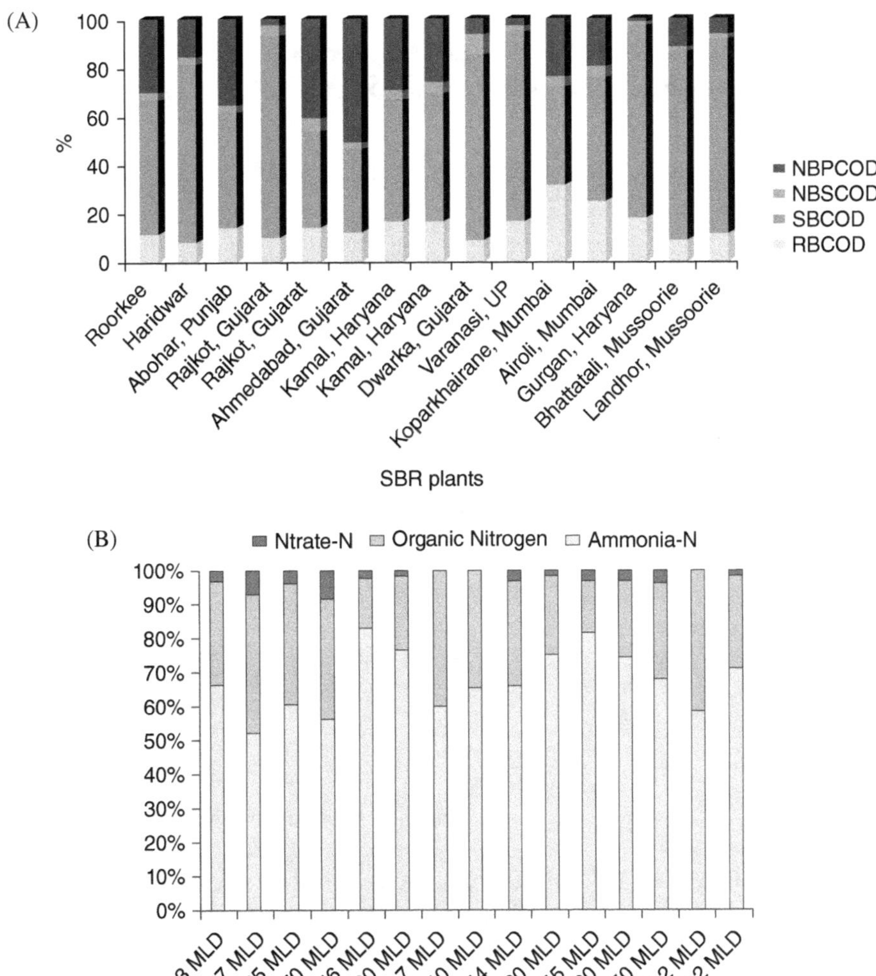

FIGURE 11.2 (A) COD fractions in SBR Plants. (B) TN and TP fractions in SBR Plants. (C-D) Relationship between rbCOD/COD (%) on EBPR % of plants.

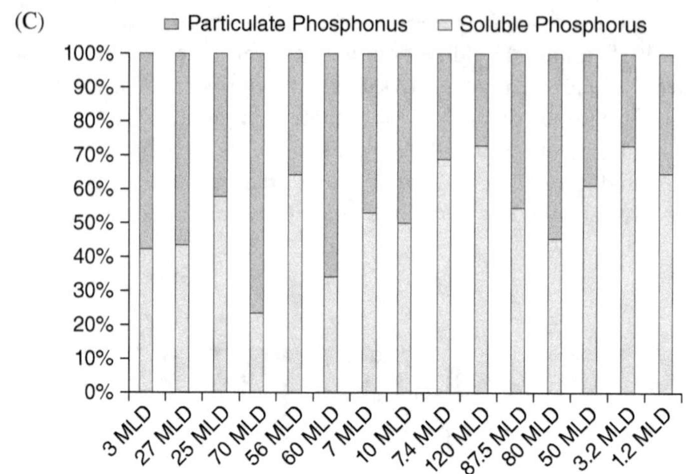

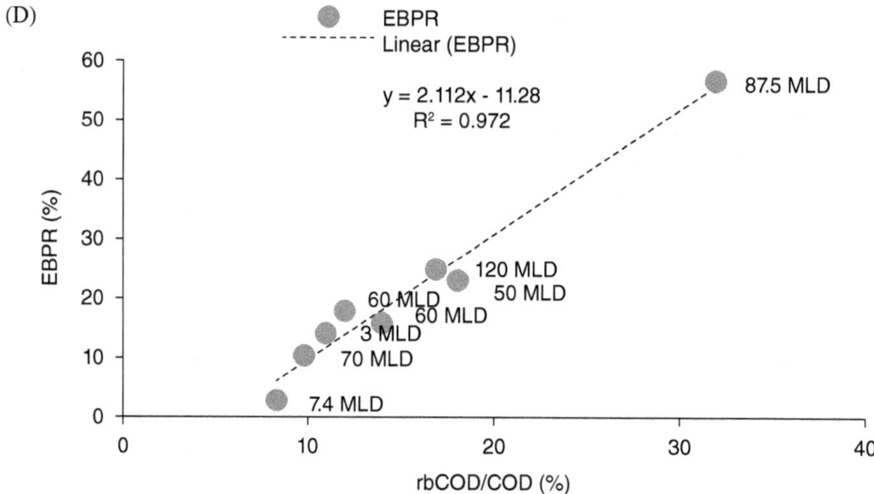

FIGURE 11.2 (Continued)

fluorescence in situ hybridization, and sequencing, which are the future scopes and objectives.

The other important and eminent role played by the bacteria is intracellular storage products formation, as (a) poly-hydroxy butyrates (PHB) for behaving as an excellent, potent, slow-degrading electron-donor, used as a carbon source during the absence of exogenous BOD for denitrification and phosphorus removal processes and (b) polyphosphate (poly-P) globules during biological phosphorus removal mechanism, considering orthophosphate release in the anaerobic zone and poly-P uptake in the subsequent aerobic zone by PAOs. Sudan black B staining was performed for the identification of poly-hydroxy butyrates (PHB), while Neisser staining, for determining polyphosphate (poly-P) globules in the selector and aerobic tank sludge of treatment plants (Figure 11.3 A).

11.4 ROLE OF COD FRACTIONATION IN BIOLOGICAL NUTRIENT-REMOVAL EFFICIENCIES

(a) **Effect of rbCOD/TN and COD/TKN on nitrate-N reduction rates and rbCOD/COD fraction on SND efficiency in SBR plants.**
The relationship between rbCOD/TN and COD/TKN on denitrification rates of the plants showed by Jimenez et al. (2010) is represented in this study in Figure 11.3 (A) and (B). Also, a typical relation between SND and rbCOD% is shown in Figure 11.3 (C). A positive, increasing linear trend justifies that rbCOD% in COD has a greater effect on SND% in SBR plants.

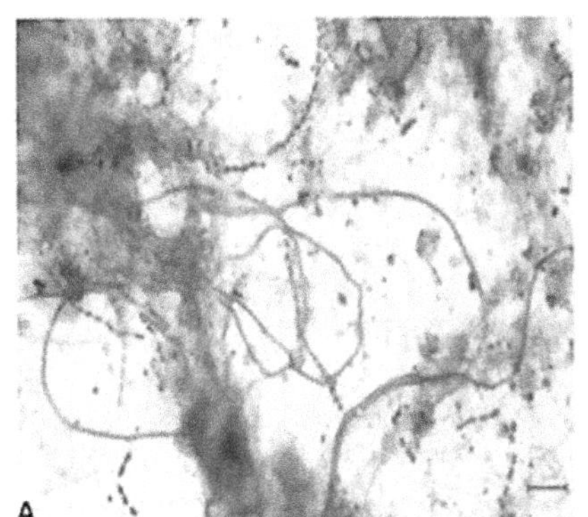

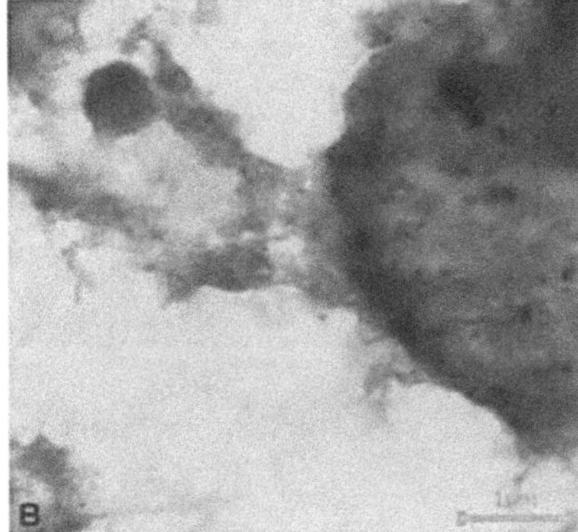

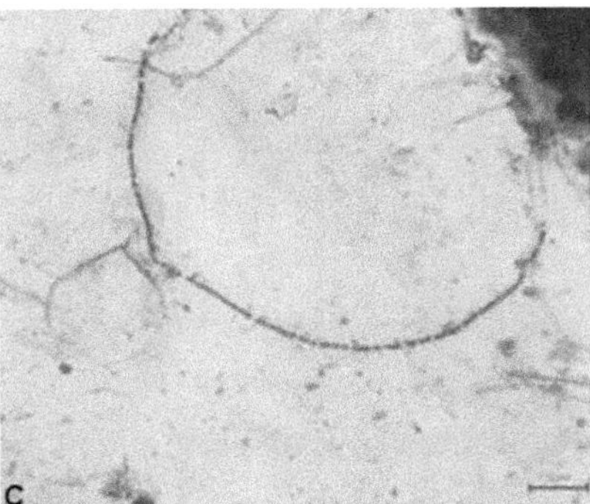

FIGURE 11.3 PHB granules in filaments and flocs of SBR sludge of (A) 87.5 MLD Koparkhairane (Mumbai) and (B) 1.2 MLD Landhor, Mussoorie (Uttarakhand) (blue-black granules as PHB (+) and red-pink blobs as PHB (-)) (C) poly-P globules in the aerobic sludge filaments indicating the presence of PAOs at 87.5 MLD SBR, Koparkhairane (Mumbai) (purple-blue-black granules as Poly-P (+) and yellow-brown blobs as Poly-P (-)).

(A)

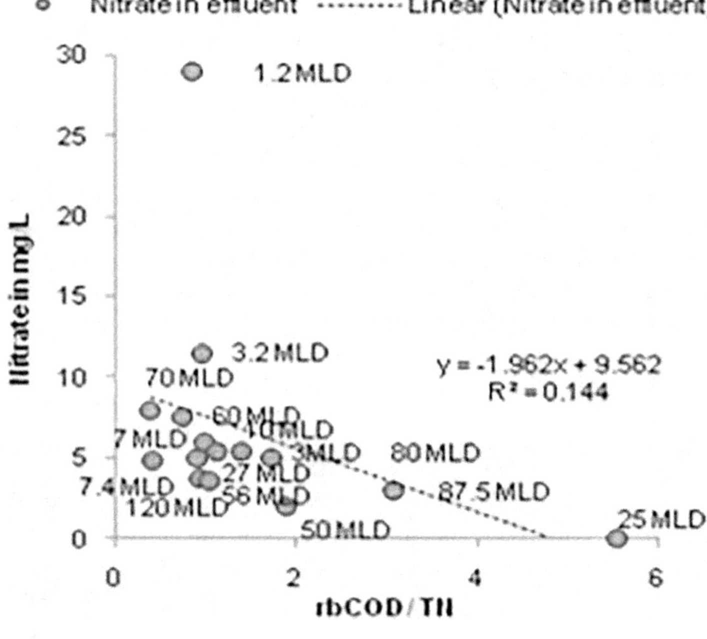

(B)

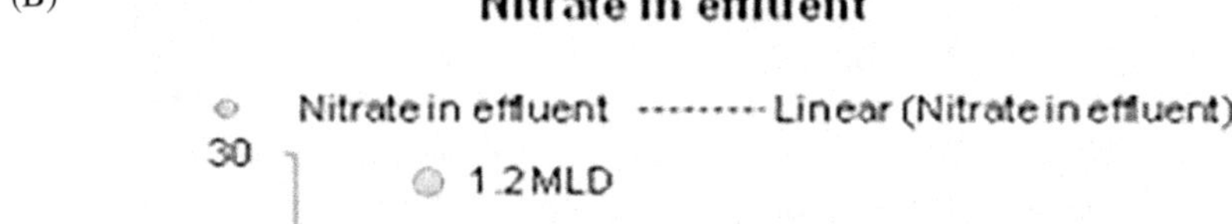

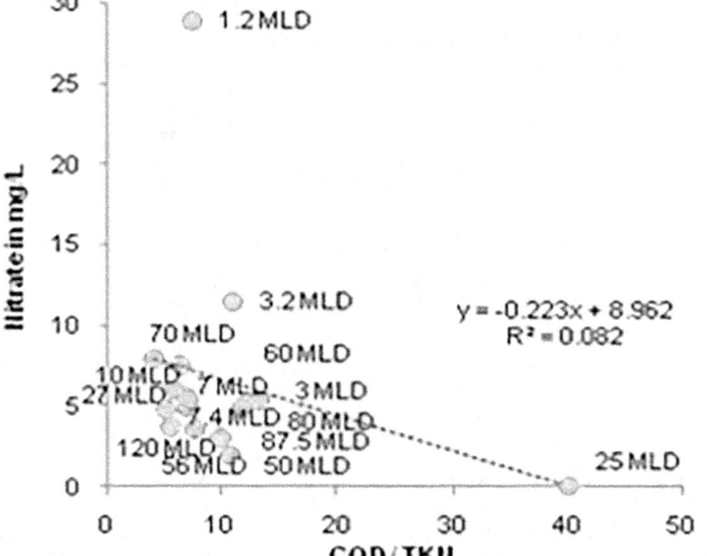

FIGURE 11.4 (A and B)—relationship between rbCOD/TN and COD/TKN on denitrification rates of the plants; (C)—a typical relation between SND and rbCOD/COD%.

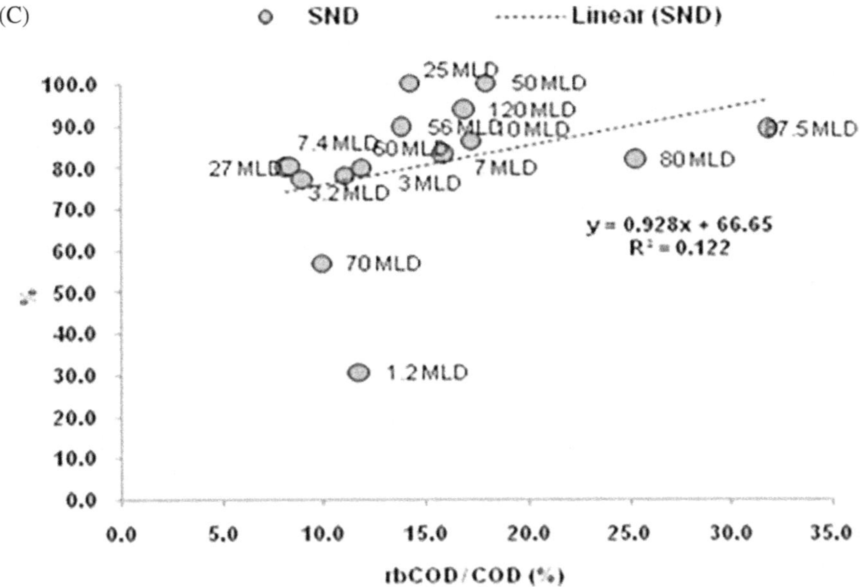

FIGURE 11.4 (Continued)

(b) Influence of rbCOD/TP on biological phosphorus removal and EBPR. The effluent concentration of TP and PO_4-P along with rbCOD/TP is shown in Table 11.3 for all the SBR plants. Figure 11.4C shows the relationship of rbCOD/COD (%) on EBPR% of plants (in which EBPR is occurring).

11.5 BIOLOGICAL SLUDGE CHARACTERIZATION IN SBR PLANTS

Sludge management and treatment are fetching the chief concerns in working with STPs. Sludge collection, treatment, management, and disposal techniques consist of essential units such as sludge tanks, gravity thickeners, and sludge dewatering process, which are comprised of centrifuge, belt press, screw press, filter press, or sludge drying beds. Treatment strategies further include anaerobic digestion of sludge and composting. The elevated prices for conduct and disposal of surplus sludge have augmented up to more than a half of the entire operation rates (Wei et al., 2003), and it frequently affects the appropriate discarding of sludge, although it has been considered an unnoticed subject in wastewater treatment plants (Wei et al., 2003; Srivastava et al., 2021). Table 11.4 shows an estimated biological sludge production and specific sludge production in all the SBR plants. The SBR plant at Gurgaon, Haryana has 0.6 Kg TSS/Kg $BOD_{removed}$ analogous to the characteristic value of sludge production—i.e., 0.65 Kg TSS/Kg $BOD_{removed}$—in extended aeration systems, as shown in Table 11.4 (USEPA, 1984; Srivastava et al., 2021).

TABLE 11.3
Relationship between rbCOD/TP and Phosphorus Removal in SBR Plants

SBR plants capacity	rbCOD/ TP	PO_4-P in effluent	PO_4-P Removal %	TP in effluent	TP removal %
70 MLD	2.3	0.7	27.0	3.1	25.0
60 MLD	5.9	1.6	23.8	4.1	32.0
56 MLD	6.4	2.1	42.3	3.9	31.6
27 MLD	6.8	1.4	17.6	2.4	38.5
7.4 MLD	6.9	1.0	44.4	1.4	46.2
3 MLD	7.3	1.8	30.8	3.6	41.0
1.2 MLD	7.8	4.6	0	5.9	11.9
10 MLD	8.4	3.4	17.1	6.9	16.0
3.2 MLD	10.3	5.9	3.3	7.4	10.5
7 MLD	10.9	2.4	22.6	4.7	19.0
80 MLD	12.2	1.7	15.0	3.7	15.9
120 MLD	12.3	0.6	81.6	1.1	76.5
87.5 MLD	14.4	1.0	71.4	1.5	76.6
50 MLD	18.6	0.3	90.8	0.9	84.7
25 MLD	50.9	2.2	41	3.9	39.0

For organic compost as per the FCO 2009 guidelines, the pH should lie in the range of 6.5–7.5; moisture content, 15–25%; total organic carbon (percent by weight), ≥12%; total nitrogen (as N) (percent by weight), ≥0.8%; total phosphate (as P_2O_5) (percent by weight), ≥0.4; and carbon-nitrogen ratio (C/N),<20. The dewatered sludge produced from SBR plants at Roorkee (3 MLD), Rajkot (56 MLD), Varanasi (120 MLD), and Mumbai (87.5 MLD and 80 MLD) have satisfied the desired limits for pH, organic carbon, TN, TP, and C/N ratio to form organic compost as per the FCO guidelines 2009 and World Health Organization (1989) (Table 11.5).

11.6 FURTHER SCOPE OF WASTEWATER AND SLUDGE CHARACTERIZATION IN DESIGNING EFFECTIVE SBR PLANTS

The scope of this study is in the optimization of SBR-based wastewater treatment plants concentrating on the aspects of nitrogen and phosphorus removal along with lower SVIs. The conclusion is that the selector-attached SBRs perform better than conventional SBRs for simultaneous nutrient removal (N and P). The sludge characterization is also an important parameter to judge the potential of dewatered sludge for organic composting. The present study can help in making insights of the key points, like COD fractions, and sludge characterization can be included in proposed guidelines and also suggest engineers construct and improve WWTPs in the most promising way.

TABLE 11.4
Specific Sludge Production in SBR Plants

SBR plants	Total estimated biological sludge produced (as TSS in Kg/d)	Total sludge produced (as VSS in Kg/d)	Specific sludge production			
			Kg TSS/Kg BOD removed	Kg TSS/Kg COD removed	Kg VSS/Kg BOD removed	Kg VSS/Kg COD removed
Roorkee (3 MLD)	411.8	221.7	1.5	0.6	0.8	0.3
Karnal (7 MLD)	750.7	387.1	1.6	0.7	0.8	0.4
Karnal (10 MLD)	545.5	381.4	1.3	0.6	0.9	0.4
Haridwar (27 MLD)	5514.8	3116.1	1.3	0.6	0.7	0.4
Rajkot (56 MLD)	2495.9	1363.7	1.5	0.5	0.8	0.3
Ahmedabad (60 MLD)	5092.1	3160.6	1.4	0.4	0.9	0.3
Koparkhairane (87.5 MLD)	5787.4	2893.7	1.4	0.6	0.7	0.3
Airoli (80 MLD)	3635.3	1911.7	1.2	0.6	0.7	0.3
Varanasi (120 MLD)	5361.2	2955.5	1.1	0.8	0.6	0.4
Rajkot (70 MLD)	2407.8	1375.9	0.8	0.6	0.5	0.3
Abohar (25 MLD)	22365.4	10606.7	1.4	0.5	0.6	0.3
Dwarka (7.4 MLD)	91.0	46.2	1.2	0.7	0.6	0.3
Gurgaon (50 MLD)	10809.7	8647.8	0.6	0.4	0.5	0.3
Bhattafall (3.2 MLD)	625.9	384.9	1.2	0.7	0.8	0.4
Landhor (1.2 MLD)	223.7	132.5	0.9	0.5	0.6	0.3

TABLE 11.5
Sludge Characterization in SBR Plants

	Aeration tank sludge			Dewatered sludge						
SBR plants	**MLSS & MLVSS**	**SVI**	**SRT**	**pH**	**TN (as N), percent by weight**	**TP (as P_2O_5), percent by weight**	**Organic carbon (%)**	**Moisture content (%)**	**Organic matter (%)**	**C/N ratio**
3 MLD	7177 & 3522	41	15.2	7.1	1.6–2.7	2.5–3.5	31	79	56	11–20
27 MLD	2789 & 1583	75	13.5	–	–	–	–	–	–	–
25 MLD	5820 & 2270	64	10.4	7.6	1.4	0.6	14.6	63	26	10.3
70 MLD	7749 & 2029	13	12	7.4	1.4	0.6	11.2	60	20	7.8
56 MLD	4549 & 2289	55	12.4	7.5	4.7	0.6	20.3	32	37	4.3
60 MLD	5874 & 2030	26	14.5	7.6	2.4	0.7	17.2	71	31	7.2
7 MLD	6546 & 3727	32	13.5	8.0	1.1	0.5	16.3	4.1	30	9.9
10 MLD	16485 & 7180	25	13.2	7.5	1.4	0.8	17.2	6.8	31	12.2
7.4 MLD	1575 & 801	37	12.2	–	–	–	–	–	–	–
120 MLD	3778 & 2140	34	10.4	7.3	1.6–3.5	4–6	16.5	66.4	29.8	4–11
87.5 MLD	2170 & 1323	83	12	7.1	3.6–4.6	5–7	32	87.4	57	7.7
80 MLD	1583 & 873	70	12	7.0	3.8	2.5	30	87	54	7.8
50 MLD	3260 & 2543	100	10	7.0	1–3	4–5	–	–	–	–
3.2 MLD	9555 & 4598	89	–	–	–	–	–	–	–	–
1.2 MLD	1390 & 825	58	–	–	–	–	–	–	–	–

REFERENCES

Alalayah, W., Demirbas, A., & Edris, G. (2017). Sludge production from municipal wastewater treatment in sewage treatment plant. *Energy Sources, Part A: Recovery, Utilization and Environmental Effects*, *39*(2).

Broughton, A., Pratt, S., & Shilton, A. (2008). Enhanced biological phosphorus removal for high strength wastewater with a low rbCOD:P ratio. *Bioresource Technology*, *99*(5), 1236–1241. https://doi.org/10.1016/j.biortech.2007.02.013

Drewnowski, J., Remiszewska-Skwarek, A., Fudala-Ksiazek, S., Luczkiewicz, A., Kumari, S., & Bux, F. (2019). The evaluation of COD fractionation and modeling as a key factor for appropriate optimization and monitoring of modern cost-effective activated sludge systems. *Journal of Environmental Science and Health, Part A*, *54*(8), 736–744. https://doi.org/10.1080/10934529.2019.1592531

Jimenez, J., Dursun, D., Dold, P., Bratby, J., Keller, J., & Parker, D. (2010). *Simultaneous nitrification-denitrification to meet low effluent nitrogen limits: Modeling, performance and reliability*. Conference Paper in Proceedings of the Water Environment Federation (WEFTEC).

Metcalf & Eddy, Inc. (2003). *Wastewater engineering treatment, disposal and reuse* (4th ed.). McGraw-Hill, Indian Edition.

Mino, T., Tsuzuki, Y., & Matsuo, T. (1987). Effect of phosphorus accumulation on acetate metabolism in the biological phosphorus removal process. In R. Ramadori (Ed.), *Biological phosphate removal from wastewaters* (pp. 27–38). Pergamon Press.

Płuciennik-Koropczuk, E., & Myszograj, S. (2019). New approach in COD fractionation methods. *Water*, *11*, 1484. https://doi.org/10.3390/w11071484.

Srivastava, G., Gani, K. M., Rajpal, A., Singh, J., & Kazmi, A. A. (2021). Reduced sludge production in a polyvinyl alcohol gel-based oxic-anoxic-oxic configured pilot-scale plant. *Case Studies in Chemical and Environmental Engineering*, *4*(1), 100136.

Srivastava, G., & Kazmi, A. A. (2022) Perspectives of intracellular polymers in functional evaluation of the microbes for EBPR. *Water Practice and Technology*, *17*(1), 112–127.

Srivastava, G., Rajpal, A., Khursheed, A., Nadda, A. K., Tyagi, V. K., & Kazmi, A. A. (2023). Influence of variations in wastewater on simultaneous nutrient removal in pre-anoxic selector attached full-scale sequencing batch reactor. *International Journal of Environmental Science and Technology*, *20*, 4355–4372. https://doi.org/10.1007/s13762-022-04052-8

Third, K. A., Burnett, N., & Cord-Ruwisch, R. (2003). Simultaneous nitrification and denitrification using stored substrate (PHB) as the electron donor in an SBR. *Wiley Periodicals, Inc., Biotechnology and Bioengineering*, *83*(6), 706–720.

U.S. Environmental Protection Agency. (1984). *Handbook improving POTW performance using the composite correction program approach* (p. 32). Centre for Environmental Research Information, Table, 3–7(16–69).

Wei, Y., Houten, R. T. V., Borger, A. R., Eikelboom, D. H., & Fan, Y. (2003). Minimization of excess sludge production for biological wastewater treatment. *Water Research*, *37*, 4453–4467. https://doi.org/10.1016/S0043-1354(03)00441-X

Wentzel, M. C., Mbewe, A., Lakay, M. T., & Ekama, G. A. (2000). *Evaluation of a modified flocculation filtration method to determine wastewater readily biodegradable COD*. WISA, Biennial Conference, Sun City, South Africa. http://citeseerx.ist.psu.edu/viewdoc/summary?doi=10.1.1.517.5571

World Health Organization. (1989). *Health guidelines for the use of wastewater in agriculture and aquaculture*. Technical Report Series 778. WHO.

12 Advances in Nanoribbons-Structured Nanomaterials for Wastewater Treatment

Himadri Tanaya Das, Swapnamoy Dutta, Aniruddha Mondal, and Payaswini Das
Nigamananda Das

12.1 INTRODUCTION

Owing to the rising population and industrial needs, the demand of freshwater is rising each and every day. So, depending just on the remaining resources, this will certainly push us towards an inevitable disaster. As an alternative approach, reclamation or reuse of wastewater came up as a suitable and greatly scalable idea, which can prevent the upcoming water-scarcity catastrophe (T. E. Balaji et al., 2021). Water pollution has been greatly influenced by the unrestricted discharge from paper, pigments, paint, rubber, and textile industries. Unwanted released compounds such as toxic dyes and heavy metals have badly contaminated water sources, which directly harms the ecosystems and subsequently affects human health also. Damaging liver, kidneys, skin irritation, dermatitis, cancer, and dysfunction of the reproductive system are the impacts observed in the human body due to the consumption of the contaminated water (Al-Degs et al., 2008; Jaiswal et al., 2022). However, researchers have given serious consideration to this very particular aspect for the last several decades. As water pollution is created not only due to industrial impact but also the effect of arsenic-like heavy metals seems to be a problematic factor. So, considering the overall difficulties, several ideas were developed by researchers such as reverses osmosis, conventional coagulation, ion exchange, chemical precipitation, adsorption, electrodialysis, electrolysis, and photoactive degradation. These concepts have been well explored in the past few decades in order to identify the best possible ones to employ for the large-scale purposes by considering the significant factors such as efficiency, eco-friendliness, cheapness, etc. In recent studies, photocatalytic degradation and adsorption are the two techniques found to be the highly suggestive and efficient ones (Sivakumar et al., 2010). The involvement of less expensive materials in these techniques has lifted up the chances of making these techniques sufficiently steady and durable. In addition, modification in materials such as making hybrid composites for better efficiency, making them magnetic for easy retrieving and reusing, etc., kept these adsorption and photodegradation technologies move forward to

 DOI: 10.1201/9781003258377-12

achieve valuable recognition (Das et al., 2017). Moreover, few bio-derived, non-toxic materials or bio-mediated, low-cost, synthesized materials exhibited great hope in accomplishing successful results in removal of hazardous compounds from wastewater (Dutta et al., 2021). Although these techniques look fascinating to be installed on a commercial scale, still, a few aspects still need to be taken into account. Additional generated compounds during the adsorption or degradation operation, longevity of the materials used, adequate availability of the illumination source for photodegradation and their possible cost expenses, interruption in the process due to fluctuation in pH, etc., are the parameters or situations to be considered before completely relying on these technologies. Studies are still underway to ponder and resolve these problems, and we expect we will be able to accomplish a nearly flawless system for real-life application.

Various nanostructures have been already developed to study the adsorption and photocatalytic degradation. Efficiency, toxicity, reusability, bandgap (especially for photocatalytic activity), etc., are the factors of materials which have always been considered in the study to accomplish the development of a commercial-scale setup for wastewater cleaning application. Morphologies like nanorods, nanotubes, nanosheets, nanowires, etc., are investigated to make improvements in the material properties and their respective performances (Bhardwaj et al., 2022; Duraisamy et al., 2018; Uddin et al., 2018). However, particularly in this chapter, we will focus on the various nanoribbons, especially carbonaceous, metal oxide, or sulphide nanoribbons for the wastewater cleaning application. Nanoribbons are often referred to as the quasi-one-dimensional nanostructures having size range less than 50 nm (Duraisamy et al., 2017). Mainly owing to the high surface area and suitable porosity, nanoribbons exhibit great applicability in eliminating heavy metals, radioactive ions, and also toxic dyes. Due to the development of mesopores and nanocavities on the nanoribbon's surfaces, trapping hazardous molecules became an easy activity; hence, adsorption performance was found to be enhanced (Li et al., 2008). On the other hand, metal oxide nanoribbons are reported to be acting like a useful semiconductor with an adequate band gap, which makes them fascinating for exploring in further studies. Nevertheless, a few studies have found other morphologies of the same materials, providing more improved outcomes, compared to their nanoribbon counterpart. For an example, Tian et al. reported about the superior photocatalytic performance of cobalt-nickel oxide nanoribbons, compared to its flowerlike nanostructures and dandelion-like spheres, whereas, in the same study, the other structures such as cobalt-nickel oxide nanorods, rose-like and balsam-like morphologies exhibited better performance than the nanoribbon (Tian et al., 2017). Similarly, in another study, due to having better light-harvesting capability, TiO_2 nanoneedle spheres were found to be depicting a more excellent photocatalytic ability than the TiO_2 nanoribbon spheres (Jie et al., 2012). Likewise, we will try to elaborate a few studies on different nanoribbons with specific approaches like morphology controlling, surface modification, and coupling with other compounds, etc., in this chapter, and also discuss their unique impacts on the applications. Additionally, basic working notions and the advantages and drawbacks will be discussed, with particular examples. Apart from that, synthesis of different types of nanoribbons will be discussed in individual sections with minute details.

12.2 DIFFERENT TYPES OF NANORIBBONS

12.2.1 Carbon-Based Nanoribbons

Carbonaceous nanoribbons are well investigated in different studies for different application purposes. Among various candidates, the most popular utilization was found to be done of graphene nanoribbons due to their excellent surface area and numerous edge sites. Compared to multi-wall nanotubes, graphene oxide nanoribbons have depicted superior catalytic performance due to their edge construction and high water-solubility (Mani et al., 2017). Based upon their termination, zigzag and armchair are the two specific structures of graphene nanoribbons which have been thoroughly focused on in studies. Through using the number of dimer lines (Na) across the ribbon, armchair nanoribbon's width is determined, whereas, based on the number of zigzag chains across the ribbon, the width of zigzag nanoribbons has been ascertained (Koskinen et al., 2008). Behavioural changes of these two separate structural configurations have also been traced using theoretical studies; for example, graphene nanoribbons with armchair formation of different widths behaved as direct band-gap semiconductors having a band-gap range from 0.5 to 3 eV, but the graphene nanoribbons with zigzag construction have exhibited metallic characteristics (Shinde et al., 2022; Vo et al., 2014). However, tight-binding theory provides insightful details about the variable (either semiconductor or metallic) behaviour of the armchair structure. Besides focusing on the structural configuration, studies have also concentrated on their individual optical characteristics, as they are helpful to understand the adsorption or degradation impact of the nanoribbons. This crucial understanding is useful to evaluate the redox property of nanoribbons for water splitting, as it is directly involved in the mechanism of elimination of unwanted elements from water or aqueous systems. The optical property affecting factors like edge shape, size, and functional group (electron-donating groups CHO and COOH) have been described elaborately in reports (Feng et al., 2021). The armchair structure has shown more bandgap and absorption edge compared to the graphene-nanoribbon zigzag shape.

There are several methods which are used in preparing graphene nanoribbons. The most frequent and recommend method to prepare graphene nanoribbon is from multi-walled carbon nanotubes (MWCNTs) via using a chemical-synthesis approach involving oxidation/reduction reactions. Compared to that, longitudinal unzipping of MWCNTs is an improved method later prescribed to minimize the defects/holes on the basal plane (James et al., 2012). This process was performed as follows: firstly, suspending MWCNTs in conc. H_2SO_4 and using K_2MnO_4 for subsequent treatment at room temperature for 60 mins and for another 60 minutes at a temperature of 55–70°C to finally attain the oxidized nanoribbons. However, later studies have modified the method (such as different time, temperature, filtration process) to improve the structure and size of nanoribbons (Sadeghi et al., 2020). Utilization of atomic hydrogen treatment (850°C, 25 Torr, 4 hours) is also used to prepare graphene nanoribbon from the vertically arranged, single-walled carbon nanotubes grown on silicon substrate (Gao et al., 2020). In recent studies, natural polymers have also been taken into account as eligible choices to make the composite environmental friendly as well as work efficiently; chitosan is one such abundantly tested material (Hu et al.,

2020). Preparation of chitosan and graphene nanoribbon is found to be almost similar to the other nanoribbon composite synthesis techniques, but here, utilization of higher temperature (more than 60°C) is not at all recommended, as it can damage the valuable constituents of the natural polymer.

12.2.2 Metal Oxides/Chalcogenides-Based Nanoribbons

Parallel to graphene nanoribbons, $AgVO_3$, TiO_2 nanoribbons have also been focused on to employ for different applications. Different synthesizing techniques have been used to prepare these nanoribbons; however, the frequent one is the hydrothermal or modified hydrothermal method. For example, Zhao et al. used a facile hydrothermal process to prepare $AgVO_3$ nanoribbons. The synthesis process involves usage of precursors like silver nitrate and ammonium metavanadate, which were thoroughly mixed and autoclaved at 180°C for 24 hours (Zhao et al., 2015). Additionally, the author prepared $Ag/AgVO_3/C_3N_4$ ternary compound by adding thermally poly-condensed melamine or g-C_3N_4 and the composite preparation was completed via using a one-step in-situ hydrothermal technique with the similar process parameter employed during the making of $AgVO_3$ nanoribbons. Finally, ethanol and deionized water was used for washing purposes, and then vacuum drying was applied at 80°C temperature. Zhen-dong Lei and his group (Lei et al., 2016) have attempted a similar preparation approach of synthesizing of $AgVO_3$ nanoribbons but recommended ultrasonication of the silver nitrate and ammonium metavanadate mixture (500 W, 40 kHz) for 30 min before autoclaving it. With an almost similar approach, Wang et al. synthesized $AgVO_3/Ba_5Ta_4O_{15}$ heterojunction photocatalysts (Wang et al., 2018). The difference here with the previous work is the preparation of $Ba_5Ta_4O_{15}$, which was done by autoclaving barium hydroxide and ammonium metavanadate for 48 hours at 270°C. Later on, a different mass ratio (1:1, 1:2, 2:1, 4:1) of $AgVO_3$:$Ba_5Ta_4O_{15}$ was taken for the composite preparation, using a hydrothermal-synthesis method. Here, the sample having 2:1 ratio ($AgVO_3$:$Ba_5Ta_4O_{15}$) exhibited superior photocatalytic performance, as the prepared heterojunction provided more active sites due to its enhanced surface area (8.36 $m^2\ g^{-1}$). Interestingly, introducing $Ba_5Ta_4O_{15}$ resulted in uniform growth of $AgVO_3$ nanoribbons compared to its pure form. Alkaline hydrothermal synthesis is used for the preparation of TiO_2 nanoribbons (Shaban et al., 2018). Firstly, sodium hydroxide solution, along with TiO_2 powder, has been stirred and autoclaved in a Teflon-lined stainless autoclave at 180°C heating temperature for a 24-hour duration. Afterwards, it has been centrifuged and separated, cleansed (using hydrochloric acid), dried (for 4 h at 80°C), and calcinated (for 2 h 450°C) carefully to obtain the TiO_2 nanoribbons. The synthesized nanoribbons are found to have an average of 5.74 nm, which was further reduced (5 nm) in the case of TiO_2 nanoribbons/CNT composite synthesized in the same study. Other than that, a $Ti(OC_4H_9)_4$ precursor has also been used in preparing TiO_2 nanoribbon frameworks using a three-step process (Cao et al., 2014). Firstly, preparation of the titanate nanoribbon framework, then utilizing the topotactic thermal conversion to TiO_2 from titanate, and finally, growing bismuth oxohalide nanoplates via using bismuth(III) nitrate precursor. Usage of bismuth(III) nitrate has also been done, along with using ammonium metavanadate to prepare monoclinic bismuth vanadate by employing a controlled heat-treatment process and varying the ethanol to water (1:0, 0:1, 1:3, 1:1, 3:1) solvent ratio (Ahmed et al., 2018). The mixture here was autoclaved at 100°C for 24 hours, and afterwards, cooled down, separated, washed, and vacuum dried at 60°C. Here, varying solvent ratios resulted in changes in width, surface area, band width, and also application efficiency. The most efficacious ethanol-to-water ratio was 3:1, which exhibited more surface area, optimized band gap, and application efficiency. Other than these oxides, precursors like nickel(II) acetate and cobalt(II) acetate were used; cobalt-nickel-oxide nanoribbon was used while using almost analogous preparation process as discussed earlier (Tian et al., 2017). In conclusion, we have summarized the hydrothermal technique as the most frequent technique applied in preparing oxides; however, studies are still under development to profoundly showcase the elaboration of properties like those which we have mentioned for graphene nanoribbons.

12.2.3 Other Nanoribbons

Other than graphene or oxide nanoribbons, a few other nanoribbons have also been studied in recent studies; however, still, there are well-hidden ideas remaining unexplored. Metal sulfide semiconductors are one of the promising materials, which have shown potential workability due to their narrow band-gap and ability of absorption in the visible light region. Bismuth sulfide and sodium bismuth sulfide nanoribbons are such examples, which were synthesized through a single-step hydrothermal technique. Utilization of bismuth nitrate, sodium sulfide nonahydrate, sodium tungstate, etc., precursors are popularly done in these studies (Koutavarapu et al., 2019; Koutavarapu et al., 2021; Xu et al., 2019). The preferred autoclaving temperature was found to be around 120°C for 24–48 hours of reaction duration, and afterwards, prepared-sample cleaning was done with the help of ethanol and deionized water, and finally, dried at 60–80°C for 8–12 h for attaining the ultimate sample. The as-prepared $NaBiS_2$ samples seemed to have band-gap of around 1.20 eV to 1.24 eV, whereas the Bi_2S_3 exhibited 1.68 eV. Another crucial factor found in case of $NaBiS_2$ is that it already poses a decent surface area of 10.1852 $m^2\ g^{-1}$, but it can be doubled by compositing with ZnO, which overall impacted the photocatalytic application in a very positive way (Koutavarapu et al., 2021). This recent finding has created a wider scope in the potential combination of nanoribbons with ZnO. Apart from these, boron nitride is another suitable material found to be effective in water-cleaning application, which is prepared through using precursors like melamine and boric acid (Li et al., 2022). The as-synthesized, porous structure had an excellent surface area of 1597 $m^2\ g^{-1}$ which promoted its pollutant adsorption capabilities.

12.3 WASTE TREATMENT BY REMOVING POLLUTANTS VIA DIFFERENT METHODS

To facilitate the pollutant degradation process, a photocatalytic system consists of two modules: a catalytically active site and a light-harvesting centre (Das et al., 2021) (Figure 12.1). The interactions of both components are dependent on charge kinetics upon photoexcitation. In a heterogeneous photocatalytic scheme, the

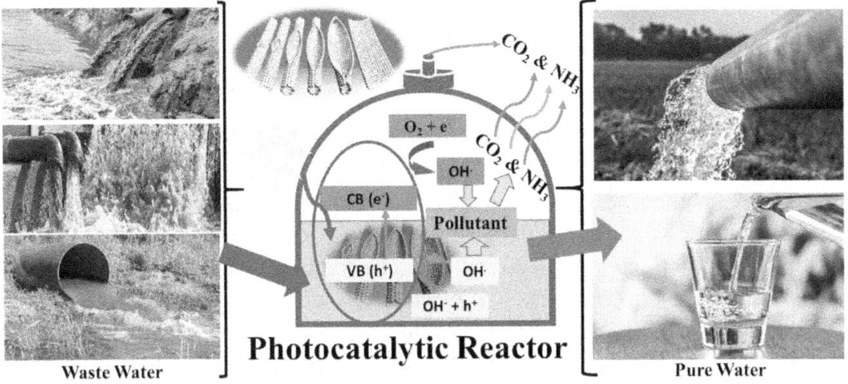

FIGURE 12.1 Mechanism of wastewater treatment by nanoribbons.

light-harvesting centre is predominantly a semiconductor. The photocatalytic process involves a succession of reductive and oxidative processes, which have been widely suggested.

Electrons (e^-) and holes (h^+) are created under appropriately simulated light irradiation with photon energy larger than or equal to the semiconductor band-gap energy, as demonstrated in Eq. (1.1).

$$\text{Photocatalyst} \quad \blacktriangledown \quad hv\ h^+ + e- \qquad (12.1)$$

Electrons move from the valence band to the conduction band at the semiconductor surface, generating holes (h^+) in the valence band. The superoxide radical anion (O_2•), hydroxyl (HO•) radical, and h^+ are the three primary active species engaged in photocatalytic processes, with the HO• radical being the principal oxidant involved in the photodegradation of pollutant in an aqueous medium. The band-edge locations of a photocatalyst are used to determine the thermodynamics of photocatalytic reduction and oxidation processes. As a result, the reduction and oxidation potentials of the substrate as well as various other intermediary processes must be considered throughout the photodegradation cycle. When the reaction is thermodynamically favourable, the electrons join with the oxygen (O_2) molecule to make O_2•, which then produces OH- radicals (Eqs. 1.2, 1.3, and 1.5).

$$e- + O_2 \quad \blacktriangledown \quad O_2^{-\bullet} \qquad (12.2)$$
$$O_2^{-\bullet} + H^+ \quad \blacktriangledown \quad HOO^\bullet \qquad (12.3)$$
$$2HOO^\bullet \quad \blacktriangledown \quad O_2 + H_2O_2 \qquad (12.4)$$
$$H_2O_2 + hv \quad \blacktriangledown \quad 2HO^\bullet \qquad (12.5)$$

Simultaneously, the h^+ interacts with H_2O or hydroxyl ion (OH^-) to generate the HO• radicals (Eqs. 1.6 and 1.7).

$$H^+ + HO^- \quad \blacktriangledown \quad HO^\bullet \qquad (12.6)$$
$$H^+ + H_2O \quad \blacktriangledown \quad HO^\bullet + H^+ \qquad (12.7)$$

The resultant HO• radical acts as a powerful oxidizing agent, combining with organic contaminants in water matrices to generate an intermediate product that produces CO_2, H_2O and other products (Eq. 1.8).

$$\text{Pollutant} + \text{HO}\bullet \blacktriangledown \quad CO_2 + H_2O + \text{Degradation product(s)} \tag{12.8}$$

Furthermore, photogenerated holes are often recognized as an oxidant to directly destroy organic contaminants, depending on the oxidation circumstances and catalyst type. The photoinduced electrons can so recombine with holes, following their creation in the absence of a hole or electron scavenger. As a result, the presence of specialized scavengers is critical for limiting the rate of charge recombination and improving photocatalytic performance. To create a semiconductor photocatalyst material that can effectively use a wider range of solar energy, the semiconductor must have the following characteristics: (1) a smaller band-gap to absorb a wider range of electromagnetic spectrum, (2) a favourable VB edge-position for the generation of h^+ and OH^-, (3) high stability, and (4) efficient charge-separation and migration.

For example, thin ribbons of graphene monolayers are a new kind of graphene that is now attracting a lot of attention in the scientific community. The majority of research into these materials has been on the thin, elongated graphene monolayer strips that can be made by "unzipping" CNTs (Sinitskii et al., 2010). Graphene nanoribbons are commonly described as a 1D sp2 -hybridized carbon strip of finite length with defined edges and non-3-coordinated carbon atoms. Depending on the edge termination, three forms of graphene nanoribbons are now recognized: (i) armchair, (ii) zigzag, (iii) and chiral nanoribbons. The width of an armchair-graphene nanoribbon is determined by the number of dimer CC lines (Na) that cross the nanoribbon, whereas the width of a graphene nanoribbon with zigzag edges is determined by the number of zigzag chains (Nz) that span the nanoribbon. Edge reconstruction may occur because the edge carbon atoms are not bonded or saturated. Edge reconstructions are predicted for zigzag graphene nanoribbons at increased temperatures, but the edge pattern of the armchair graphene nanoribbon is stable due to the existence of strong dangling bonds. Hydrogen saturation is widely used to stabilize the edge structure. Other edge profiles incorporating pentagonal and heptagonal carbon rings have been seen; however, such edge reconstructions are extremely rare. Hybrid-graphene nanoribbons are made up of graphene nanoribbons that have a variety of various edges, most commonly with heterojunctions of armchair and zigzag patterns. Graphene nanoribbons, like graphene, can have bi-layered or few-layered configurations; the design is called a graphitic nanoribbon when multiple layers of finite graphene strips are placed together. Due to the increased number of defect sites and higher edge-to-surface ratios of graphene nanoribbons generated from MWNTs, their Raman spectra have a stronger D band than those of the beginning nanomaterials (Georgakilas et al., 2015).

The fabrication of zein-nanoribbons and usage in the removal of Pb(II) from contaminated water are described in this work. The smooth and continuous synthesis of zein-nanoribbons was achieved using a modified coaxial electrospinning technique. To avoid protein–metal interactions and spinneret clogging, the technique used a sodium lauryl sulfate ethanol aqueous solution as the sheath fluid. The zein-nanoribbons made utilizing the modified procedures exhibited a flatter and narrower morphology than equivalent materials made using a single-fluid technique, according to scanning electron microscope pictures. These zein-nanoribbon mats were utilized to remediate Pb(II)-polluted water by using the favourable interactions between metal and protein. For Pb(II) solutions with starting concentrations of 100, 150, and 200 mg L^{-1}, adsorption studies showed that equilibrium was reached in 60 minutes. The pseudo-second-order model may be used to describe the process. The Langmuir isotherm model suited the isotherm data well, with a maximum adsorption of 89.37 mg g^{-1}. The adsorption capacity can stay up to 82.3% after five cycles of re-use, according to desorption data. Both the synthesis and application of zein-nanoribbons have been proposed as mechanisms (Wen et al., 2016). Similarly, Amin et al. has successfully proposed this work: water-intensive sectors must adhere to strict environmental standards and increasing regulatory frameworks, which necessitate the development of novel water-recycling technology. Polymeric membranes have provided an operative solution for improving water recycling; however, ionic and molecular sieving requires finely tuned pore size and surface chemistry to be successful. Fouling is also a significant issue that restricts the practical use of membranes in water recycling in these sectors. Using a non-solvent-induced phase-separation (NIPS) approach, four distinct graphene oxide (GO) derivatives were integrated into a polyethersulfone (PES) matrix. The GO derivatives utilized had diverse forms (nanosheets versus nanoribbons) and oxidation states (C/O = 1.05–8.01), all of which have the potential to improve water flow and reduce membrane fouling by controlling pore size, hydrophilicity, and surface charge. A water sample from the Athabasca oil sands in Alberta was used to test the permeation capabilities of the PES/GO membranes. The contact angle and streaming potential measurements show that the PES/GO nanocomposite membranes are more hydrophilic and negatively charged. When compared to the unmodified PES membrane, all graphene-based nanocomposite membranes showed improved water flow and organic matter rejection. Fouling was slowed as a result of improved membrane surface qualities, according to the findings of the fouling measurements. At an optimal loading of 0.1 wt.%, longitudinally unzipped graphene oxide nanoribbons (GONR-L) produced the greatest water flow (70 LMH at 60 pressure), organic matter rejection (59%), and antifouling qualities (30% improvement compared to pristine PES membrane). Flux recovery-ratio tests revealed that PES/GO nanocomposite membranes had a significant improvement in fouling resistance. (Karkooti et al., 2018)

12.4 NANORIBBONS FOR WASTEWATER-TREATMENT APPLICATION (POLLUTANT REMOVAL BY VARIOUS PROCESSES)

12.4.1 Carbon-Based Nanoribbons

Starting from energy production and storage to drug delivery, carbonaceous materials have a wide range of usages. Eliminating unnecessary substances like heavy-toxic metals, hazardous dye is one of the prominent applications. Carbon nanotubes, nanocarbon, graphene, or graphene oxide-like compounds are thermally and mechanically well-stable and also have good radiation resistance, due to which they have been frequently used in the adsorption of radioactive elements such as thorium, uranium, or the heavy metals such as arsenic, mercury, etc., from the contaminated water resources. For instance, Yun Wang and his group have extensively studied the graphene oxide nanoribbons and presented a set of possible combinations of graphene oxide nanoribbons, either with functionalizing or modifying it, using other compounds to enhance the adsorption process (Wang et al., 2015; Wu et al., 2019; Xiu et al., 2019). Initially, utilizing individual graphene oxide nanoribbons exhibited adequate uranium sorption capacity (394.1 mg g^{-1}) from watery solutions due to the existence of oxygen-containing functional groups (Wang et al., 2015). The effect of pH was found to be a crucial factor where acidic pH levels were recommended to achieve higher efficiency, and also, the sorption process on the prepared nanoribbons were sensitive towards the ionic strength, which indicates that the prevailing mechanisms are either outer-surface complexation or ion exchange. The prepared nanoribbons were employed for sorption/desorption and reused for five successive cycles, where the sorption dropped to certain level (126.5 mg g^{-1}). In order to achieve better recovery and reusability of the sorbent, utilization of magnetic graphene-oxide nanoribbons proved to be a fascinating option (Wu et al., 2019). After performing sorption or adsorption process in the aqueous or watery solution, a magnetic separation and recovery technique is quite helpful to retrieve the adsorbent material without much quantitative loss. Similar to graphene oxide nanoribbons, its magnetic counterpart was also found to be working well in the acidic pH levels. The main reason behind its superior activity in the acidic pH region is because the adsorbent surface was positively charged, which created electrostatic repulsion; thus, adsorption of thorium ions favourably takes place. Compositing manganese oxide with graphene oxide nanoribbons has also shown satisfactory adsorption capacity (166.11 mg g^{-1}) for thorium ion removal. Manganese oxide is a stable, less expensive material, having a large surface area, and works perfectly under acidic conditions, just like graphene nanoribbons, so compositing both was anticipated to be working better than the individual nanoribbon. However, in a previous study, individual graphene nanoribbons were examined for the removal of thorium ion using where the adsorption capacity was much higher (273 mg g^{-1}) than the nanoribbon/manganese oxide composite (Xiu et al., 2019). Various kinds of ligands have also been used with graphene oxide nanoribbons aerogel to improve selectivity and adsorption capacity (Wang, Chen et al., 2021). Graphene oxide nanoribbons aerogel have excellent physical and chemical characteristics, which make them suitable for eliminating or adsorbing organic pollutants like applications. On top of that, the added ligands such as amidoxime, phosphorus-containing ligands (vinylphosphonic acid), etc., exhibited great impact on the adsorbing activity of the composites as well as depicted excellent durability and simple solid-liquid separation procedure. For instance, amidoxime is already a proven compound and has been effectively utilized to eradicate uranium ions from salt water or seawater due to their high selectivity and fast adsorption rate (Wang et al., 2014). Similarly, employing phosphonic acid groups facilitates easy trapping of thorium ions, as the acid groups are suitable in developing stable complexes with the ions on the surface of the composite-adsorbent material (Wang, Chen et al., 2021).

Yun Wang and his group have also composited natural polymer (chitosan), along with the graphene oxide nanoribbons, which depicted adsorption capacity of 320 mg g^{-1} (Hu et al., 2020). Apart from cellulose, chitin-derived chitosan is a naturally abundant linear polysaccharide which has shown great activity in removal of heavy-metal ions from aqueous solutions (Aizat et al., 2019; Sun et al., 2018). Predicting its viable utilization due to the presence of hydroxyl and amino functional groups turned out to be quite successful, as the graphene nanoribbons/chitosan composite membrane has shown beneficial adsorption of uranium (Hu et al., 2020). Additionally, low toxic nature is another crucial factor behind its adsorption application, which almost nullifies further formation of hazardous compounds during the adsorption process. Recently, Wang, Wang et al. (2021) synthesized 3-D graphene nanoribbons, which have exhibited excellent specific area of 597.4 m^2 g^{-1}, and as a result, improvement on the adsorbing activity was observed, along with improved solid–liquid separation of adsorbent and the toxic compound-containing solution. By utilizing the as-prepared 3-D nanoribbons rich in oxygen groups, adsorption capacity of 380.4 and 430.6 mg g^{-1} was observed for thorium and uranium ions, respectively. The prepared adsorbent was also found to be mechanically stable and highly porous, which effectively helped in achieving suitable adsorption performance. Removal of arsenic and mercury ions using chemical adsorption is also a favourable and evidently scalable process. Like the thorium and uranium ions adsorption, 1-D graphene oxide nanoribbons are also found effectual in adsorbing the arsenic and mercury ions (Sadeghi et al., 2020). Formation of numerous active sites because of graphene-oxide nanoribbon oxygen containing functional groups on the basal planes and the edges helped in removal of these heavy metals. Other than these toxic compounds, graphene nanoribbons are also used for adsorbing hazardous dyes like anionic orange II and methylene blue (Gao et al., 2020). However, particularly, graphene nanoribbons have not yet been explored very much in studies, and hopefully, in upcoming times, these individual nanoribbons or compositing with other materials will be investigated with a competent, high-efficiency adsorption approach.

As we discussed in this section about the carbonaceous nanoribbons, it is very clear that reports have explored their capability for removal of toxic metal ions and dyes via utilizing adsorption process, which came up as an effective, cost-efficient and scalable technology. However, a few studies have considered their usage as photocatalyst, but mostly, reports have focused on using them individually, rather doped or by compositing the nanoribbons with other suitable materials. Tunable band-gap and excellent conductivity of graphene nanoribbons resulted the composite materials to achieve superior photocatalytic activity. For example, Zhang et al. (Zhang et al., 2016) reported the visible-light photoconductivity response of vanadium oxide, composited with graphene nanoribbons, which clearly indicates the sufficient capability of the prepared composite as a promising photocatalyst. Vanadium oxide is a well-verified photocatalyst which has shown excellent ability in degrading azo dyes (Sajid et al., 2020). However, their addition has further improved high mobility and charge-transfer characteristics which can enhance the photodegradation process to facilitate the dye containing water cleaning in a more precise manner. In a similar way, the author has also proposed the viability and potential usage of cerium oxide/graphene nanoribbons as photocatalysts (Zhang et al., 2015). Similar to vanadium oxide, cerium oxide also poses adequate capacity of degrading harmful dyes under the visible-light influence. Properties like photostability, high oxygen storage ability, and non-toxic nature made the cerium oxide utilization more alluring (Ma et al., 2019). Likewise, the oxide catalyst ZnS, CdS-like sulfide materials, are also found to be interesting photocatalysts to explore the elimination of azo dyes (Lee et al., 2017; Liu et al., 2021). However, ternary $Zn_xCd_{1-x}S$ has been proven more efficacious than the ZnS and CdS focusing, which S. Lou et al. proposed facile preparation to synthesize hybrid photocatalyst via immobilizing $Zn_xCd_{1-x}S$ nanocrystals on the graphene nanoribbons, which completely degraded the organic dye (methyl orange) in almost ~2h 30 min (Lou et al., 2016). This unique hybrid structure of ternary semiconductor/graphene nanoribbon has shown larger specific surface-area along with a porous structure, which was favourable for the photodegradation. Doping graphene nanoribbons with boron or nitrogen have shown immense exploring opportunities because the combination offers controllable and predictable charge transport characteristics, which assist well to improve the electronic conductivity as well as to achieve better the photoactivity (Tang et al., 2012; Wang et al., 2009). For instance, under UV light illumination, boron-doped graphene nanoribbons effectively degraded Rhodamine-B dye (Xing et al., 2014). Such outcomes confirm the efficiency of the synthesized doped-graphene nanoribbons and also its possible utilization in graphene-based photoelectric devices.

12.4.2 Metal Oxides/Chalcogenides-Based Nanoribbons (Pollutant Removal by Various Processes)

Among the mostly employed metal oxide semiconductors, TiO_2 is the most preferably employed photocatalytic material due to its adequate band-gap feature, which demonstrates its excellent activity in the degradation of hazardous pollutants from water or aqueous solutions. Mainly, the photocatalytic activity of TiO_2 relies on their respective particle dimensions, morphology, phase state, and surface area. For an instance, Tao et al. synthesized TiO_2 nanoribbon sphere (having a diameter of 50 and 40 μm) to study the photocatalytic degradation of methyl orange dye, where the as-synthesized nanoribbon seemed to be performing better compared to the nanoparticle TiO_2 (Tao et al., 2012). The nanoribbon's structural viabilities such as better light harvesting abilities, enhanced surface area, etc., boosted the electron transfer rate and reduced the charge-pair recombination rate, which augmented the photocatalytic efficiency. However, most of the recent years' studies have attempted to improve the properties and morphologies of TiO_2 via modifications like doping or compositing with other compounds. These helped in achieving swift mass transfer, a larger number of reaction sites via improving specific surface area and upgrade the light absorption capability, as well as slowing down the photoactivated charge pair recombination rate. So, as a result, studies are focusing on the surface tailoring or compositing the TiO_2 with narrow-gap semiconductors (such as CdSe, CdS, CdTe, etc.) or metals (such as Ag, Au, Pt, Cu, etc.) rather than employing it in its individual form. For instance, Cao et al. combined the bismuth oxohalides (BiOI, BiOBr, and BiOCl) and TiO_2 nanoribbon framework to prepare unique, hybrid structures for degrading rhodamine B, methyl orange, and salicylic acid under sunlight illumination (Cao et al., 2014). Owing to having a flat and wide surface built of negatively charged octahedral, TiO_2 nanoribbon provides suitable nucleation and growth conditions for bismuth oxohalides containing oppositely charged [Bi_2O_2] units. The as-developed mesoporous architecture seems to be enhancing mass transport as well as reducing the incident light reflection, which occurred due to intense scattering; therefore, light absorption effectively improves and is able to harvest more photons, by which the degradation process gets improved. Moreover, the formed interconnections in the hybrid structure of TiO_2 and bismuth oxohalides offers rapid electron-transport, which weakens the electron-hole recombination rate. Owing to properties such as enhanced electroconductivity, physical, chemical, and mechanical features, CNTs are always reflected as an efficient material to produce hybrid materials. Shaban et al. reported preparation of composite via combining carbon nanotubes (CNTs) with TiO_2 nanoribbons, which was employed for studying degradation of MB dye under natural sunlight illumination (Shaban et al., 2018). Compared to composite nanoribbons (TiO_2/CNTs) individual TiO_2 nanoribbons needed more irradiation duration for completely degrading the dye. Similarly, with increasing catalyst dosage, composite nanoribbons reacted at a quicker pace than the individual TiO_2 nanoribbons. Other than as individual photocatalysts, TiO_2 nanoribbons have also been employed in the membrane system for dye removal. Incorporating these nanoribbons with the membrane prevented it from having worsened permeate quality, flux reduction, and reduced membrane-lifespan-like drawbacks. Besides, it also enhanced the filtration ability of prepared membranes. Bai et al. mentioned about the preparation of TiO_2 nanoribbon spheres, where long, one-dimensional (1D) nanoribbons were used to develop three-dimensional (3D) dendritic TiO_2 nanospheres (Bai et al., 2013). To develop an efficient water-purification system, a high-performing, concurrent photocatalytic-membrane was formed via integrating commercially available cellulose acetate (CA) membrane with the as-synthesized TiO_2 nanoribbon spheres. The porous, functional layer on the CA membrane was found to be useful in eliminating Rhodamine-B and acid-orange-like dye pollutant containing water compared to the individual CA membrane. The addition of the TiO_2 nanoribbon spheres in the membrane meaningfully uplifted the removal amount of the dye and enhanced the membrane lifespan with almost-diminished fouling problems. Similarly, X. Cao et al. developed highly porous (~97%) titanate-nanoribbon membrane with excellent mechanical properties via using a supersaturation-directing self-assembly approach (Cao et al., 2013). The as-prepared membrane is found to have the ability to discriminate smaller molecules and also has nanosolid-capture abilities. The author clearly stated that this developed membrane is well suitable for water purification and environmental-remediation application and can be constructed for large-scale usage. Likewise, the TiO_2, titanate-nanostructured materials are also a fascinating choice for pollutant-removal applications (Banerjee & Choudhury, 2022; Fan et al., 2018). Degradation of acetaminophen using bi-modified titanate nanoribbon was observed to occur more than 88% in 3 hours of irradiation time. Contrasting to TiO_2, bismuth titanate has lower energy band-gap, which makes it favourable to make better use of the visible-light range and improved photocatalytic activity. The unique configuration of the core–shell heterostructure nanoribbon-photocatalysts having high specific-surface presented more active surfaces, which particularly enhanced the carrier mobility and reduced the charge-carrier recombination rate.

Silver vanadium oxide ($AgVO_3$) is an eminent visible-light-responsive photocatalyst which has grabbed significant consideration for its narrow band gap and highly dispersed valence bands, but its individual application has been poorly restricted due to its inadequate capacity to split the electron-hole pairs, and as a result, a higher degree of photocatalytic performance does not seem to be achievable. So, to improve the electron trapping capability and increase the electron-hole pair splitting capacity, modifying with other photocatalysts or compositing with highly conductive material are the viable options. Owing to the localized surface-plasmon resonance (SPR) effect, the combination of $AgVO_3$ nanoribbons with Ag and graphitic carbon nitride has exhibited extended visible-light response. The as-synthesized ternary plasmonic photocatalyst was successfully used to study the degradation of the basic cationic dye (Zhao et al., 2015). Previously, the same group has approached the theoretical study using density functional theory and 3-D finite-difference time-domain technique to study the impact of Ag nanoparticles on the electronic and band structure of $AgVO_3$ nanoribbon and also to summarize the influence of Ag nanoparticles on the improvement of the electric field and SPR of the $Ag/AgVO_3$ (Wei et al., 2014). With the purpose of gaining structural stability benefits as well as catalytic performance enhancement, quantum dots with $AgVO_3$ nanoribbons were observed as favourable candidates. For instance, developing heterojunction via utilization of graphene quantum dots (GQDs) along with $AgVO_3$ nanoribbons has evidently shown excellent photogenerated-charge pair-splitting efficiency, which raised overall photocatalytic efficiency (Lei et al., 2016). Various concentrations of GQD (0.5, 1, 3, 5, 7 wt%) were experimented with to identify the optimized one. In degrading Ibuprofen, particularly the $AgVO_3$/3 wt% GQD exhibited the highest degradation efficiency (> 90%). Furthermore, the author mentioned that the prepared catalyst was stable and showed no noticeable deactivation up to the recycling study up to four successive cycles and steadily performed the degradation as well. Similarly, by compositing SnO_2 quantum dots along with graphitic carbon nitride ($g\text{-}C_3N_4$) nanosheets and $AgVO_3$ nanoribbons, a suitable heterostructure was developed, using which prominent degradation activity was achieved (Koyyada et al., 2021). Compared to the $AgVO_3/g\text{-}C_3N_4$ composite, the as-prepared ternary composite has demonstrated much higher degradation efficiency. Under solar light, almost 80% Rhodamine-B dye was degraded using the prepared ternary composite in less than an hour, and additionally, the catalytic performance was found to be stable up to five cycles. Individual $g\text{-}C_3N_4$ nanoribbon has also been studied as a photocatalyst for degrading methyl-orange- and methylene-blue-like dye in recent years, but their further beneficial developments are still underway (Liu et al., 2020). Other than these heterojunction photocatalysts, the heterojunction formed using $AgVO_3$ nanoribbons with $Ba_5Ta_4O_{15}$ nanosheets was effective in degrading acid-red dye (Wang et al., 2018). Behind the successful photoactivity of the synthesized heterojunction photocatalysts, the major reason was the weakened photogenerated-charge pair recombination that was caused due to the holes-dominated charge-transfer process and localized surface-plasmon resonance. Considering eco-friendliness, less cost, and abundance-like factors, binary cobalt-nickel oxides ($NiCo_2O_4/CoNi_3O_4$) have also shown great impact in removal of unwanted pollutants from aqueous solutions, and also, they are easily recoverable. Studying adsorption abilities of $NiCo_2O_4$ nanoribbons using methyl orange dye was found to be effective; however, the same study depicted that the nanorod morphology of the same cobalt-nickel oxide has higher adsorbing capacity (360.5 mg g^{-1}) with the similar dye of equal concentration (Tian et al., 2017). In a recent study, degradation of atrazine was performed, using the highly porous and hierarchical structure of cobalt-nickel oxides ($CoNi_3O_4$) nanoribbon/diatomite hybrid, which has revealed greater adsorption capacity (Dong et al., 2020). The as-prepared hybrid offered numerous active sites and open diffusion channels, which, promisingly, assured efficient removal of atrazine. Other than these metal oxides, Zn_2GeO_4 has also shown an efficacious outcome in removal of lead and copper-like toxic ions from water, along with augmented material properties such as good thermal stability, high selectivity, and recyclability; however, this specific combination has not been studied to a wide extent due to high-cost issues (Yu et al., 2011).

12.4.3 Other Nanomaterial-Based Nanoribbons (Pollutant Removal by Various Processes)

As we discussed in the earlier sections, the carbonaceous (mainly graphene) and metal oxide nanoribbons have been widely employed in wastewater cleaning, either by using chemical adsorption or by using photoactivated degradation-techniques. However, a few other nanoribbons have also shown prominent scope for similar applications. Bismuth-based materials—specifically, bismuth sulfide and sodium bismuth sulfide nanoribbons—are such photocatalyst materials which have been probed by several researchers. Under a visible-light range, the unique, layered structure of bismuth-based photocatalysts has exhibited easy splitting of photo-activated charge pairs, which resulted in superior water oxidation activity and photodegradation performance (He et al., 2018; Ye et al., 2019). Relating to those factors and due to having a narrow band-gap (around 1.25 eV) and photostability, sodium bismuth sulfide ($NaBiS_2$) has clearly shown promising effects in eliminating toxic compounds (like dyes and heavy metal) under visible-light influence. However, to make further improvements, ZnO nanosheets were anchored on $NaBiS_2$ nanoribbon to prepare the hybrid composites, which delivered a highly efficacious outcome in degrading tetracycline and Rhodamine B under solar-light irradiation compared to their individual counterparts (Koutavarapu et al., 2019; Koutavarapu et al., 2021). From the studies, it was clearly observed that combining ZnO reduces the charge pair recombination rate; however, over-increasing the quantity of ZnO nanosheets agglomerated on the $NaBiS_2$ nanoribbon surfaces in an undesirable manner and developing some additional junction sites which influence the recombination results in the photocatalytic activity being hampered. So, the recommended rate of ZnO proposed in the study was less than 10 mg (Koutavarapu et al., 2019). Also, the degradation process was influenced by the existing mesoporous channels in the synthesized composite, as they facilitate ion transport in a swift way. Utilization of bismuth sulfide nanoribbon has also been reported by Xu et al., but the author recommended its usage as co-catalyst heterojunction with bismuth tungstate to get a favourable outcome (Xu et al., 2019). Apart from the several detailed nanoribbons, photodegradation, in very recent studies, utilization of boron nitride, and ZnSe/CdSe core-shell nanoribbons have also been mentioned to eliminate pollutants like bisphenol A, methyl orange, rhodamine B, and methylene blue via adsorption and photodegradation technique, although these studies need further extensive analysis to derive the best possible approaches out of it (Li et al., 2022; Sun et al., 2020).

12.5 PROS AND CONS OF NANORIBBONS IN WASTEWATER TREATMENT

In search of the most suitable material, several compounds are been investigated, where TiO_2 was found to be one of the prominent materials, but still, many aspects remained vague, due to which still various materials and their different morphologies are being examined. In this chapter, we have elaborately discussed different types of nanoribbons and their existing application in contamination eradication from water or aqueous solutions. We have aimed here to emphasize significantly the properties and factors which make the nanoribbons more efficient compared to other structural configurations like nanorods, nanospheres, nanowires, nanosheets, etc. For example, under visible illumination, bi-modified titanate nanoribbons were found to be efficient in degrading acetaminophen compared to the bi-modified titanate nanosheets and nano-bulks (Fan et al., 2018). However, a few studies have evidently shown contrary results as well. For an instance, nanorods, balsam-like and rose-like structures of nickel-cobalt oxide seemed to be performing better than their nanoribbon counterpart (Tian et al., 2017). Similarly, TiO_2 nanoneedles were found to be more capable in performing photocatalytic degradation of methyl orange than the nanoribbons (Akpan & Hameed, 2009). Two reasons were mainly identified behind the behaviour; firstly, the electron transferred swiftly through appropriate conventional channels, increased efficiency, and declined the charge recombination. Secondly, the nanoneedle structure has the capacity of harvesting a larger amount of light due to its structural benefits. On the other hand, in the same study, it was pinpointed that the synthesized nanoribbons were having 30 nm of thickness, so incident light was able to transmit the nanoribbon due to its inadequate thickness, and as a result, its light harvesting ability got weakened; hence, its photocatalytic ability dropped. After having these pitfalls, also, nanoribbons have exhibited a few structural advantages. The armchair and zigzag-edges of the MoS_2 and graphene nanoribbons have revealed abundant potential (Shinde et al., 2022). In the previous sections, we have summarized their correlation with optical characteristics and their overall impact on the adsorption and photocatalytic degradation. Doping and compositing the nanoribbons also have a promising future. Like in studying boron-doped graphene nanoribbons, it was closely observed and noted that the semiconductor characteristics were improved, so the armchair and zigzag-edges were enriched as well as the band gap (Xing et al., 2014). Owing to nanoribbon's plentiful surface area, it is favourable to prepare composites, as it provides more active edge sites and stronger interfacial interaction. Several studies have successfully explored various composites with different nanoribbons, heterostructure formation, edge functionalization, and tuning functional groups, which we have minutely explained in the previous section. Modifying nanoribbons using organic frameworks is a recent approach which can also bring hopeful developments in the future, as it has already been proven useful for various applications (Liu et al., 2017; Wang et al., 2019). Still, further detailed research is required to investigate the structural aspects like defects, intercalation, and vacancy aspects and to make their applicative understanding more prominent and more easily understandable. Synthesis approaches like controlling twists, flatness, and orientation throughout their growth also needs to be considered for further advancements. Two other concerning issues have also been identified in reports: the involvement of lower cost in overall process and the reusability of the materials. To tackle the first aspect, several novel synthesis methods are being developed as well as, in many cases, utilization of natural materials are also being taken account of. Applying such ideas will not only help in cutting the total cost but also will help us in eliminating usage of unnecessary and toxic chemicals. Secondly, the reusability of the materials has been tried by many researchers. In general, it was observed that the material acts stable in five to 10 successive cycles; afterwards, the efficiency is reduced. As an alternative solution, preparing magnetic materials came up in recent studies. It has been able to perform the separation, retrieval, and reusable studies properly, to some extent. Wu et al. have already investigated magnetic graphene-oxide nanoribbons in thorium adsorption, which was well workable but in need of more modifications in specifying the morphological analysis and the application prospective (Wu et al., 2019). However, the reusable samples must be reanalyzed at the microscopic level for more conclusive details and also the possible danger of secondary contamination needs to be considered to avoid unwanted failures.

12.6 FUTURE SCOPE OF NANORIBBONS

Recently, materials with various morphologies and sizes have been explored to study their compatibility with various ranges of applications. Subtle analysis of these studies will probably lead us to commercial-scale application, and hence, towards a better future. Among several tested designs, nanoribbon architectures are one such which exhibited great potential. In this chapter, we have illustrated different types of nanoribbons, with added examples, and also indicated their advantages and drawbacks to make their upcoming prospects clearer. Owing to their unique structure, nanoribbons are able to easily offer highly exposed, numerous active sites on the edges, which is very favourable for electron transfer or movement and therefore beneficial for application purposes like sensors, photocatalysts, etc. Mainly through analyzing their bandgap, crystal structure, and other characteristics such as (electrical, optical, magnetic, etc.), it has been evidently clear that this particular morphology has a good response against adsorbance performance or catalytic performance. However, day-to-day efforts are continuously pushing the limits to achieve better efficiency of the nanoribbons. Doping, composite/heterojunction formation, morphological alteration, and phase and strain engineering have brought more efficacious results in eradication of heavy metal ions or toxic dyes from aqueous solutions. For instance, via doping, reaction activity can be boosted, or forming heterojunction provides more edges or active sites, which enhances the catalytic applications. As of now, several theoretical predictions have depicted various possible improvements in properties of nanoribbons, which requires experimental confirmation. Modifications in properties or characteristics are not the only aspects focused on in studies, but also, recent studies have approached several facile, low-cost preparation methods to make them usable commercially. Optimal control in synthesis is also required to maintain the size, morphology, and desired properties, as this will help in developing large-scale preparation strategies. Hence, we believe that, with further improvements in research and sincere considerations to eliminate the drawbacks as well as optimizing the synthesis and application processes, these will be very useful to extend the usage of nanoribbons from lab-scale to real-life operations.

Acknowledgement: HTD thanks RUSA, Utkal University, Bhubaneswar for funding.

REFERENCES

Ahmed, T., Zhang, H.-l., Gao, Y.-Y., Xu, H.-b., & Zhang, Y. (2018). Surfactant-free synthesis of m-$BiVO_4$ nanoribbons and enhanced visible-light photocatalytic properties. *Materials Research Bulletin*, *99*, 298–305. https://doi.org/10.1016/j.materresbull.2017.11.029

Aizat, M. A., & Aziz, F. (2019). 12 — Chitosan nanocomposite application in wastewater treatments. In A. Ahsan & A. F. Ismail (Eds.), *Nanotechnology in water and wastewater treatment* (pp. 243–265). Elsevier.

Akpan, U. G., & Hameed, B. H. (2009). Parameters affecting the photocatalytic degradation of dyes using TiO_2-based photocatalysts: A review. *Journal of Hazardous Materials*, *170*(2–3), 520–529.

Al-Degs, Y. S., El-Barghouthi, M. I., El-Sheikh, A. H., & Walker, G. M. (2008). Effect of solution pH, ionic strength, and temperature on adsorption behavior of reactive dyes on activated carbon. *Dyes and Pigments*, *77*(1), 16–23.

Bai, H., Liu, L., Liu, Z., & Sun, D. D. (2013). Hierarchical 3D dendritic TiO_2 nanospheres building with ultralong 1D nanoribbon/wires for high performance concurrent photocatalytic membrane water purification. *Water Research*, *47*(12), 4126–4138. https://doi.org/10.1016/j.watres.2012.09.059

Balaji, T. E., Das, H. T., & Maiyalagan, T. (2021). Recent trends in Bimetallic Oxides and their composites as electrode materials for supercapacitor applications. *ChemElectroChem*, *8*, 1723. https://doi.org/10.1002/celc.202100098

Banerjee, A., & Choudhury, M. (2022). Green technology for wastewater treatment and remediation. In C. Baskar, S. Ramakrishna, & A. Daniela La Rosa (Eds.), *Encyclopedia of green materials*. Springer. https://doi.org/10.1007/978-981-16-4921-9_30-1

Bhardwaj, L. K., Rath, P., & Choudhury, M. (2022). A comprehensive review on the classification, uses, sources of nanoparticles (NPs) and their toxicity on health. *Aerosol Science and Engineering*, *7*(1), 69–86. https://doi.org/10.1007/s41810-022-00163-4

Cao, X., Lu, Z., Zhu, L., Yang, L., Gu, L., Cai, L., & Chen, J. (2014). A new family of sunlight-driven bifunctional photocatalysts based on TiO_2 nanoribbon frameworks and bismuth oxohalide nanoplates. *Nanoscale*, *6*(3), 1434–1444.

Cao, X., Zhou, Y., Wu, J., Tang, Y., Zhu, L., & Gu, L. (2013). Self-assembled, robust titanate nanoribbon membranes for highly efficient nanosolid capture and molecule discrimination. *Nanoscale*, *5*(8), 3486–3495. https://doi.org/10.1039/C3NR00347G

Das, H. T., Mahendraprabhu, K., Maiyalagan, T., & Elumalai, P. (2017). Performance of solid-state hybrid energy-storage device using reduced graphene-oxide anchored sol-gel derived Ni/NiO nanocomposite. *Scientific reports*, *7*(1), 1–14. https://doi.org/10.1038/s41598-017-15444-z

Das, H. T., Vinoth, S., Thirumoorthi, M., Alshahrani, T., Hegazy, H. H., Somaily, H. H., Shkir, M., & Aifaify, S. (2021). Tuning the optical, electrical, and optoelectronic properties of cuo thin lms fabricated by facile SILAR dipcoating technique for photosensing applications. *Journal of Inorganic and Organometallic Polymers and Materials, 31*(6), 2606–2614. https://doi.org/10.1007/s10904-021-01928-z

Dong, X., Ren, B., Zhang, X., Liu, X., Sun, Z., Li, C., Tan, Y., Yang, S., Zheng, S., & Dionysiou, D. D. (2020). Diatomite supported hierarchical 2D $CoNi_3O_4$ nanoribbons as highly efcient peroxymonosulfate catalyst for atrazine degradation. *Applied Catalysis B: Environmental, 272*, 118971. https://doi.org/10.1016/j.apcatb.2020.118971

Duraisamy, E., Das, H. T., Selva Sharma, A., & Elumalai, P. (2018). Supercapacitor and photocatalytic performances of hydrothermally-derived Co_3O_4/CoO@carbon nanocomposite. *New Journal of Chemistry*, *42*(8), 6114–6124. https://doi.org/10.1039/c7nj04638c

Duraisamy, E., Gurunathan, P., Das, H. T., Ramesha, K., & Elumalai, P. (2017). [Co (salen)] derived Co/Co_3O_4 nanoparticle@ carbon matrix as high-performance electrode for energy storage applications. *Journal of Power Sources*, *344*, 103–110. https://doi.org/10.1016/j.jpowsour.2017.01.100

Dutta, S., Pohrmen, C. B., Banerjee, I., Trivedi, A., Verma, R., & Dubey, S. (2021). Bio-derived metal and metal oxide incorporated biopolymer nanocomposites for dye degradation applications: A review. *Octa Journal of Biosciences*, *9*(1).

Fan, G., Peng, H., Zhang, J., Zheng, X., Zhu, G., Wang, S., & Hong, L. (2018). Degradation of acetaminophen in aqueous solution under visible light irradiation by Bi-modified titanate nanomaterials: Morphology effect, kinetics and mechanism. *Catalysis Science & Technology*, *8*(22), 5906–5919.

Feng, J., Mao, X., Zhu, H., Cui, M., Ma, Y., Zhang, D., & Bi, S. (2021). How size, edge shape, functional groups and embeddedness influence the electronic structure and partial optical properties of graphene nanoribbons. *Physical Chemistry Chemical Physics, 23*(36), 20695–20701.

Gao, Y., Liang, X., Han, S., Wu, L., Zhang, G., Qin, C., Bao, S., Wang, Q., Qi, L., & Xiao, L. (2020). High-efficiency adsorption for both cationic and anionic dyes using graphene nanoribbons formed by atomic-hydrogen induced single-walled carbon nanotube carpets. *Carbon Letters, 30*(2), 123–132. https://doi.org/10.1007/s42823-019-00089-x

Georgakilas, V., Perman, J. A., Tucek, J., & Zboril, R. (2015). Broad family of carbon nanoallotropes: Classification, chemistry, and applications of fullerenes, carbon dots, nanotubes, graphene, nanodiamonds, and combined superstructures. *Chemical Reviews*, *115*(11), 4744–4822. https://doi.org/10.1021/cr500304f

He, R., Xu, D., Cheng, B., Yu, J., & Ho, W. (2018). Review on nanoscale Bi-based photocatalysts. *Nanoscale Horizons*, *3*(5), 464–504. https://doi.org/10.1039/C8NH00062J

Hu, X., Wang, Y., Yang, J. O., Li, Y., Wu, P., Zhang, H., Yuan, D., Liu, Y., Wu, Z., & Liu, Z. (2020). Synthesis of graphene oxide nanoribbons/chitosan composite membranes for the removal of uranium from aqueous solutions. *Frontiers of Chemical Science and Engineering, 14*(6), 1029–1038. https://doi.org/10.1007/s11705-019-1898-9

Jaiswal, K. K., Dutta, S., Banerjee, I., Pohrmen, C. B., Singh, R. K., Das, H. T., . . . Kumar, V. (2022). Impact of aquatic microplastics and nanoplastics pollution on ecological systems and sustainable remediation strategies of biodegradation and photodegradation. *Science of the Total Environment*, *806*, 151358.

James, D. K., & Tour, J. M. (2012). The chemical synthesis of graphene nanoribbons—a tutorial review. *Macromolecular Chemistry and Physics*, *213*(10–11), 1033–1050.

Karkooti, A., Yazdi, A. Z., Chen, P., McGregor, M., Nazemifard, N., & Sadrzadeh, M. (2018). Development of advanced nanocomposite membranes using graphene nanoribbons and nanosheets for water treatment. *Journal of Membrane Science*, *560*, 97–107. https://doi.org/10.1016/j.memsci.2018.04.034

Koskinen, P., Malola, S., & Häkkinen, H. (2008). Self-passivating edge reconstructions of graphene. *Physical Review Letters*, *101*(11), 115502. https://doi.org/10.1103/PhysRevLett.101.115502

Koutavarapu, R., Lee, G., Babu, B., Yoo, K., & Shim, J. (2019). Visible-light-driven photocatalytic activity of tiny ZnO nanosheets anchored on $NaBiS_2$ nanoribbons via hydrothermal synthesis. *Journal of Materials Science: Materials in Electronics*, *30*(11), 10900–10911. https://doi.org/10.1007/s10854-019-01434-6

Koutavarapu, R., Reddy, C. V., Syed, K., Reddy, K. R., Shetti, N. P., Aminabhavi, T. M., & Shim, J. (2021). Ultra-small zinc oxide nanosheets anchored onto sodium bismuth sulfide nanoribbons as solar-driven photocatalysts for removal of toxic pollutants and phtoto-electrocatalytic water oxidation. *Chemosphere*, *267*, 128559. https://doi.org/10.1016/j.chemosphere.2020.128559

Koyyada, G., Siva Kumar, N., Al-Ghurabi, E. H., Asif, M., & Mallikarjuna, K. (2021). Enhanced solar-driven photocatalytic performance of a ternary composite of SnO_2 quantum dots//$AgVO_3$ nanoribbons//g-C_3N_4 nanosheets (0D/1D/2D) structures for hydrogen production and dye degradation. *Environmental Science and Pollution Research*, *28*(24), 31585–31595. https://doi.org/10.1007/s11356-021-12962-2

Lee, G.-J., & Wu, J. J. (2017). Recent developments in ZnS photocatalysts from synthesis to photocatalytic applications—A review. *Powder Technology*, *318*, 8–22. https://doi.org/10.1016/j.powtec.2017.05.022

Lei, Z.-d., Wang, J.-j., Wang, L., Yang, X.-y., Xu, G., & Tang, L. (2016). Efficient photocatalytic degradation of ibuprofen in aqueous solution using novel visible-light responsive graphene quantum dot/$AgVO_3$ nanoribbons. *Journal of Hazardous materials*, *312*, 298–306. https://doi.org/10.1016/j.jhazmat.2016.03.044

Li, Q., Zhang, J., Liu, B., Li, M., Liu, R., Li, X., . . . Zou, Y. (2008). Synthesis of high-density nanocavities inside TiO_2– B nanoribbons and their enhanced electrochemical lithium storage properties. *Inorganic Chemistry*, *47*(21), 9870–9873.

Li, Z., Zhao, X., Hong, X., Yang, H., Fang, D., Wang, Y., & Lei, M. (2022). Hierarchically porous boron nitride nanoribbon for safe and high-performance bisphenol A adsorption. *Materials Letters*, *307*, 131022. https://doi.org/10.1016/j.matlet.2021.131022

Liu, W., Dai, X., Bai, Z., Wang, Y., Yang, Z., Zhang, L., . . . Wang, S. (2017). Highly sensitive and selective uranium detection in natural water systems using a luminescent mesoporous metal–organic framework equipped with abundant Lewis basic sites: A combined batch, X-ray absorption spectroscopy, and first principles simulation investigation. *Environmental Science & Technology*, *51*(7), 3911–3921. https://doi.org/10.1021/acs.est.6b06305

Liu, X., Sayed, M., Bie, C., Cheng, B., Hu, B., Yu, J., & Zhang, L. (2021). Hollow CdS-based photocatalysts. *Journal of Materiomics*, *7*(3), 419–439. https://doi.org/10.1016/j.jmat.2020.10.010

Liu, Y., Guo, X., Chen, Z., Zhang, W., Wang, Y., Zheng, Y., . . . Huang, Y. (2020). Microwave-synthesis of g-C_3N_4 nanoribbons assembled seaweed-like architecture with enhanced photocatalytic property. *Applied Catalysis B: Environmental*, *266*, 118624. https://doi.org/10.1016/j.apcatb.2020.118624

Lou, S., Wang, W., Jia, X., Wang, Y., & Zhou, S. (2016). A unique nanoporous graphene-ZnxCd1−xS hybrid nanocomposite for enhanced photocatalytic degradation of water pollutants. *Ceramics International*, *42*(15), 16775–16781. https://doi.org/10.1016/j.ceramint.2016.07.161

Ma, R., Zhang, S., Wen, T., Gu, P., Li, L., Zhao, G., . . . Wang, X. (2019). A critical review on visible-light-response CeO_2-based photocatalysts with enhanced photooxidation of organic pollutants. *Catalysis Today*, *335*, 20–30. https://doi.org/10.1016/j.cattod.2018.11.016

Mani, V., Govindasamy, M., Chen, S.-M., Chen, T.-W., Kumar, A. S., & Huang, S.-T. (2017). Core-shell heterostructured multiwalled carbon nanotubes@reduced graphene oxide nanoribbons/chitosan, a robust nanobiocomposite for enzymatic biosensing of hydrogen peroxide and nitrite. *Scientific reports*, *7*(1), 11910. https://doi.org/10.1038/s41598-017-12050-x

Sadeghi, M. H., Tofighy, M. A., & Mohammadi, T. (2020). One-dimensional graphene for efficient aqueous heavy metal adsorption: Rapid removal of arsenic and mercury ions by graphene oxide nanoribbons (GONRs). *Chemosphere*, *253*, 126647. https://doi.org/10.1016/j.chemosphere.2020.126647

Sajid, M. M., Shad, N. A., Javed, Y., Khan, S. B., Zhang, Z., Amin, N., & Zhai, H. (2020). Preparation and characterization of Vanadium pentoxide (V_2O_5) for photocatalytic degradation of monoazo and diazo dyes. *Surfaces and Interfaces*, *19*, 100502. https://doi.org/10.1016/j.surfin.2020.100502

Shaban, M., Ashraf, A. M., & Abukhadra, M. R. (2018). TiO_2 nanoribbons/carbon nanotubes composite with enhanced photocatalytic activity; fabrication, characterization, and application. *Scientific reports*, *8*(1), 1–17.

Shinde, P. V., Tripathi, A., Thapa, R., & Sekhar Rout, C. (2022). Nanoribbons of 2D materials: A review on emerging trends, recent developments and future perspectives. *Coordination Chemistry Reviews*, *453*, 214335. https://doi.org/10.1016/j.ccr.2021.214335

Sinitskii, A., Dimiev, A., Kosynkin, D. V., & Tour, J. M. (2010). Graphene nanoribbon devices produced by oxidative unzipping of carbon nanotubes. *ACS Nano*, *4*(9), 5405–5413. https://doi.org/10.1021/nn101019h

Sivakumar, M., Towata, A., Yasui, K., Tuziuti, T., Kozuka, T., & Iida, Y. (2010). Dependence of sonochemical parameters on the platinization of rutile titania–an observation of a pronounced increase in photocatalytic efficiencies. *Ultrasonics Sonochemistry*, *17*(3), 621–627.

Sun, C., Li, T., Wen, W., Luo, X., & Zhao, L. (2020). ZnSe/CdSe core–shell nanoribbon arrays for photocatalytic applications. *CrystEngComm*, *22*(5), 895–904.

Sun, Z., Chen, D., Chen, B., Kong, L., & Su, M. (2018). Enhanced uranium(VI) adsorption by chitosan modified phosphate rock. *Colloids and Surfaces A: Physicochemical and Engineering Aspects*, *547*, 141–147. https://doi.org/10.1016/j.colsurfa.2018.02.043

Tang, Y.-B., Yin, L.-C., Yang, Y., Bo, X.-H., Cao, Y.-L., Wang, H.-E., . . . Lee, C.-S. (2012). Tunable band gaps and p-type transport properties of boron-doped graphenes by controllable ion doping using reactive microwave plasma. *ACS Nano*, *6*(3), 1970–1978. https://doi.org/10.1021/nn3005262

Tao, J., Deng, J., Dong, X., Zhu, H., & Tao, H.-j. (2012). Enhanced photocatalytic properties of hierarchical nanostructured TiO_2 spheres synthesized with titanium powders. *Transactions of Nonferrous Metals Society of China*, *22*(8), 2049–2056. https://doi.org/10.1016/S1003-6326(11)61427-1

Tian, Y., Li, H., Ruan, Z., Cui, G., & Yan, S. (2017). Synthesis of $NiCo_2O_4$ nanostructures with different morphologies for the removal of methyl orange. *Applied Surface Science*, *393*, 434–440. https://doi.org/10.1016/j.apsusc.2016.10.053

Uddin, M. S., Das, H. T., Maiyalagan, T., & Elumalai, P. (2018). Influence of designed electrode surfaces on double layer capacitance in aqueous electrolyte: Insights from standard models. *Applied Surface Science*, *449*, 445–453. https://doi.org/10.1016/j.apsusc.2017.12.088

Vo, T. H., Shekhirev, M., Kunkel, D. A., Morton, M. D., Berglund, E., Kong, L., . . . Sinitskii, A. (2014). Large-scale solution synthesis of narrow graphene nanoribbons. *Nature Communications*, *5*(1), 3189. https://doi.org/10.1038/ncomms4189

Wang, B., Zhao, M., Li, L., Huang, Y., Zhang, X., Guo, C., . . . Zhang, H. (2019). Ultra-thin metal-organic framework nanoribbons. *National Science Review*, *7*(1), 46–52. https://doi.org/10.1093/nsr/nwz118

Wang, C.-Z., Lan, J.-H., Wu, Q.-Y., Luo, Q., Zhao, Y.-L., Wang, X.-K., . . . Shi, W.-Q. (2014). Theoretical insights on the interaction of uranium with amidoxime and carboxyl groups. *Inorganic Chemistry*, *53*(18), 9466–9476. https://doi.org/10.1021/ic500202g

Wang, H., Wang, H. S., Ma, C., Chen, L., Jiang, C., Chen, C., . . . Wang, X. (2021). Graphene nanoribbons for quantum electronics. *Nature Reviews Physics*, *3*(12), 791–802.

Wang, K., Wu, X., Zhang, G., Li, J., & Li, Y. (2018). $Ba_5Ta_4O_{15}$ nanosheet/$AgVO_3$ nanoribbon heterojunctions with enhanced photocatalytic oxidation performance: Hole dominated charge transfer path and plasmonic effect insight. *ACS Sustainable Chemistry & Engineering*, *6*(5), 6682–6692. https://doi.org/10.1021/acssuschemeng.8b00477

Wang, X., Li, X., Zhang, L., Yoon, Y., Weber, P. K., Wang, H., . . . Dai, H. (2009). N-doping of graphene through electrothermal reactions with ammonia. *Science*, *324*(5928), 768–771. https://doi.org/10.1126/science.1170335

Wang, Y., Chen, X., Hu, X., Wu, P., Lan, T., Li, Y., . . . Chew, J. W. (2021). Synthesis and characterization of poly(TRIM/VPA) functionalized graphene oxide nanoribbons aerogel for highly efficient capture of thorium(IV) from aqueous solutions. *Applied Surface Science*, *536*, 147829. https://doi.org/10.1016/j.apsusc.2020.147829

Wang, Y., Wang, Z., Ang, R., Yang, J., Liu, N., Liao, J., . . . Tang, J. (2015). Synthesis of amidoximated graphene oxide nanoribbons from unzipping of multiwalled carbon nanotubes for selective separation of uranium(vi). *RSC Advances*, *5*(108), 89309–89318. https://doi.org/10.1039/C5RA15977F

Wei, Z., Feng, L., Zhi-Ming, J., Xiao-Bo, S., Peng-Hui, Y., Xue-Ren, W., . . . Liang-Sheng, L. (2014). Efficient plasmonic photocatalytic activity on silver-nanoparticle-decorated $AgVO_3$ nanoribbons. *Journal of Materials Chemistry A*, *2*(33), 13226–13231.

Wen, H.-F., Yang, C., Yu, D.-G., Li, X.-Y., & Zhang, D.-F. (2016). Electrospun zein nanoribbons for treatment of lead-contained wastewater. *Chemical Engineering Journal*, *290*, 263–272. https://doi.org/10.1016/j.cej.2016.01.055

Wu, P., Wang, Y., Hu, X., Yuan, D., Liu, Y., & Liu, Z. (2019). Synthesis of magnetic graphene oxide nanoribbons composite for the removal of Th(IV) from aqueous solutions. *Journal of Radioanalytical and Nuclear Chemistry*, *319*(3), 1111–1118. https://doi.org/10.1007/s10967-018-6375-2

Xing, M., Fang, W., Yang, X., Tian, B., & Zhang, J. (2014). Highly-dispersed boron-doped graphene nanoribbons with enhanced conductibility and photocatalysis. *Chemical Communications*, *50*(50), 6637–6640. https://doi.org/10.1039/C4CC01341G

Xiu, T., Liu, Z., Wang, Y., Wu, P., Du, Y., & Cai, Z. (2019). Thorium adsorption on graphene oxide nanoribbons/manganese dioxide composite material. *Journal of Radioanalytical and Nuclear Chemistry*, *319*(3), 1059–1067. https://doi.org/10.1007/s10967-019-06417-9

Xu, X., Meng, L., Li, Y., Sun, C., Yang, S., & He, H. (2019). Bi_2S_3 nanoribbons-hybridized {001} facets exposed Bi_2WO_6 ultrathin nanosheets with enhanced visible light photocatalytic activity. *Applied Surface Science*, *479*, 410–422. https://doi.org/10.1016/j.apsusc.2019.02.086

Ye, L., Deng, Y., Wang, L., Xie, H., & Su, F. (2019). Bismuth-based photocatalysts for solar photocatalytic carbon dioxide conversion. *ChemSusChem*, *12*(16), 3671–3701. https://doi.org/10.1002/cssc.201901196

Yu, L., Zou, R., Zhang, Z., Song, G., Chen, Z., Yang, J., & Hu, J. (2011). A Zn_2GeO_4-ethylenediamine hybrid nanoribbon membrane as a recyclable adsorbent for the highly efficient removal of heavy metals from contaminated water. *Chemical Communications*, *47*(38), 10719–10721.

Zhang, B., He, X. C., Gao, M. J., Ma, X. F., & Li, G. (2015). Entanglement of CeO_2 nanorods and graphene nanoribbons and their properties studies of nanocomposites. *Materials Science Forum*, *814*, 153–160. https://doi.org/10.4028/www.scientific.net/MSF.814.153

Zhang, B., Zheng, J., Ma, L. W., Guo, B., He, X. C., Gao, M. J., . . . Li, G. (2016). Charge behavior of low-dimensional V_2O_5/graphene nanoribbons oxides nanocomposites under irradiation of visible light and its application. *Materials Science Forum*, *847*, 203–210. https://doi.org/10.4028/www.scientific.net/MSF.847.203

Zhao, W., Guo, Y., Wang, S., He, H., Sun, C., & Yang, S. (2015). A novel ternary plasmonic photocatalyst: Ultrathin g-C_3N_4 nanosheet hybrided by Ag/$AgVO_3$ nanoribbons with enhanced visible-light photocatalytic performance. *Applied Catalysis B: Environmental*, *165*, 335–343. https://doi.org/10.1016/j.apcatb.2014.10.016

13 Beneficial Bacteria in Restoring the Soil Contamination Caused due to Intense Agrochemicals

Dhayalan Vaithiyanathan, Karuppasamy Sudalaimuthu, and Rajchandar Padmanaban

13.1 INTRODUCTION

Globally, lands are classed into three different groups; namely, agricultural, forestry and other lands. A global statistical report reveals that about 54% of Asian land has been utilized for agricultural practices: 45% in Oceania, 38% in Africa, 30% in America and 21% in European countries. Forestry lands are converted to agricultural lands for food sustainability in Africa and America. Asia and European countries still retain the agricultural land without any expansion. Thus, it is apparent from the global statistics that most of land globally has been utilized for agriculture and agro-based activities. Table 13.1 reveals that the usage of agrochemicals has been tremendously increased since the year 2000. This adversely affects the soil fertility, causing the land to be unsuitable for cultivation, which is reflected in the breakage of the continuous supply of the human food chain (Giller et al., 2021). Consumption of chemical fertilizers globally increased in order to attain crop productivity and yield in a short stretch of time, where the percentage of using organic fertilizers gets reduced, as represented in Figure 13.1. It is specifically identified that three major chemical fertilizers have been utilized in agricultural lands globally in the forms of N for nitrogen nutrients, P_2O_5 for Phosphorous and K_2O for potassium nutrients. A corrective measure to overcome these issues is to find a better alternative to existing agricultural practices; more specifically, utilization of agrochemicals.

Long-term utilization of agrochemicals destroys the physical and biological characteristics of soil, causing severe contamination that, in turn, affects agricultural activity and human health (Lal, 2020). Agrochemicals tremendously improve crop productivity and yield, which meets the needs of raising global population; nevertheless, its adverse effects disrupting the entire soil ecosystem is a long-term concern (Arévalo-Gardini et al., 2015). Extensive works in agrochemicals disclose the adverse effects of using them as fertilizers. Effects of those fertilizers are not found promptly by virtue

DOI: 10.1201/9781003258377-13

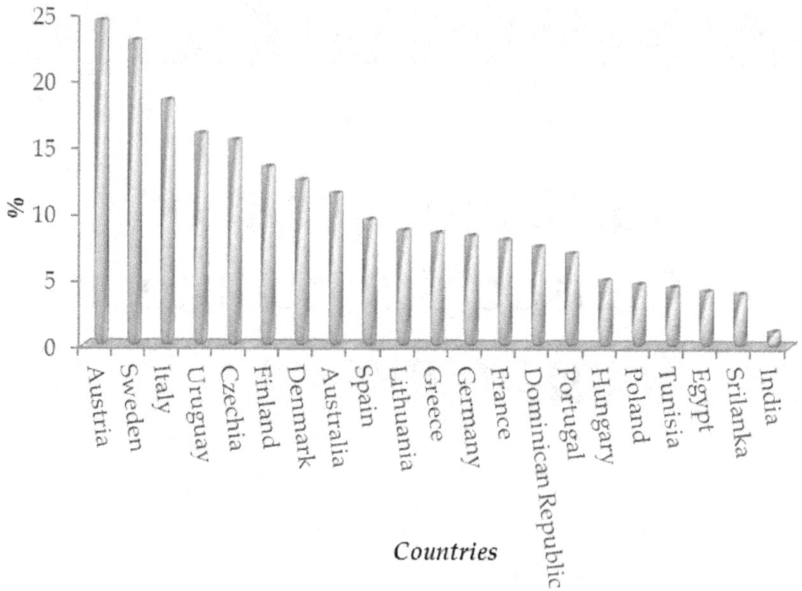

FIGURE 13.1 Land utilized for agricultural practices in top countries (2018).

of robust buffering competency of soil constituents (Dorney, 2012). Such a pivotal soil structure gets deteriorated by high-level, concentrated chemical fertilizers, peculiarly fertilizers consisting potassium and sodium at a high dosage. Certain agrochemicals induce acid formation that leads to a decrease in pH level (Jeschke, 2016). Incessant utilization of these types of chemical fertilizers initially multiplies the crop yield and causes unforeseen drops in crop cultivation as well as in soil quality. Accumulation of salt components of agrochemicals results in gradual magnification of soil contamination sites. Hence, there is a great need to reclaim and restore the contaminated soil.

Bacterial biodiversity plays a major part in soil that gives a pathway for the nutrient intake by plants. Preserving the soil bacterial diversity is highly imperative (Dubey et al., 2019). This has been an increasing problem globally that makes tremendous changes in bacterial flora present in the soil. The dehydrogenase activity and biomass of the bacterial community present in soil eventually decreases due to continuous utilization of agrochemicals (Muñoz-Leoz et al., 2011). Multiplication in the bacterial community tremendously restores the contaminated site. Thus, the bacterial community and soil health are firmly interdependent on each other. Toxicity induced by the agrochemicals can be remediated through employing microorganism as a bioremediator (Castro et al., 2017). Malignancy of chemical fertilizers can be reduced through lowering the biological elements present in the agrochemicals by the process of bacterial remediation. Bacterial remediation includes a coterie of beneficiary bacteria that reduces the intensity of pollutants. Soil ecology will remain undisturbed through utilization of a bacterial remediation process in view of the fact that the remedial employment of microbes causes no harm to the environment. Bacterial remediation is the best-suited recommendation for restoration of contaminated soil sites rather than choosing other technical restoring methods (Hashim et al., 2011). A well-balanced,

TABLE 13.1
World Agrochemical Usage from 2000 to 2010 and Its Increase Percentage

Country	Year	Nitrogen as N	Increase %	Phosphorus as P_2O_5	Increase %	Potassium as K_2O	Increase %
World	2000	80,727.6	34.5	32 375.5	25.5	21 626.9	79.6
	2018	1,08,658.0		40 647.9		38 854.0	
Africa	2000	2,444.4	75.8	918.9	61.9	458.9	91.8
	2018	4,298.9		1 487.9		880.6	
Americas	2000	17,414.5	415.1	8 142.2	58.4	8 229.4	63.6
	2018	24 573.8		12 903.3		13 466.9	
Asia	2000	46,468.5	35.6	17 643.7	19.6	7 8115.1	100
	2018	63,053.0		21 113.4		19 678.8	
Europe	2000	13,210.5	12.2	4 0915.1	0 (-5.3)	4 778.4	0 (-8.7)
	2018	14,827.3		3 873.4		4 359.9	
Oceania	2000	1,189.7	60.1	1 579.5	0 (-19.6)	349.0	34
	2018	1,904.9		1 269.9		467.8	

augmented mixture of organic manure nourishes the bacterial community that, in turn, uplifts plant productivity through various hormone productions. Although the utilization of organic manure is a sluggish process in attaining complete crop productivity, it does completely preserve the soil from prolonged contamination. The breakdown of organic pollutants, effortless techniques, low cost, low manpower and natural remedy are the key functionalities of the bacterial remediation technique (Banerjee & Choudhury, 2022). Prime stages of bacterial remediation include isolation of bacteria, culturing the bacteria, sequencing identification of bacteria, formulation of identified bacteria and sprinkling the bio-fertilizers consisting of the bacteria. Beneficial bacteria accumulate in the soil and trigger their hormones and efficiency by multiplying themselves against pollutants that are the root cause of soil contamination (Choudhary et al., 2018). This chapter details the effects of agrochemicals in soil contamination, various beneficial bacterias' role in inducing soil health and its activities in restoring the contaminated soil.

13.2 SOIL CONTAMINATION

Soil contamination is generally explained as the presence of chemicals or other particles at a more excessive rate than the usual concentration that poses serious threats to human and environmental safety (Ajibade et al., 2021). The chemical particles are either organic or inorganic compounds, but some of the emerging chemical contaminants are beyond human categorizations. Based on the detailed study of soil contaminants, the intrinsic characteristics of soil contamination and its adverse effects can be assessed. Determining the point source of soil contamination is a pivotal factor to understand the contamination pattern so as to make remediation measures. Anthropogenic elements and compounds that naturally occur in parent rock is normally known as inorganic soil contaminants. This includes trace elements, radio nuclides and asbestos. The trace elements are generally found at very low concentrations in the environment but are highly toxic to the organism associated with them. This also includes the heavy metals (copper (Cu), zinc (Zn), mercury (Hg), tin (Sn), lead (Pb), cadmium (Cd), cobalt (Co), nickel (Ni) and chromium (Cr)) and non-metals (selenium (Se), antimony (Sb) and arsenic (As)). The aforementioned trace elements are incapable of being deteriorated by the metabolic process. Naturally, the trace-element contaminants are in the form of salts, oxides, ions and sulfides. Other sources of contaminants contribute equally to the inorganic such as agricultural sources, industrial sources, urban sources, sewer sources, mining and smelting sources, nuclear sources, deforestation, biological agents, acid rain and so on.

Agricultural practices that involve the usage of inorganic compounds on crops and fields include pesticides, agrochemical fertilizers, herbicides and fungicides. The chemical compounds used to prepare the herbicides, pesticides and other insecticides are non-biodegradable in nature. Thus, it settles in the soil in the form of salts or reacts with the soil molecules and gets shaped as highly toxic soil contaminants. These contaminants readily collapse the entire soil structure, composition and pH. Indiscriminate utilization of agrochemical fertilizers due to the negligence of farmers contaminates the soil. For example, agrochemical fertilizers are manufactured having rock phosphate minerals as the main constituent, which possesses traces of lead, cadmium and asbestos. Such toxic elements accumulate substantially in the soil and later become a great threat to the environment and human health when the accumulated toxic contaminants get a response from the atmosphere. In addition to agricultural sources, industrial sources in various forms contribute to soil contamination. It may be a gas, liquid or solid substance. The industrial by-products, when combined with other point sources of soil contamination, get highly toxic in nature. Following the industrial source, developed and developing countries, due to incredible development in technology, dump the waste that is the subject of a separate, huge study by the upcoming researchers on soil contamination. Waste disposal that includes all types of waste such as garbage, household waste, industrial waste, commercial waste, agro-based waste, e-waste and so on contributes in huge part to soil contamination. Specifically, when it is dumped in the agricultural sites, the soil gets contaminated, and the agricultural land becomes unfit for cultivation any more. Moving to the sewer source of soil contamination, the sewer sludge disposed from commercial and domestic sites accumulates in the soil. The pollutants present in the sludge in various forms interact with the soil, thus posing the risk of soil contamination. Mineral deposits during soil erosion, harmful dust particles, heavy metal release from the mining sites, exposure of heavy radiation due to radionuclides such as radon-222 and radium-226, uranium, thorium, isotopes of potassium and carbon to the soil increase the toxic nature of soil; the mixture of sulfur dioxide and nitrogen oxide in the form of acid rain contaminates the soil; biological agents consisting of the excreta of birds and deforestation are major contributors to soil contamination.

13.2.1 Agricultural Soil

Throughout the world, agriculture is the paramount to all activities. Soil is the salient element that forms the basis for agricultural activities. The nutrients enriched in the soil help the crops to grow. Nutrients flow through the soil to the plants and nourish the activities of plant growth. Indirectly, the animals, birds, human beings and other organisms that make use of plants get the nutrients of the soil in different forms. Thus, soil and its nutrients either directly or indirectly play a pivotal role in the biotic environment. It is evident from the aforementioned statement that the healthiest soil feeds the healthy nutrients to the world population. Over the years, the nutrients in the soil have been found to be declining continuously. This is because of the ceaseless salt deposits in the soil, which is commonly termed as salinization. Water holding salt content gets deposited in the soil and produces salinization. Water having salt content is due to various natural as well as man-made activities. Besides, the chemical constituents that are capable of being accumulated for a long period start contaminating the soil from the day of their reaction with the environment. Specifically, in agricultural fields, the soil comes in contact with agro-chemicals in the form of fertilizers, pesticides, herbicides and so on, which makes the soil an accumulated site of toxic chemicals. Contaminated sites get lowers in soil health, thus causing a deficiency in mass production of nutritious foods. By nature, the soil nutrients get decomposed by the microorganisms sustained in the roots of the plants, thus inducing the nutrients' synthesis and uptake by the plant. Hence, the nutritional cycle is highly influenced by the beneficial microorganisms existing in the soil. Therefore, the survival of microorganisms in the soil is directly proportional to the existence of the nutrient level of the soil. Applying continuous chemical fertilizers makes the nutrient levels decrease at a certain point due to the unfavorable conditions that are created for microbial survival. Keeping the microbes undisturbed retains the soil nutrients. For agricultural global elongation, organic farming is essential that helps preserve the soil in years to come.

13.2.2 Agrochemicals in Soil

Although there are various sources contributing to soil contamination, the major source is found to be agro-based chemicals, since global statistics reveal that more than 50% of total global land was occupied for agriculture and its dependent activities. Modern agricultural practices and changes in civilizations lead to indulging in the utilization of hybrid seeds, chemical fertilizers, herbicides and pesticides. Although their employment in agricultural practices increases the productivity in the short term, it poses the threat of soil contamination and leads to an unfavorable environment. Agrochemicals are man-made compounds that consist of large nutrient dosages in various forms that are toxic to soil ecology. Basically, the agrochemicals are introduced with more restrictions and guidelines in global market. But due to the negligence of farmers and other consumers, it leads to excess utilization that contaminates the soil and its environment. Nowadays, fertilizers and other insecticides are unavoidable in agricultural practices. Nevertheless, if this persists, agricultural sustainability will become a great question mark. Although the current agrochemicals fulfill the global need for crop productivity, future soil quality and its capability of crop production is less feasible. Chemical fertilizer usage is gradually increased in many developing countries in order to fulfill their food needs, as represented in Table 13.2.

TABLE 13.2
World Chemical Fertilizers Usage (Top Countries)

Country	Chemical fertilizers (Kg/ha)
Netherlands	665.6
Egypt	624.8
Japan	373.2
China	3015.5
Britain	287.5
Germany	205.4
France	180.1
USA	160.8
Italy	126.4
India	1215.4
Greece	115.4
Indonesia	106.9
Turkey	100.4

Agrochemicals are primarily classed as nitrogenous, phosphorous and potassium fertilizers. In nitrogenous fertilizers, ammonium sulfate, ammonium chloride and amide nitrogen are present in the terms of urea (Table 13.3). The nitrate form as well as ammoniacal forms exist in the nitrogenous fertilizer that helps to meet the nitrogen deficiency, whereas concerning phosphorous fertilizers, the phosphorous nutrients required for the plants will exist in the form of available phosphate. For potassium fertilizers, the plant-required potassium nutrients will be fulfilled in the form of potassium chloride, commonly termed as muriate and potash sulfate. Chemical fertilizers possess the nutrients' availability, and the plants take in the nutrients through various chemical forms. China ranks first after to USA in fertilizer production, where India ranks third. As per the FAO analysis in between 2015 and 2030, there will be a 138-million-ton increment in fertilizer consumption with an annual increment of 1%.

In addition to three primary chemical fertilizers, the agrochemicals also include pesticides. Pesticides are the fusion of two or more chemical compounds that are widely utilized for controlling pests in agricultural activities. These pesticides come in various forms such as fungicides, herbicides, rodenticides, nematicides and insecticides that indulge in the mass production, processing, storage, transportation and marketing of the products. Globally, there are about 1000 varieties of pesticides commercially that consist of about 800 activated chemical ingredients, and the numbers are still increasing. Increasing utilization of pesticides leads to bio-accumulation of

TABLE 13.3
Nutrients and Their Chemical Forms Absorbed by Plants

Element	Chemical forms of elements absorbed by plants
Primary nutrients	
Nitrogen (N)	NO_3 –, NH_4 +
Phosphorus (P)	$H_2PO_4^-$, HPO_4^{3-}, PO_4^{3-}
Potassium (K)	K^+
Secondary nutrients	
Magnesium (Mg)	Mg^{2+}
Calcium (Ca)	Ca^{2+}
Sulfur (S)	SO_4^{2-}
Micronutrients	
Boron (B)	BO_3^{3-}
Chlorine (Cl)	Cl^-
Copper (Cu)	Cu+, Cu^{2+}
Iron (Fe)	Fe^{3+}
Manganese (Mn)	Mn^{2+}
Molybdenum (Mo)	MoO_4^{2-}
Nickel (Ni)	Ni^{2+}
Zinc (Zn)	Zn^{2+}

pesticides residually that leads to soil contamination to a greater extent. Globally, the utilization of pesticides increases in order to tackle different varieties of pests that attack plants. Thus, day by day deposition of trace elements and other chemical compounds grouped as pollutants give raise to soil contamination.

13.2.3 Heavy Metals in Agrochemicals

Agrochemicals encompass chemical compounds for the nutritional support and pest control of plants. The agrochemical production industries indulge in the use of heavy metals during the production of fertilizers and pesticides for more impact on the soil (Table 13.4). In general, the major heavy metals identified in fertilizers are arsenic (As), cadmium (Cd), copper (Cu), lead (Pb), mercury (Hg) and nickel (Ni). Concerning pesticides, the major heavy metals found are copper, indium, iron, manganese, molybdenum, nickel and cobalt. It has been estimated that 0.002 mg of pesticides are consumed by the plants, on average. The consumed pesticides indirectly affect the products associated with the plant.

13.2.4 Accumulation of Agrochemicals

Heavy metal traces identified in agrochemicals fertilizers accumulate in the plant's tissues. Although some of the microelements such as iron, copper, zinc, nickel, cobalt, selenium, manganese and molybdenum are required by plant growth, at higher concentrations, they turn toxic, leading to contamination of the soil. Cobalt,

TABLE 13.4
Average Intake of 15 Pesticide Chemicals by the Soil

Pesticide	Average (mg)	Range (mg)
TDE	0.012	0.005–0.018
DDT	0.027	0.016–0.041
DDE	0.018	0.011–0.028
Endrin	< 0.001	< 0.001–0.001
PeP	0.002	< 0.001–0.006
Kelthane	0.007	0.002–0.012
Dieldrin	0.005	0.004–0.007
Lindane	0.003	0.001–0.005
Heptachlor epoxide	0.002	0.001–0.003
Diazinon	< 0.001	< 0.001–0.001
2,4-D	0.002	< 0.001–0.005
Aldrin	0.001	< 0.001–0.002
BHC	0.002	0.001–0.004
Malathion	0.007	0.003–0.012
Carbaryl	0.037	None-0.150

copper, nickel and zinc elements are considerably less toxic in nature to plants, and there is less chance of there being soil contamination when they are found at a lower range; once the ranges of the aforementioned heavy metal traces go higher, there is a greater risk of soil being contaminated. A high dosage of these heavy metals reacts with each other and diminishes the microbes present in the root of the plants; thus, the microbial load in the soil gets reduced. This leads to failure in soil structure and composition, leading to heavy contamination. During the absorption phase, heavy-metal ion-throng enters the food chain of soil microbes, causing reduction in soil quality. The prolonged usage of agrochemicals leads to soil contamination that takes more time to restore. Their existence in the soil leads to a variety of transformations or a degradation process. Microbial degradation, photolysis, volatilization, leaching, chemical hydrolysis, surface runoff and chemical conversion will occur due to the exposure of agrochemicals to the soil structure. In addition to that, through the water and air medium, the pesticides and fertilizers accumulate in the soil, which leads to soil contamination.

13.2.5 Effects of Agrochemicals in Soil Contamination

From the day of agrochemical introduction on the global market, the practice of farming collapsed, which led to depletion of soil health in numerous aspects. Soil contamination present at the agricultural site is due to the extensive utilization of chemical fertilizers and pesticides. Since soil health is strongly associated with air and water quality, soil contamination leads to water as well as air pollution, to some extent. Thus, the effect of agrochemicals goes beyond soil contamination. Agrochemical utilization results in the failure of the entire endeavor of soil conservation. Soil degradation shows adverse effects on various physiochemical properties of the soil. Agricultural practices have been modernized through quick productivity and reduced cropping span by the utilization of agrochemicals (Agnolucci et al., 2020). The inorganic fertilizers in the majority hold salts of ammonium, nitrate, phosphate and potassium, usually commercialized through fertilizer industries. The industries extract the salts from the radionuclide such as uranium, thorium and heavy metals, which includes mercury, cadmium, lead, copper and nickel. Industrialization and agro-modernization altogether paved the way for rapid utilization of agrochemicals. In spite of the quick productivity achieved from these fertilizers, the salts accumulate in the soil and cause major soil contamination, making the land not fit for agricultural practices over the years. Sometimes, the contamination may give a major lead to water and air pollution; since agricultural land is subjected to irrigation, water acts as an excellent carrier of agrochemical ingredients, bringing extensive depth to the land and also evaporating the salts into the atmosphere. In general, plants utilize only 50% of the fertilizers, and about 25% react with soil compounds, whereas 10% react with surface as well as underground water; the remaining 15% evaporates, on average. Since the percentage of the fertilizer's reaction with the soil compound is in excess, this leads to massive soil contamination (Savci, 2012). Therefore, there is a great requirement to collapse this inorganic agricultural-practice chain. Finding the best-suited alternatives to agrochemicals is essential in order to restore the contaminated agricultural land sites. Nevertheless, the crop yield and productivity span need to be considered in order to meet global food needs. Biofertilizers are an ideal solution to meet the aforementioned challenges. Biofertilizers, having soil-utilitarian bacteria, generally known as Plant Growth Promoting Rhizobacteria (PGPR), multiply enormously in the soil and defend the factors responsible for soil contamination (Khatoon et al., 2020).

13.3 SOIL MICROBES

In this section, soil microbes and the importance of bacterial utilization in reducing soil contamination is discussed in detail. Soil microbes have assorted utilities in agricultural activities. Microbes are classified as pathogen and non-pathogen microbes based on their characteristics. There are numerous pathogenic microbes that contribute to soil degradation through diminishing plant growth. Non-pathogenic microbes are not harmful to the soil ecosystem or for agricultural practices. Certain beneficial microbes help in the restoration of soil to its normal nature through its idiosyncratic nature of producing natural hormones. Beneficial microbes—more specifically, utilitarian bacteria in restoring contaminated soil—are discussed in forthcoming sections.

13.3.1 Beneficial Microbes

Toxicity intensity gets excessive due to the inorganic pollutants existing in the soil. These types of inorganic contaminants do not undergo any decomposition, either chemically or biologically. Thus, inorganic contaminants are found to be jeopardizing. It is ascertained that certain beneficial microbes accumulate the inorganic pollutants and transform to detox in nature. Because of the fact that, heavy metals cannot be decomposed, but they can be transformed from one oxidation state to another form; through biological processes, the microbes have numerous impacts on restoration of contaminated soil. Repercussions of employing the microbes at the contaminated site result in the conversion of heavy metals more water soluble, less toxic in nature get precipitated, thus reducing their toxicity levels; and volatility thus can be removed from the site. Indulging the microbe depends on bioavailability of microbes, to a greater extent. The heavy metals have the tendency of destroying the microbes in the soil because of their toxicity. Thus, mass colonies of microbes having an intensity greater than the intensity level of heavy metal are essential. Intensity of heavy metal at the contaminated site needs to be assessed in order to load the microbes in huge colonies through mass culturing. By loading the microbes in the specific, contaminated sites, the metal detoxification process gets triggered, thus removing the toxicity from the heavy metals. Among various microbes, bacteria play a significant role in reducing the toxic nature of soil. Researchers insist that the utilization of bacteria is the best-suited restoration agent rather than employing other microbes such as fungi, viruses, algae and protozoa. Other than bacteria, these microbes, although having the tendency to restore the soil at a contaminated site, may cause dangerous repercussions to plants and human beings.

13.3.2 Beneficial Bacteria

Soil-beneficial bacteria detoxify the inorganic contaminants to greater extent through the bio-remediation process. Many bacterial species such as *Pseudomonas putida, Enterobacter cloacae, Bacillus subtilis, Achromobacter xylosoxidans, Acinetobacter baumannii, Acinetobacter calcoaceticus, Aeromonas hydrophila, Arthroder macuniculi, Aspergillus niger, Bacillus aerius, Bacillus altitudinis, Bacillus amyloliquefaciens, Bacillus licheniformis, Bacillus megaterium, Bacillus mucilaginous, Bacillus subtilis, Bacillus thuringenesis, Burkholderia cepacia, Burkholderia gladioli, Enterococcus casseliflavus, Enterococcus gallinarum, Fusarium proliferatum, Lecanicilliump salliotae, Paenibacillus favisporus, Paenibacillus taichungensis, Pseudomonas entomophila, Pseudomonas azotoformans, Bacillus licheniformis, Paenibacillus* spp, *Bacillus circulans, Bacillus edaphicus, Acidithiobacillus ferrooxidans* and *Burkholderia* play a significant part in detoxifying the soil contaminants and as well stabilize the soil health.

13.3.3 Beneficial Bacterial Role in Inducing Soil Health

Soil-utilitarian bacteria accumulate the inorganic substances present in the soil through leaching, thus improving the soil structures (Davison, 1988; Brierley, 1985; Ehrlich, 1990). The utilitarian bacteria are generally referred to as plant-growth-promoting rhizhobacteria, which are responsible for plant-microbe interactions (Burr et al., 1984). Based on their residing characteristics, free-living rhizhobacteria and symbiotic bacteria are the major classifications of PGPR (Khan, 2005). Basically, symbiotic bacteria sustain and produce nodules inside the plant cells with specific structures, whereas free-living rhizhobacteria induce the plant growth from outside the plant without producing any nodules (Gray & Smith, 2005). A species of rhizobium is well-known symbiotic bacteria, which are globally used to fix nitrogen to the soil, especially in leguminous crops (Bottomley & Maggard, 1990; Bottomley & Dughri, 1989). *Klebsiella, Azospirillum* and *Azotobacter* are some of the free living rhizhobacteria that tend to fix the nitrogen in most of the soil on a large scale (Lynch, 1983). A species of bacillus is highly recommended for phosphorous enhancement in the soil (Brown, 1974). *Achromobacter xylosoxidans, Acinetobacter baumannii, Acinetobacter calcoaceticus, Aeromonas hydrophila, Arthroderma cuniculi, Aspergillus niger, Bacillus aerius, Bacillus altitudinis, Bacillus amyloliquefaciens, Bacillus licheniformis, Bacillus megaterium, Bacillus mucilaginous, Bacillus subtilis, Bacillus thuringenesis, Burkholderia cepacia, Burkholderia gladioli, Enterococcus casseliflavus, Enterococcus gallinarum, Fusarium proliferatum, Lecanicillium psalliotae, Paenibacillus favisporus, Paenibacillus taichungensis, Pseudomonas entomophila, Pseudomonas koreensis, Pseudomonas luteola, Pseudomonas simiae, Serratia marcescens, Serratia nematodiphila, Sphingomonas paucimobilis* are some of the bacterial species inducing phosphorous solubilization in the soil (Dhayalan & Sudalaimuthu, 2021). *Pseudomonas azotoformans, Bacillus licheniformis, Paenibacillus* spp, *Bacillus circulans, Bacillus edaphicus, Acidithiobacillus ferrooxidans* and *Burkholderia* are the beneficial bacteria that induce about 30% of potassium solubility to the plant and enrich the soil health through active synthesis of complexolysis, chelation, polysaccharides and acidolysis (Archana et al., 2013; Meena et al., 2015; Saha et al., 2016; Prajapati & Modi, 2016; Subhashini, 2015). Leaching of soil nutrients is one of the primary reasons for contamination. The aforementioned bacterial species tremendously increase the soil-nutrient status, thus reducing the contamination. Plant hormones such as auxins, gibberellins (GA), abscisic acid (ABA), cytokinins (CK), salicylic acid (SA), ethylene (ET), jasmonates (JA), brassinosteroids (BR) and peptides are induced by the bacterial species to build a self-immune system to fight against the pathogenic microbes that tremendously degrade the soil structure and agricultural practices (Yang et al., 2013). Thus, the beneficial bacteria take immense responsibilities in inducing the soil health.

13.4 POTENTIAL OF SOIL BACTERIA IN RESTORING CONTAMINATED SOIL

This section discusses the potential of soil bacteria in restoring contaminated sites (Table 13.5). In comparison to other restoration techniques, the bacterial remediation of soil contamination has been proven as an economically viable and eco-friendly technology. This is because of low initialization cost, less requirements of labor and equipment. Bio transformation, bioaccumulation, biosorption and bioremediation are different mechanisms inhibited by soil bacteria due to their complex structure. The mechanism of cellular metabolism is classed as active method or dependent method and passive method or non-dependent method. Intracellular accumulation of soil contaminants occurs when the pollutants migrate through the cell membrane. In passive methods, the compound or ion of heavy metal interacts with the cell wall ligands of bacteria by physiochemical interactions, thus achieving the removal of soil contaminants. Polysaccharides, lipids and proteins consisting many functional groups such as phosphate, sulphate, carboxyl and amino greatly help in binding with the toxic components. The binding mechanism involves various processes such as ion-exchange, complexation, precipitation and so on that makes the soil contamination-free. In an aqueous medium, physical adsorption takes place through the effects of Vander Waals' forces, in which the metal ions (As^{3+}, Pb^{2+}, Cr^{6+}, Cd^{2+}, Zn^{2+}, Cu^{2+}) exchange occurs with bacterial cellular ions (Mg^{2+}, Ca^{2+}, Na^{+}, K^{+}), thus contributing for bioremediation. Precipitation also takes place during the metabolism that makes the pollutants be released from the soil.

13.4.1 Bacterial Functions in Soil Restoration

Bacterial activities such as urease, acid phosphatase, dehydrogenases, catalase and alkaline phosphates act as the resister to soil contaminants. Soil organic compounds get destroyed due to the increased utilization of superphosphate fertilizers, dithiothreitol (DDT), endosulfan, heptachlor and lindane in the soil (Carvalho et al., 2017; Jayaraj et al., 2016). Eventually, the volatile nature of these chemical fertilizers causes tremendous air pollution, in addition to soil contamination (Bhattacharyya et al., 2020). Moreover, heavy metals make up a major part, along with fertilizers, of soil contamination caused by industrialization and urban growth (Shinwari et al., 2015). Since heavy metals and other inorganic chemicals are non-biodegradable in nature, it makes the soil possess arduous, non-degradable characteristics, leading to contamination. Biofertilizers containing active beneficial bacteria characterized by biodegrading the non-biodegradable heavy metals and inorganic compounds find the best solution (Akhtar et al., 2013; Lim et al., 2014). *Pseudomonas, Achromobacter, Brevibacillus, Kluyvera, Mesorhizobium, Azotobacter, Bacillus, Bradyrhizobium, Rhizobium* and *Xanthomonas* are some of the bacterial genera capable of synthesizing 1-aminocyclopropane-1-carboxylatedeaminase, thus reducing the heavy metal stress in the soil and restoring the contamination sites (Kumar & Puri, 2012). The enzymes such as hydrolases, glutathione and esterases, synthesized by the bacterial genera including *Enterobacter, Gordonia, Klebsiella, Paenibacillus, Azotobacter, Bacillus, Serratia* and *Pseudomonas*, intensively degrade the heavy metals and pesticides that exist in the soil (Shaheen & Sundari, 2013). Bacteriocins, siderophores and antibiotics produced by the bacteria tend to boost the antagonist mechanism to destroy the accumulation of heavy metals and inorganic salts, thus assiduously restoring the soil. Many of the soil-utilitarian bacteria produce kinetins, gibberellins and Indole-3-acetic acid for effective restoration of soil that has been contaminated by plastic and agrochemicals (Ikhwan & Nurcholis, 2020). Certain peculiar characterizations of bacteria in the soil tremendously ameliorate the soil condition and restore the contaminated sites, as represented in Table 13.5.

13.4.2 Formulation of Beneficial Bacteria

Beneficial microbes need to be formulated so as to carry out the bacterial functions in the contaminated sites. Initially, the bacterial genus curing specific heavy metals needs to be identified. Identified beneficial bacteria need to be cultured based on the intensity of heavy metal. If the intensity of heavy metals is found to be massive, the identified bacterial species colonies need to be cultured at mass production. Produced bacteria need to be blind with soils in order to undergo enzymatic activities to restore the soil from contamination (Sahu & Brahmaprakash, 2016). Formulation of biofertlizers with active bacterial species is usually carried out for inducing adequate soil health (Arora et al., 2016). Although there exist different states of formulations, liquid formulations are found to be very effective (Kour et al., 2020). Growth media is essential for formulating the bacteria. Natural growth media such as whey, sludge, compost, etc., can be used for formulation, since they are cost effective and environmental friendly. Apart from growth media, proper inoculation is essential for binding the beneficial bacteria with soil contaminants. In addition, the carrier is essential for transferring the bacterial species in a live state. Perlite, peats, vermiculture, saw dust, compost, coal, clays, soybean and wheat bran are the most widely utilized carriers in bioformulations of

TABLE 13.5
Bacterial Strains Utilized in Restoring Contaminated Soil

Soil contamination factors	Beneficial bacteria	Bacterial activity	Reference
Heavy metal	*Achromobacter* *Azotobacter* *Bacillus* *Bradyrhizobium* *Brevibacillus* *Kluyvera* *Mesorhizobium* *Ochrobactrum* *Pseudomonas* *Psycrobacter* *Ralstonia* *Rhizobium* *Sinorhizobium* *Variovox* *Xanthomonas*	synthesizing 1-aminocyclopropane-1-carboxylate deaminase, hydrolases, glutathione and esterases	Shinwari et al., 2015
pesticides	*Azospirillum* *Azotobacter* *Bacillus* *Enterobacter* *Gordonia* *Klebsiella* *Paenibacillus* *Pseudomonas* *Serratia*	production of reactive oxygen species, phytoalexins, phenolic compounds or pathogenesis-related proteins	Shaheen and Sundari, 2013
Plastics and other non-biodegradable elements	*Azospirillum* *Azotobacter* *Bacillus* *Enterobacter* *Ralstonia* *Rhizobium* *Sinorhizobium*	Synthesize of Kinetins, gibberellins and Indole-3-acetic acid	Ikhwan and Nurcholis, 2020

beneficial bacteria. The bioformulations of beneficial bacteria help to transfer the bacterial species in an active state. Active bacteria get blind with the soil through proper sprinkling over the contaminated sites at a regular interval of time. After two to three weeks, the bacteria get blind with the soil and start interacting with the contaminants as well as with other particles. As a result, the toxic qualities of soil contaminants get reduced due to the enzymatic activities of bacterial species. Formulation plays a significant part in restoration of soil-contaminated sites, as represented in Figure 13.2.

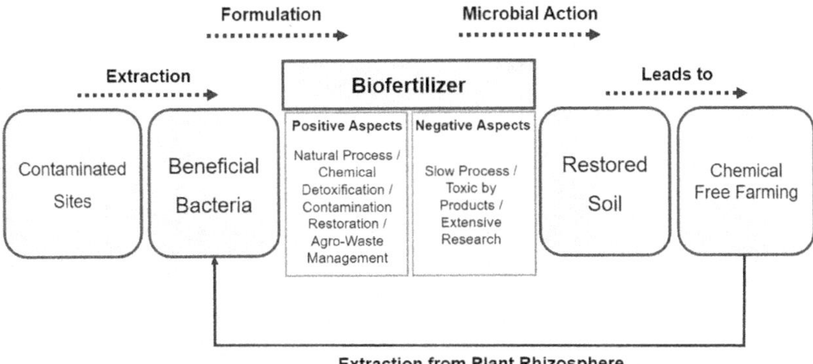

FIGURE 13.2 Beneficial bacteria in restoring contaminated soil.

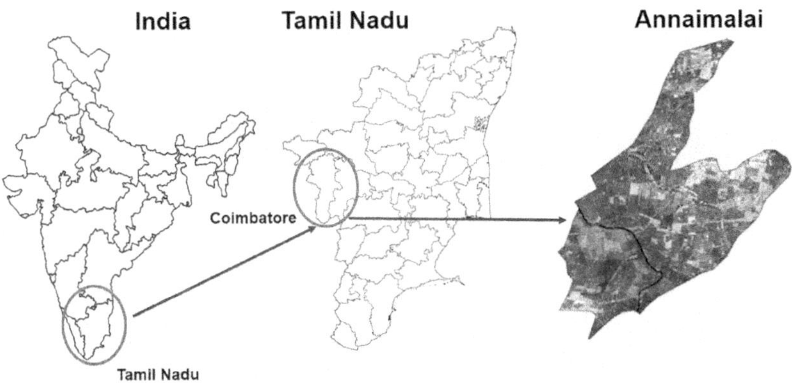

FIGURE 13.3 Experimental study at Anaimalai location of Coimbatore district, Tamilnadu, India.

13.4.3 Investigation of Beneficiary Bacteria in Restoring the Soil Suitably for Agriculture—An Illustrative Example of Anaimalai Block, Coimbatore, India

For practical validation of aforementioned theoretic claims, a small part of a soil-contaminated agricultural site is taken into consideration. Practical implementation on the concept of restoration of the contaminated site by utilizing bacteria is carried out in Anaimalai block having the geographical extent of 10°39' N and 76°49' E, positioned 40 km to the south of Coimbatore district of Tamilnadu, India, which is represented in Figure 13.3. Soil in that area is contaminated by various agrochemicals and became nitrogen-deficient soil. Local farmers suffered a lot, and at one end, the multiple-cropping agro-lands were turned to mono-cropping. Subsequently, the agricultural soils encountered by the pesticides and other agro-chemicals were identified. Since the area is supported by an

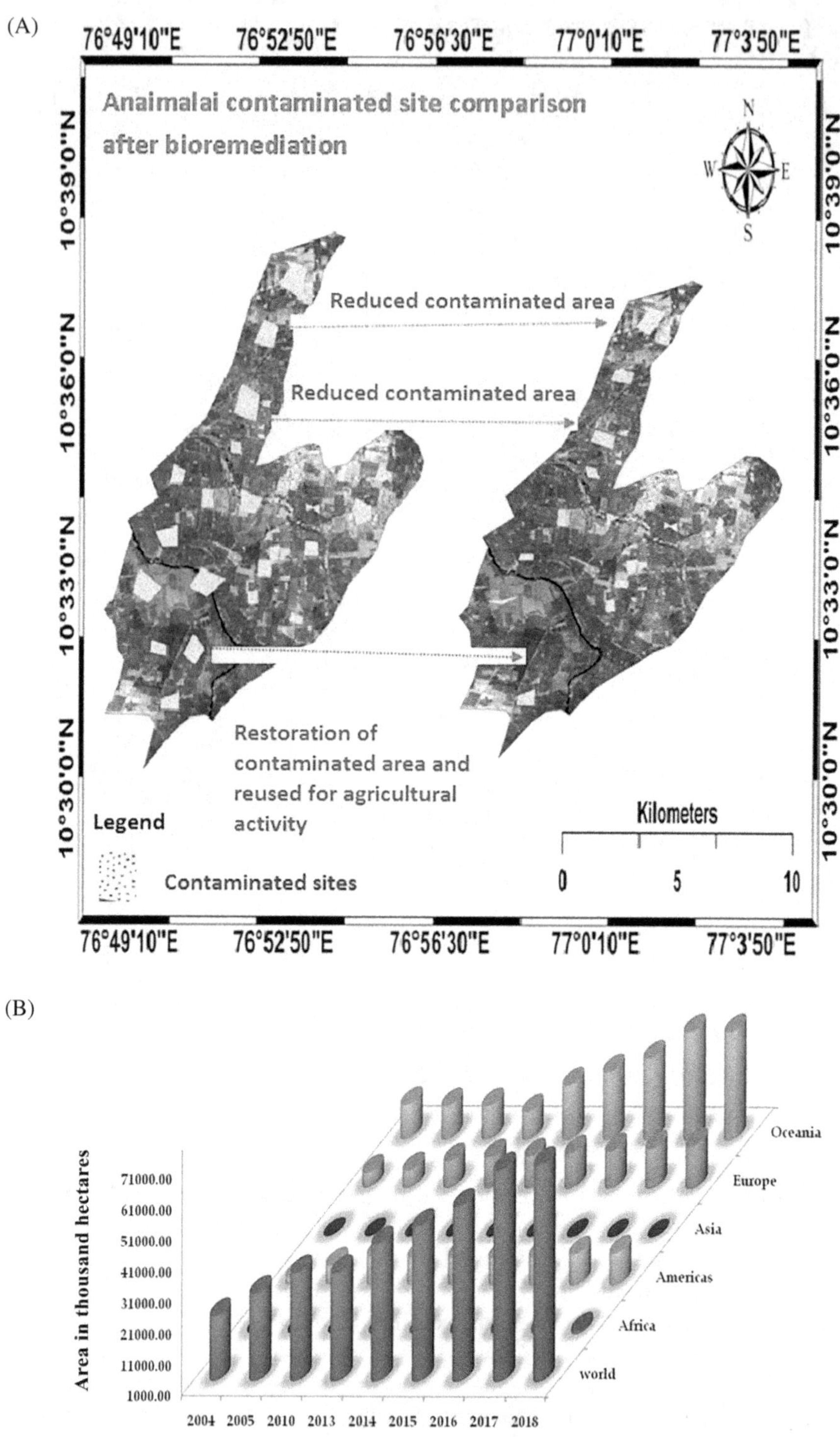

FIGURE 13.4 (A)—Contaminated sites before and after the bioremediation process. (B)—Area under organic agriculture (one thousand ha).

irrigation system, the water carries the agrochemical residuals to each individual parcel of lands.

In order to restore those contaminated agro-lands, bacterial strains such as *Arthrobacter, Aspergillus, Alcaligenes, Azotobacter, Bacillus, Corynebacterium, Flavobacterium, Methosinus, Methanogens, Mycobacterium, Nocardia, Pseudomonas, Pleurotus, Phormidium, Rhodococcus, Stereum* and *Rhizopus* were cultured separately and formulated as individual biofertilizers carrying aforementioned bacterial species. The organic bio-fertilizers consisting of the aforementioned bacterial species were sprinkled over various agro-soils, which were contaminated. Sprinkling dosage is taken as 2 tonne/ha, and the sprinkling time-period is two months at the frequency of twice in a week. The sprinkling was done during spring season under a nominal climate. It is made sure that there is no rainfall during the time of sprinkling. The soil sample was indulged with bacteria was taken. The results were compared and spatially assessed through a kriging technique, which implies that the bacterial species enormously contributes for the restoration of soil-contaminated sites. Contaminated sites underwent a bioremediation process through bacteria sprinkling. The site was compared after six months, as represented in Figure 13.4-A.

13.5 GLOBAL VIEW ON ORGANIC BIO-FERTILIZERS

In order bring quality and sustainability to agro-based products, global countries turned to organic farming in order to avoid chemical-fertilizer consumption. From the global statistical record in 2018, agricultural areas covering 715.1 million hectares were converted to organic farms globally, as revealed in Figure 13.4-B. Twenty-four countries commercially engaged in manufacturing organic bio-fertilizers with bacterial constituents. Almost 78% of the global market was encroached on by the rhizobial inoculants. It is expected that the annual growth rate of utilization of bio-fertilizers will increase by 10% during the period of 2020–2025. The substantial utilization of biofertilizers intensifies the agricultural soil health, thus restoring all the agro-soil-contaminated sites. Although it is a time-intensive process, utilization of biofertilizers increase the soil health and help in restoration of contaminated sites. Asia-Pacific nations attained maximum growth in commercializing the utilization of bacteria that leads to restoration of contaminated sites.

13.6 FUTURE RESEARCH DIRECTIONS

The world looks to the best substitute for agro-chemicals that can restore the contaminated agro-sites and also maintain existing soil health. Future research is on utilization of consortium multi-tarit PGPR strains, which enhance the soil health and protect the soil components from contaminants (Vassilev et al., 2015; Khan et al., 2017). Globally, it is expected to introduce biofilms in protecting the bio inoculation so as to produce mass bacterial colonies (Shaikh et al., 2017; El-Ramady et al., 2018). According to the global research in agro-contamination, the restoration future will be in utilization of nano-particles consisting of interactive bacteria that may termed as nanofertilizers. Nanoencapsulation will be a versatile technology in the future, consisting of bacterial functionalities in restoring the contaminated soil.

ACKNOWLEDGMENT

The authors would like to thank the editors for giving us a great chance to produce our contributions. We the authors are grateful for support in the form of a fellowship and encouragement received from SRM Institute of Science and Technology, Kattankulathur, India. It is pleasure to extend the acknowledgment to Mr. Vickram Muthu Rathinasabari for their valuable field assistance.

REFERENCES

Agnolucci, P., Rapti, C., Alexander, P., De Lipsis, V., Holland, R. A., Eigenbrod, F., & Ekins, P. (2020). Impacts of rising temperatures and farm management practices on global yields of 18 crops. *Nature Food*, *1*(9), 562–571.

Ajibade, F. O., Adelodun, B., Lasisi, K. H., Fadare, O. O., Ajibade, T. F., Nwogwu, N. A., . . . Wang, A. (2021). Environmental pollution and their socioeconomic impacts. In *Microbe mediated remediation of environmental contaminants* (pp. 321–354). Woodhead Publishing.

Akhtar, M. S., Chali, B., & Azam, T. (2013). Bioremediation of arsenic and lead by plants and microbes from contaminated soil. *Journal of Plant Science and Research*, *1*, 68–73.

Archana, D. S., Nandish, M. S., Savalagi, V. P., & Alagawadi, A. R. (2013). Characterization of potassium solubilizing bacteria (KSB) from rhizosphere soil. *Bioinfolet-A Quarterly Journal of Life Sciences*, *10*(1b), 248–257.

Arévalo-Gardini, E., Canto, M., Alegre, J., Loli, O., Julca, A., & Baligar, V. (2015). Changes in soil physical and chemical properties in long term improved natural and traditional agroforestry management systems of cacao genotypes in Peruvian Amazon. *PLoS One*, *10*(7), e0132147.

Arora, N. K., & Mishra, J. (2016). Prospecting the roles of metabolites and additives in future bioformulations for sustainable agriculture. *Applied Soil Ecology*, *107*, 405–407.

Banerjee, A., & Choudhury, M. (2022). Green technology for wastewater treatment and remediation. In C. Baskar, S. Ramakrishna, & A. Daniela La Rosa (Eds.), *Encyclopedia of green materials*. Springer. https://doi.org/10.1007/978-981-16-4921-9_30-1

Bhattacharyya, C., Roy, R., Tribedi, P., Ghosh, A., & Ghosh, A. (2020). Biofertilizers as substitute to commercial agrochemicals. In *Agrochemicals detection, treatment and remediation* (pp. 263–290). Butterworth-Heinemann.

Bottomley, P. J., & Dughri, M. H. (1989). Population size and distribution of Rhizobium leguminosarum bv. trifolii in relation to total soil bacteria and soil depth. *Applied and Environmental Microbiology*, *55*(4), 959–964.

Bottomley, P. J., & Maggard, S. P. (1990). Determination of viability within serotypes of a soil population of Rhizobium leguminosarum bv. trifolii. *Applied and Environmental Microbiology*, *56*(2), 533–540.

Brierley, J. A. (1985, June). Use of microorganisms for mining metals. In H. Halvorson et al. (Eds.), *Engineered organisms in the environment: Scientific issues* (pp. 10–13). Proceedings of a Cross-Disciplinary Symposium Held in Philadelphia, PA.

Brown, M. E. (1974). Seed and root bacterization. *Annual Review of Phytopathology*, *12*(1), 181–197.

Burr, T. J., Caesar, A., & Schrolh, M. N. (1984). Beneficial plant bacteria. *Critical Reviews in Plant Sciences*, *2*(1), 1–20.

Carvalho, T. L. G., Balsemão-Pires, E., Saraiva, R. M., Ferreira, P. C. G., & Hemerly, A. S. (2014). Nitrogen signalling in plant interactions with associative and endophytic diazotrophic bacteria. *Journal of Experimental Botany*, *65*(19), 5631–5642.

Castro, A. A. D., Prandi, I. G., Kuca, K., & Ramalho, T. C. (2017). Organophosphorus degrading enzymes: Molecular basis and perspectives for enzymatic bioremediation of agrochemicals. *Ciencia e Agrotecnologia*, *41*(5), 471–482.

Choudhary, M., Ghasal, P. C., Yadav, R. P., Meena, V. S., Mondal, T., & Bisht, J. K. (2018). Towards plant-beneficiary rhizobacteria and agricultural sustainability. In *Role of rhizospheric microbes in soil: Volume 2: nutrient management and crop improvement* (pp. 1–46). Springer.

Davison, J. (1988). Plant beneficial bacteria. *Bio/technology*, *6*(3), 282–286.

Dhayalan, V., & Sudalaimuthu, K. (2021). Plant growth promoting rhizobacteria in promoting sustainable agriculture. *Global Journal of Environmental Science and Management*, *7*(3), 401–418.

Dorney, R. S. (2012). *The professional practice of environmental management*. Springer Science & Business Media.

Dubey, A., Malla, M. A., Khan, F., Chowdhary, K., Yadav, S., Kumar, A., . . . Khan, M. L. (2019). Soil microbiome: A key player for conservation of soil health under changing climate. *Biodiversity and Conservation*, *28*(8), 2405–2429.

Ehrlich, H. L. (1990). *Geomicrobiology* (2nd ed.). Dekker.

El-Ramady, H., El-Ghamry, A., Mosa, A., & Alshaal, T. (2018). Nanofertilizers vs. biofertilizers: New insights. *The Environment, Biodiversity & Soil Security*, *2*, 40–50.

Giller, K. E., Hijbeek, R., Andersson, J. A., & Sumberg, J. (2021). Regenerative agriculture: An agronomic perspective. *Outlook on Agriculture*, *50*(1), 13–25.

Gray, E. J., & Smith, D. L. (2005). Intracellular and extracellular PGPR: Commonalities and distinctions in the plant–bacterium signaling processes. *Soil Biology and Biochemistry*, *37*(3), 395–412.

Hashim, M. A., Mukhopadhyay, S., Sahu, J. N., & Sengupta, B. (2011). Remediation technologies for heavy metal contaminated groundwater. *Journal of Environmental Management*, *92*(10), 2355–2388.

Ikhwan, A., & Nurcholis, M. (2020, February). Bacteria analysis as plastic biodegradation agent and biofertilizer. In *IOP conference series: Earth and environmental science* (Vol. 458, No. 1, p. 012017). IOP Publishing.

Jayaraj, R., Megha, P., & Sreedev, P. (2016). Organochlorine pesticides, their toxic effects on living organisms and their fate in the environment. *Interdisciplinary Toxicology*, *9*(3–4), 90.

Jeschke, P. (2016). Propesticides and their use as agrochemicals. *Pest Management Science*, *72*(2), 210–225.

Khan, A. G. (2005). Role of soil microbes in the rhizospheres of plants growing on trace metal contaminated soils in phytoremediation. *Journal of Trace Elements in Medicine and Biology*, *18*(4), 355–364.

Khan, W. U., Ahmad, S. R., Yasin, N. A., Ali, A., Ahmad, A., & Akram, W. (2017). Application of Bacillus megaterium MCR-8 improved phytoextraction and stress alleviation of nickel in Vinca rosea. *International Journal of Phytoremediation*, *19*(9), 813–824.

Khatoon, Z., Huang, S., Rafique, M., Fakhar, A., Kamran, M. A., & Santoyo, G. (2020). Unlocking the potential of plant growth-promoting rhizobacteria on soil health and the sustainability of agricultural systems. *Journal of Environmental Management*, *273*, 111118.

Kour, D., Rana, K. L., Yadav, A. N., Yadav, N., Kumar, M., Kumar, V., . . . Saxena, A. K. (2020). Microbial biofertilizers: Bioresources and eco-friendly technologies for agricultural and environmental sustainability. *Biocatalysis and Agricultural Biotechnology*, *23*, 101487.

Kumar, M., & Puri, A. (2012). A review of permissible limits of drinking water. *Indian Journal of Occupational and Environmental Medicine*, *16*(1), 40.

Lal, R. (Ed.). (2020). *The soil-human health-nexus*. CRC Press. https://doi.org/10.1201/9780367822736.

Lim, S. P., Pandikumar, A., Huang, N. M., Lim, H. N., Gu, G., & Ma, T. L. (2014). Promotional effect of silver nanoparticles on the performance of N-doped TiO 2 photoanode-based dye-sensitized solar cells. *RSC Advances*, *4*(89), 48236–48244.

Lynch, J. M. (1983). *Soil biotechnology: Microbiological factors in crop productivity*. Blackwell Scientific Publications.

Meena, V. S., Maurya, B. R., Verma, J. P., Aeron, A., Kumar, A., Kim, K., & Bajpai, V. K. (2015). Potassium solubilizing rhizobacteria (KSR): Isolation, identification, and K-release dynamics from waste mica. *Ecological Engineering*, *81*, 340–347.

Muñoz-Leoz, B., Ruiz-Romera, E., Antigüedad, I., & Garbisu, C. (2011). Tebuconazole application decreases soil microbial biomass and activity. *Soil Biology and Biochemistry*, *43*(10), 2176–2183.

Prajapati, K., & Modi, H. (2016). Growth promoting effect of potassium solubilizing Enterobacter hormaechei (KSB-8) on cucumber (Cucumis sativus) under hydroponic conditions. *International Journal of Advanced Research in Biological Sciences*, *3*(5), 168–173.

Saha, M., Maurya, B. R., Meena, V. S., Bahadur, I., & Kumar, A. (2016). Identification and characterization of potassium solubilizing bacteria (KSB) from Indo-Gangetic Plains of India. *Biocatalysis and Agricultural Biotechnology*, *7*, 202–209.

Sahu, P. K., & Brahmaprakash, G. P. (2016). Formulations of biofertilizers–approaches and advances. In *Microbial inoculants in sustainable agricultural productivity* (pp. 179–198). Springer.

Savci, S. (2012). An agricultural pollutant: Chemical fertilizer. *International Journal of Environmental Science and Development*, *3*(1), 73.

Shaheen, S., & Sundari, K. (2013). Exploring the applicability of PGPR to remediate residual organophosphate and carbamate pesticides used in agriculture fields. *International Journal of Agriculture and Food Science Technology*, *4*(10), 947–954.

Shaikh, S., & Saraf, M. (2017). Zinc biofortification: Strategy to conquer zinc malnutrition through zinc solubilizing PGPR's. *Biomedical Journal of Scientific & Technical Research*, *1*(1), 224–226.

Shinwari, K. I., Shah, A. U., Afridi, M. I., Zeeshan, M., Hussain, H., Hussain, J., . . . Jamil, M. (2015). Application of plant growth promoting rhizobacteria in bioremediation of heavy metal polluted soil. *Asian Journal of Multidisciplinary Research & Review*, *3*(4), 179–185.

Subhashini, D. V. (2015). Growth promotion and increased potassium uptake of tobacco by potassium-mobilizing bacterium Frateuria aurantia grown at different potassium levels in vertisols. *Communications in Soil Science and Plant Analysis*, *46*(2), 210–220.

Vassilev, N., Vassileva, M., Lopez, A., Martos, V., Reyes, A., Maksimovic, I., . . . Malusa, E. (2015). Unexploited potential of some biotechnological techniques for biofertilizer production and formulation. *Applied Microbiology and Biotechnology*, *99*(12), 4983–4996.

Yang, D. L., Yang, Y., & He, Z. (2013). Roles of plant hormones and their interplay in rice immunity. *Molecular Plant*, *6*(3), 675–685.

14 Innovations in Nanotechnology for Wastewater Treatment

Annika Durve Gupta, Vandana Gupta, Sonali Zankar Patil, and Goldie Oza

14.1 INTRODUCTION

Water is natural resource on the planet, and its supply in its purest form is critical for life itself, as the notion of existence is unfathomable without it. Water is also known as a universal solvent owing to the potential features such as solubility and power. Water pollution is currently a big concern around the world, owing to a variety of factors such as poor industrial wastes, sewage remediation, marine dumping difficulties, radioactive waste material, some agricultural viewpoints, and so on. (Zhang et al., 2016; Ahmad et al., 2015). Water contamination has a negative impact on the ecosystem, and it may also cause air pollution, which has highly serious consequences for human health. Water contamination has a negative influence on the concerned area's economic growth and social viewpoints.

According to a recent UN study, cleansed and freshwater supply is a worldwide issue that has become a concern for the entire globe in the twenty-first century, since the life of living things is endangered by contaminated water (Hodges et al., 2018; Gitis & Hankins, 2018). Water contamination appears when unwanted pollutants go to bodies of water or reservoirs, making them unsuitable for drinking as well as other purposes. There is an assortment of physical, chemical, and mechanical procedures to coping with this emerging problem. Furthermore, researchers are always developing new technologies to improve low-cost water filtering systems (Umar et al., 2013; Umar et al., 2019; and Rajpal et al., 2022).

Nanotechnology, a newly developing area, has the potential to purify water at a cheap cost, with high working-efficiency in eliminating contaminants and reusability (Banerjee et al., 2021). Nanomaterials have been effectively utilised in a variety of areas in the past, including catalysis, medical research, and so on. Experts have recently found that nanomaterials are a better option for treating wastewater because they have few unique properties, viz., larger surface area, size, strong solution mobility, highly responsive (Wu et al., 2019; Bhardwaj et al., 2022), a porous character, strong mechanical properties, dispersibility, hydrophilicity, and hydrophobicity (Daer et al., 2015; Yaqoob et al., 2019; Tang et al., 2014). Exploiting numerous nanomaterials, heavy metals such as Mo, Pb, and a few organic as well as inorganic contaminants and numerous hazardous microorganisms are being reported to be

DOI: 10.1201/9781003258377-14

effectively eliminated (Mir et al., 2012; Umar, 2018). As per the WHO, over 1.7 million people have died due to pollution of water, and 4 billion patients with different health concerns are recorded yearly because of waterborne illnesses (Briggs et al., 2016). Table 14.1 shows the many different water pollutants, their roots, and their negative consequences.

14.2 NANOMATERIALS

Nanotechnology is the study of research, engineering, and technology at nanoscale, which ranges from 1–100 nanometers. Nanotechnology is the ability to manufacture an extremely tiny substance from the bottom up or the top down. There are two approaches for creating nanoparticles: chemical and physical (Figure 14.1). The physical system employs a "top-down" approach in which force is utilised to

TABLE 14.1
Different Types of Water Contaminants, along with Their Origins and Undesirable Consequences

Water contaminants	Origins	Negative consequences	References
Infectious microorganisms	Bacteria and viruses	Waterborne diseases	Umar et al., 2019
Agricultural contaminants	Agricultural chemicals	Affects the fresh water resources	Tang et al., 2014
Suspended and sediments solids	Land farming, destruction, mining processes	Fish spawning and the aquatic habitat of insects and fishes is being harmed	Richter & Ayers, 2018
Inorganic contaminants	Heavy metals, trace elements, metal compounds, inorganic salts, mineral acids,	Aquatic fauna and flora, Community health problems	Sizmur et al., 2017
Organic contaminants	Herbicides, insecticides, detergents	Aquatic life problems, dysgenic	Wang et al., 2019
Industrial contaminants	Civic pollutant water	Caused air and water pollution	Liu et al., 2018
Radioactive toxins	Diverse isotopes	Damages teeth, skin, bones, and can result in cancers	Bayoumi & Saleh, 2018
Nutrients contaminants	Plant fragments, fertilizer	Effect on eutrophication process	Ma et al., 2018
Macroscopic contaminants	Different marine debris	A large plastic pollution	Longwane et al., 2019
Contaminated and sewage water	Domestic wastewater	Waterborne diseases	Rajasulochana & Preethy, 2016

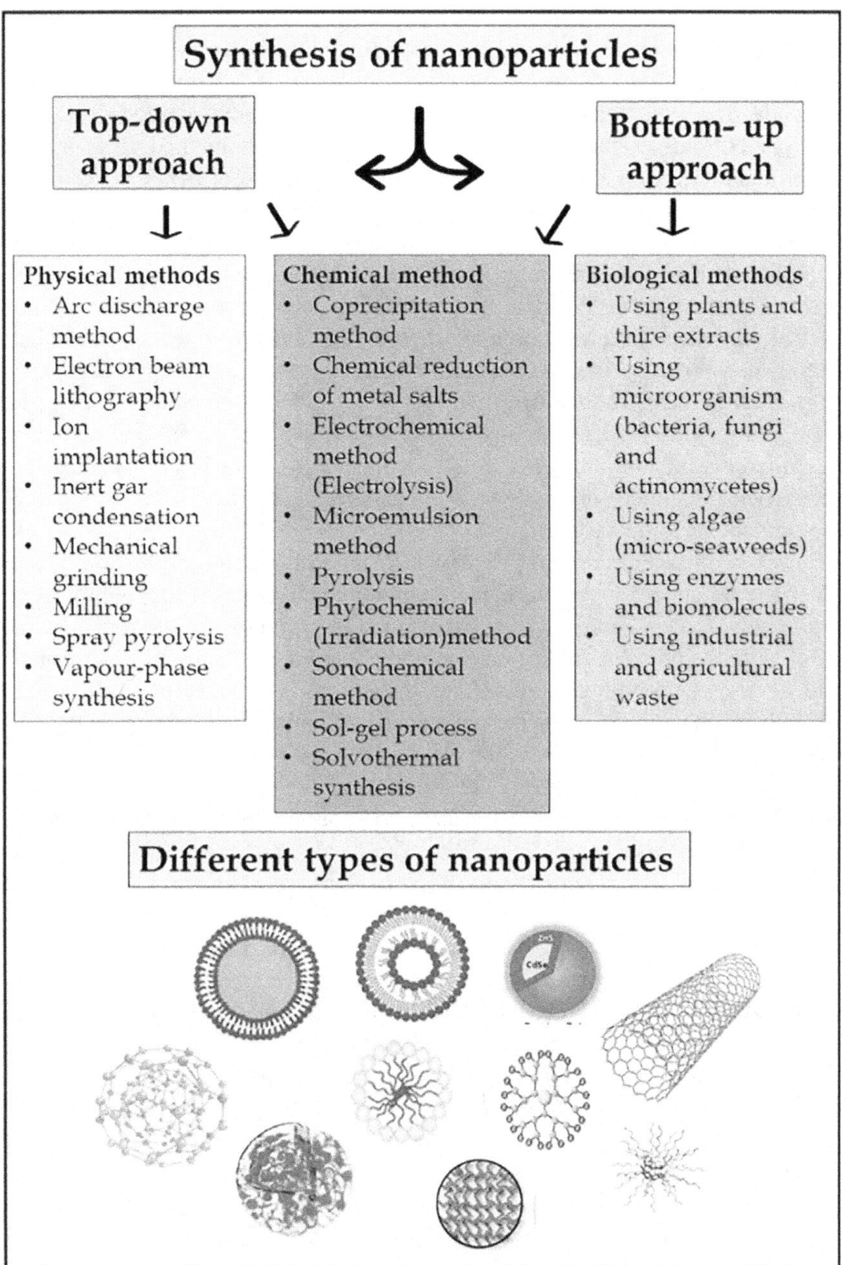

FIGURE 14.1 Different methods for synthesis of nanoparticles.

pulverise the bulk material into minute particle components. These physical techniques are adaptable methods for producing nanoparticles with larger scale, diameter, and volume that are still regulated but create surface defects, corrosion, and are costly and time-consuming. The chemical preparation methodology, which employs a bottom-up approach, is the most frequently used method. Physical methods for nanoparticle synthesis are costly and need the use of heavy materials with high hardness, strength, and other properties that may not be suited for large-scale nanoparticle processing. While chemical techniques are used in the mass production of nanoparticles, the main product generated seems to create pollution, posing a harm to terrestrial and aquatic life and adding to environmental accumulation. Physical and chemical procedures are therefore useful processes, but a biological method incorporating living creatures, bacteria, plants, and so on, also known as the green system, can resolve the problem (Gogoi & Choudhury, 2022). These green solutions include employing particles that are non-hazardous to humans, trees, microorganisms, and enzymes as reducing, capping, and stabilising agents for nanoparticle manufacturing using natural biological agents that are eco-friendly and biodegradable (Figure 14.2-A)

However, since humans first expressed a curiosity in the production of nanostructured materials for various application in many aspects of life, both the quantity of study and its popularity have expanded in an unanticipated way (Albrecht et al., 2006; Rajput, 2015). Agriculture and nutrition, biotechnology, pharmaceuticals, energy, health, environment, and even material science have all benefited greatly from the expansion of the nano sector (Kandasamy & Prema, 2015). The applications of nanomaterials in the cleanup of wastewater and water, in particular, has received a lot of interest. Nanomaterials have greater adsorption capabilities and reactivity, causing their tiny sizes and consequently huge specific surface areas. Heavy metals (Tang et al., 2014; Yan et al., 2015), organic contaminants (Liu et al., 2014; Kalhapure et al., 2015), and microorganisms (Borrego et al., 2016) have all been stated to be efficiently removed by many types of nanomaterials. Numerous studies have demonstrated that nanoparticles have great promise for usage in water and wastewater remediation. At the time, the most commonly researched nanomaterials for water and wastewater remediation are comprised of metal oxide as well as nanoparticles from zero-valent metal, nanocomposites, and carbon nanotubes (CNTs). More advanced breakthroughs in nanomaterials, such as nanomotors, nanophotocatalysts, nanosorbents, and nanomembranes, have happened recently, and certain imprinted polymers are now useful for polluted-water treatment operations. In conclusion, it is believed that research into nanomaterial uses in water filtering will be useful (Sekoai et al., 2019). We attempted to compile a comprehensive chapter of several papers stating applications of nanoparticles in wastewater remediation. Based on published research, the use of nanoparticles in remediation of wastewater may be categorised into four different categories. The latest developments and methods related with each category are attempted to be incorporated in a chronological order. Finally, we discuss the limitations of nanotechnology in wastewater remediation, as well as its conclusion and future directions.

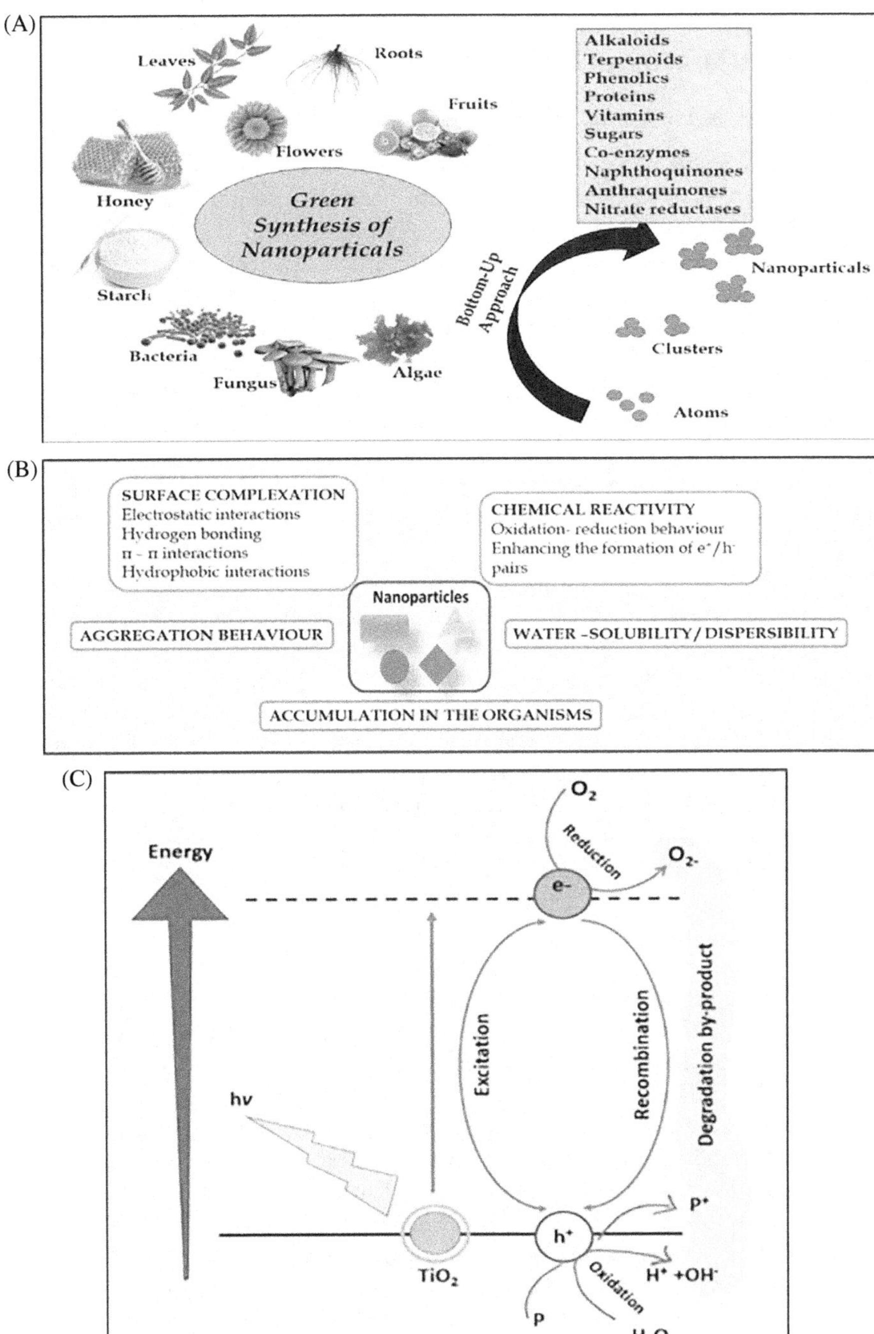

FIGURE 14.2 (A)—Green synthesis of nanoparticles. (B)—Properties and interactions of nanoparticles responsible for wastewater remediation. (C)—Redox reactions seen in TiO_2 (Chen et al., 2016).

14.3 VARIOUS METHODS FOR WASTEWATER REMEDIATION USING NANOMATERIALS

14.3.1 Clean-Up Using Nanomaterials-Adsorption Process

Adsorption is a potential process for eliminating contaminants from wastewater in which the sorbate, or water contaminants, clings to the sorbent surface due to attraction forces. Covalent (chemisorption) also non-covalent (the physiosorption) forces can be involved in adsorption procedure. Nanoparticles are incredibly competent and are being studied for adsorption characteristics because of their great surface areas as well as diffusion distance of small intraparticles (Kalfa et al., 2009). Furthermore, size-dependent surface structure and greater surface energy at nanoscale may result in extremely energetic adsorption sites, which results in greater normalised surface area adsorption capacity. Surface alteration is also straightforward, allowing for the discriminatory adsorption of certain impurities onto the nanoparticles' surface. Once pollutants have been surface adsorbed, they will be desorbed by adjusting the temperature, pH, and other parameters of the solution (Saikia et al., 2011; Saikia et al., 2013; Saha et al., 2011). Pollutant desorption from the surface allows these nanoparticles to be reused, cutting overall V costs. In addition to shape and size, the porosity of these nanoparticles may be altered. Adsorption kinetics can be altered by porous nanomaterials with structure and variable pore size like for, e.g., carbon nanofibres activated by electrospun (Shi et al., 2008). Nanostructured oxides of mixed ions, metallic and metallic oxide nanoparticles, and magnetic nanoparticles (Al_2O_3, TiO_2, MnO_2, ZrO_2, ZnO, MgO, CeO_2) are only a few examples of nanomaterials employed in nano-adsorbents (Gupta et al., 2015). This category now includes carbon nanoparticles, carbon nanosheets, and carbon nanotubes as well as a new class called carbonaceous nanomaterials (CNMs). As nano-adsorbents, various silicon-based nano-materials—e.g., silicon nanoparticles, silicon nanosheets, and silicon nanotubes—are used (Anjum et al., 2016). Nanofibres, aerogels, polymer-based nanomaterials, and nanoclays are also excellent at removal of heavy metals from polluted water. Surface chemistry, size, aggregation state, fractal dimension and shape, chemical composition, solubility, and crystal structure are all factors that influence adsorption properties. Surprisingly, a number of these nanosorbents are now accessible in water purifiers. Nanomaterials made of metal and nonmetal oxides are commonly used to eradicate injurious toxins from waste and polluted waters. Ferric oxide micro-particles are commonly utilised for a range of uses due to the natural availability of iron and small cost of manufacture (Li et al., 2006). Nanoparticles which are based on iron are commonly used to remove different colours (Saha et al., 2011; Istratie et al., 2016) and heavy metals present in wastewater due to their redox potential, high surface charge, and reusability (Li et al., 2003).

Numerous aspects impact the adsorption process, including adsorbent dose, temperature, pH, and cultivation time. To improve adsorption, the surface of Fe_2O_3 nanoparticles is often modified using compounds like amino-propyl-tri-methoxysilane (Palimi et al., 2014). Other properties of nanoscale, zero-valent iron (nZVI) include the capacity to catalyse heavy metal ions reduction to trivalent form, leading in Cr separation as insoluble hydroxides. Because of the magnetic properties of metallic iron and its quick oxidation, nZVI has a robust propensity to consolidate and produce large aggregates. As a result, the surface-active sites are blocked, the response rate is delayed, and the reaction is frequently promptly inactivated. As a result, controlling combination as well as oxidation is vital for developing low-cost nZVI applications. However, iron which is oxidised is released into the solution, and that has its set of disadvantages. A protective covering for the surface of iron nanoparticles is employed to overcome this issue. ZVI nanoparticles, Chitosan-coated, are well-known for the capacity to remove Cr(VI) from wastewater by reducing it to Cr(III) while also creating precipitation with Fe (III).

Magnetic nanoparticles (NPs), viz., Fe_3O_4 and γ-Fe_2O_3 are widely used due to their ease of separation, reusability, greater responsiveness, low cost, and ecological approachability. Magnetic NPs, on the other hand, are prone to corrosion due to their high reactivity in aqueous environments, limiting their lifetime. As a result, coatings such as porous carbon, mesoporous silica, HA, thiosalicylhydrazide, and others are routinely employed. In addition to iron oxides, zinc oxide (ZnO), magnesium oxide (MgO), and manganese oxide (MnO) are found. Heavy metals, viz., Cd, As, and Zn are eliminated far too effectively by NPs. They have greater BET surface areas along with varied nanostructures. MnO NPs are often reformulated as nanotunnel/ nanoporous manganese oxides and hydrous oxides of manganese for water remediation (HMO). Nanoplates, nano assemblies, hierarchical ZnO nano-rods, and microspheres with nano-sheets are utilised in wastewater remediation.

Because of their vast surface area also ability to interact with all pollutants via a variety of interactions—e.g., π-π interaction, covalent, hydrophobic, electrostatic interactions, and hydrogen bonding—carbon nanotubes are extremely efficient in adsorbing diverse organic pollutants. Carbon nanotubes' exterior surface, interstitial channel, inner site, and peripheral groove are all potential adsorption sites (Das et al., 2017). They're also hollowed on the inside, which aids in nanoparticle adsorption. Organic compounds with C=C bonds or benzene rings, polar aromatic substances, polycyclic aromatic hydrocarbons (PAHs), and are readily adsorbed on the surfaces of nanotubes of carbon (Yang & Xing, 2010). Similarly, as a result of hydrogen bond formation with nanotubes of graphitic carbon surface, organic molecules having -$NH_{2,}$ -OH, and -COOH, functionalities are surface adsorbed on carbon nanotubes (Yang et al., 2008). Electrostatic interaction amongst organic contaminants positively charged and carbon nanotubes electron-rich is also possible at an appropriate pH.

Carbon nanotubes are particularly efficient in adsorbing various organic contaminants due to surface area and capability to interact with pollutants via a number of interactions like π-π interaction, hydrophobic, covalent, electrostatic interactions, and hydrogen bonding. It is worth noting that the interstitial channel, inner site, external surface, and groove of peripheral carbon nanotubes are all conceivable adsorption sites (Das et al., 2017). They are hollowed on the inside, which aids in nanoparticle adsorption. Again, nanotubes of (CN) cyano-functionalised carbon are good adsorbents for water which are phenol-type contaminants, viz., 2,4-dichlorophenol (DCP), 4-chlorophenol, 2-naphthol, also 1-naphthol. The altered nanotubes of magnetic carbon have a substantial dispersion ability; also, they may be readily recovered from wastewater and/or utilized with the help of a magnet. The acid treatment, functional group/molecules, metal impregnation, and grafting greatly boosted carbon nanotube dispersion, surface charge, BET surface area, and hydrophobicity, resulting in greater adsorption potentials. To introduce functional groups like oxy on the surface, inorganic acids, viz., HNO_3, HCl, and H_2SO_4 are used. Carbon nanotubes readily eliminate heavy metal ions like Cu (II), Cd (II), Hg (II), and Pb (II), from polluted water. This, again, eliminates any impurities from the carbon nanotubes.

Because carbon nanotubes are more expensive, nanosheets of graphene oxide have developed as an alternate absorbent for both metal and organic pollutants. Graphene is the carbon allotrope which has unique properties; it makes it appropriate for a variety of environmental applications. Graphene oxide nanosheets were formed through chemical oxidation of graphene. To increase heavy metal adsorption, a few selected hydrophilic groups such as carboxylic and hydroxyl acids are induced in graphene oxide. However, no additional acid treatment is required to boost adsorption capacity. Organic water contaminants are adsorbed by hydrophobic, π -π, electrostatic, and hydrogen bonding, whereas inorganic water pollutants are adsorbed by electrostatic, bonding (anion–π and cation–π), and Lewis acid–base interactions; for example, chelating. Metal ions and graphene-based ions are being researched for the adsorption of anionic toxins. Two anions that are effectively adsorbed on graphene-based materials are perchlorate and fluoride. Graphene oxide and its derivatives are particularly good in removing cationic dyes, but graphene and its derivatives are typically competent at removing anionic dyes. The nano-polymers such as dendrimers and nano-cellulose are being investigated as wastewater-remediation adsorbents. Among the finest qualities of nano-cellulose are biocompatibility, biodegradability, and cost-effectiveness. Dendrimers have the ability to absorb both inorganic and organic impurities.

Hydrophobic organic-molecules will be adsorbed by the hydrophobic core of the dendrimer, whereas metal and/or ionic impurities are attracted to the interior and exterior branches via electrostatic and hydrogen bonding. The finest thing is that dendrimer desorption is only required when the pH of solution changes. All over, nano-adsorbents are simple to create, and high adsorption capacity as well as their ability to reuse and regenerate (after pollutant desorption) make them a viable wastewater-treatment option. Contaminants interact with and bind to nanoparticle surfaces due to a variety of surface forces (hydrogen bonding, electrostatic, etc.). The dispersibility or solubility of a nanoparticle in polluted fluids also affects its efficiency. Nanoparticle capacity to form e-/h+ pairs and the oxidation and reduction potential all play a role in nanoparticles in polluted-water remediation. Accumulation and aggregation limit the effective surface area of nanoparticles, lowering their overall effectiveness.

14.3.2 Photocatalytic Water-Remediation Using Nanoparticles

A nano-catalyst in photocatalytic wastewater-remediation uses light energy (generally UV) to break or degrade an extensive array of organic materials, viz., estrogens, dyes, pesticides, organic acids, microbes (like chlorine resistant microbes and viruses), crude oil, and inorganic molecules (nitrous oxides). The underlying idea behind photocatalysis is the oxidation of organic molecules, water microbes, and decontamination leftovers. Photocatalysts generate highly reactive transient types (e.g., OH, H_2O_2, O_3, O_2) that decompose organic molecules into readily biodegradable compounds before being converted to CO_2 and water. These are examples of Advanced Oxidation Processes (AOP) that generate extremely volatile and oxidising hydroxyl radicals (•OH), which are observed to be the principal species accountable for the breakdown of the organic impurities from polluted water (Kuwahara et al., 2010). Figure 14.2-B depicts the destiny of pollutant in the occurrence of TiO_2 nanoparticles.

Because of their exceptional band-gap features, which form electron-hole (e-/h+) pairs when exposed to light, nanostructured metal-oxide semiconductors have caught the interest of scientists working on photocatalysis and wastewater treatment technologies. Additional properties seen in the nanostructured semiconductors are: they are cheaper, very little or no harmfulness, size decrease, maximising responsive surfaces, doping, or sensitisers are examples of tunable characteristics that may be changed, and their ability to be used for longer periods of time without losing significant photocatalytic activity also makes them useful in wastewater remediation.

Zinc oxide (ZnO) and titanium dioxide (TiO_2) are the most promising metal oxide semiconductor nanostructures thus far. The size and geometry of these materials determine their sorption, (e-/h+) dynamics, and solid-phase transition. The quicker recombination of the h+ and e- reduces TiO_2 effectiveness. Recombination can be lessened by decreasing the particle size of TiO2, which increases interfacial charge-carrier movement (Zhang et al., 1998), though, in order to achieve maximum efficiency, the appropriate size within the nanometer range should be fine-tuned because lowering the size to a few nanometers generates additional surface recombination of the h+ and e-. Likewise, increasing the element size of ZnO during high-temperature calcination affects degradation efficacy of photocatalysis (Hayat et al., 2011). Due to smaller carrier-diffusion courses in tube walls and also faster mass transmission of reactants to nanotube surface, TiO2 nanotubes are confirmed to be more competent, as compared to TiO2 nanoparticles, from the degradation of organic molecules (Macak et al., 2007). A broad reaction (Figure 14.2-C), including a variety of chain redox reactions, happens at the TiO2 surface.

Changing the shape and size of the photolysis kinetics or the photolytic range may not be enough to obtain the requisite efficiency. Each nanomaterial has a band gap that corresponds to its efficient absorption range of optimal-light wavelength. The wavelength's associated energy must be larger than and/or equal to the band gap in order to activate electron and also to s photocatalysis cascade. TiO_2 has a band gap ranging from 3.2 to 3.0 eV, and light wavelength is lower than 400 nm, confining it to UV-irradiation wavelengths. Furthermore, the UV zone accounts for just 3–5% of the whole solar spectrum. As a result, the efficiency with which such photocalysts (TiO_2) use solar energy is inadequate. As a result, the need to widen TiO_2's excitation range to accept visible solar radiation has been explored. Catalyst alteration approaches include metal impurities, hybrid nanoparticles, dye sensitization, or composites based on semiconductors of narrow band-gap and anions (Ni et al., 2007). Dye sensitization is accomplished by utilising a dye molecule, which is triggered by sunlight. Following that, the excited dye molecule contributes additional electrons to TiO2 to help in the creation of e+/h- couples (Vinodgopal et al., 1996).

The usage of anions declines band-gap power, allowing the departing electron to be transferred to the semiconductor under solar radiation (Fujishima et al., 2008). Furthermore, doping anions like F, N, S, C in the TiO_2 crystalline lattice evidently alters photo-responsiveness towards visible spectrum (Ohno et al., 2004; Torres et al., 2004; Asahi et al., 2001; Hattori et al., 1998). Under visible light, N-doped TiO_2 has the highest feasibility of all these anions, despite the fact that e+/h- recombination is amplified (Torres et al., 2004). As a consequence, integrating, for example, a system with additional technologies, such as a noble metal for electron capturing, would boost photochemical efficiency in the long run (Ni et al., 2007). Crystallographic surfaces can also improve TiO_2's photocatalytic activity. By employing capping agents (fluorides), the proportion of greater energy surfaces with greater, more effective photcatalytic activity may be amplified from 10% to 89% (Han et al., 2009). All of these techniques are utilised to boost the photolytic efficiency and efficacy of numerous nanomaterials. Inclusively, the setup and operating parameters of a photocatalytic waste water remediation process have a significant impact on its efficiency. Photocatalytic water treatment methods, whether limited photocatalytic systems using artificial UV light, extensive solar treatment facilities, or pilot programmes in developing nations, are nearly mature.

14.3.3 Decontamination of Wastewater Using Nanomaterials

Decontamination is one of the greatest extensively used and efficacious ways of wastewater cleaning. However, there is a crucial need to remove the restrictions and dangers presented by standard disinfection technologies, which result in the production of toxic disinfection byproducts (DBPs). While chemical disinfectants like chlorine, chloramines, and ozone are effective in removing microbiological illnesses, they can respond with diverse components in natural water to produce DBPs. These classical disinfectants have a high oxidation proclivity, which results in the formation of many DBPs. Numerous nanoparticles have exhibited exceptional disinfectant powers without forming such DBPs, raising the durability and reliability of disinfection to unprecedented altitudes. The nano-disinfectants are water-inert oxidants that are friendlier on the environment (Adhikari et al., 2014).

Disinfectants derived from nanoparticles work in a different way than ordinary decontaminators. The nanoparticles can either infiltrate or interact with cellular membrane, interrupting electron transport or causing cell damage by generating reactive oxygen species (ROS) (Li et al., 2008). Antimicrobial, polymeric silk-peptide nanoparticles cause osmotic collapse and cell death by forming channels of nano-scale in bacterial plasma membranes (Gazit, 2007). Chitosan nanoparticles exhibit bactericidal action. According to the proposed process, positively chitosan-charged particles interact with cell membranes which are negatively charged, enhancing membrane permeability and triggering breakage and seepage of intracellular components (Qi et al., 2004). Ag-nano is presently the most extensively used bactericidal nanomaterial, among other nano-disinfectants, due to its excellent antibacterial action, larger microbicidal spectrum, and low human-toxicity (Xiu et al., 2012).

Antibacterial effects of Ag nanoparticles have also been demonstrated to be reliant on the shape and size of the nanoparticle. Small Ag nanoparticles (8 nm) were more efficient than large particle sizes (11–23 nm) (Morones et al., 2005). In addition, shortened trilateral silver nanoplates having the basal plane in the form of a lattice plane showed the maximum biocidal activity against the gram-negative bacterium *Escherichia coli*, in comparison to spherical and rod-shaped counterparts (Pal et al., 2007). Silver ion nano-release activates a number of activities, including DNA replication inhibition, protein binding and enzymatic injury, and membrane injury (Liau et al., 1997; Danilczuk et al., 2006; Kim et al., 2007;).

TiO_2 has several biocidal properties in addition to photocatalytic destruction of harmful organic and inorganic chemicals. TiO_2's bactericidal activity is connected to the creation of reactive oxygen species (ROS);—namely, peroxide and hydroxyl free radicals—when the material is exposed to UV radiation (Kikuchi et al., 1997). Additionally, Ag doping with TiO_2 permits TiO_2 to be activated by visible light, resulting in much-improved bacterial inactivation under visible light (Seery et al., 2007; Sung-Suh et al., 2004; Sökmen et al., 2001). A nanomaterial with a band-gap chemistry comparable to TiO_2 has also been broadly utilised for its antibacterial nature. ZnO nanoparticles have larger UV absorption competence and powerful bactericidal capabilities against an extensive assortment of pathogens (Jones et al., 2008). When *E. coli* was eliminated from the water, the antibacterial properties of palladium-coated ZnO nanoparticles were similarly enhanced (Khalil et al., 2011). The efficiency of ZnO nanoparticles as a broad-spectrum antibacterial is assumed to be related to intracellular nanoparticle accumulation, cell membrane breakdown, H_2O_2 production, and Zn^{2+} ion release (Huang et al., 2008; Sawai, 2003). An alternative form of nanomaterial that has been proven to have broad antibacterial activity is fullerene (a C60 derivative). Despite the fact that this sort of nanomaterial is intrinsically insoluble, it may be changed to surge its water solubility (Lyon et al., 2008). Surfactants and polymers have been employed to increase carbon-nanotube dispersion in order to better understand their biocidal response. Carbon nanotubes break cell membranes and cause oxidative stress to kill microorganisms (Vecitis et al., 2010).

14.3.4 Nanomembrane Used in Wastewater Remediation

Membrane filtration works by removing undesirable molecules from water and reusing it via the use of semi-permeable membranes. As a result, water-permeable membranes must be not porous to solutes and other particles. Membrane technology is concerned with the many resources and processes that may be utilised to make better membrane performance, minimise membrane entangling, and reduce energy usage.

Nanofiltration has the advantage of being able to operate at lower pressures than other membrane processes such as reverse osmosis, ultrafiltration, and microfiltration (Gehrke et al., 2015). As a result, nanofiltration is very effective in separating pollutants from surface water with low osmotic pressure. Heavy metals and all other organic molecules are rejected preferentially by its pore size; however, monovalent ions like NaCl pass through. Nanofiltration successfully eliminates all microbes from water while also lessening the amount of salt in the water. Reverse osmosis is very effective in removal of small organic and dissolved inorganic molecules. Electrostatic exchanges with charged membrane and charged species are more important than ion size in their passage through the membrane (Shi et al., 2008). Salts with divalent sulphate anions, on the other hand, are excluded by membranes more effectively than salts containing monovalent anion chloride. Because of these characteristics, nanofiltration membranes are particularly beneficial for selectively eliminating solutes obtained from multifaceted processes (Jagadevan et al., 2012; Qu et al., 2013).

Carbon nanotubes can also be utilised in the fabrication of the polymer composite membranes. This composite's tensile modulus, small mass density, and extraordinarily greater strength, considerable flexibility, and broad aspect ratio are all properties that boost its performance. Depending on the manufacturing procedure, the structure may be made of single-walled carbon nanotubes (SWCNTs) or multi-walled carbon nanotubes (MWCNTs). Carbon nanotube membranes are mostly hydrophobic and have a greater mechanical strength. The pore sizes of the majority of the VA-carbon nanotube membranes investigated are equivalent to ultrafiltration and microfiltration membranes. Nanofibres with effective diameters ranging from 1–200 nm are highly effective for nanofiltration. Polymers (both natural and synthetic) and inorganic materials, viz., some metal oxides are electrospun into mats of nanofiber having an intricate pore construction (Ramakrishna et al., 2006; Wegmann et al., 2008). The key benefits of this approach are increased porosity and a better surface and volume ratio. The resultant fibres are orders of magnitude thinner than those produced by standard approaches. The surface-area-to-volume ratio is determined by fibre diameter (Anjum et al., 2016), which may be changed by adjusting procedure parameters like used voltage, substance concentration, spinning distance, and surface tension. This sort of nanofibre membrane is now available for the air purification, but its potential in water treatment is largely unexplored. They are exceptionally successful at removing micron-sized nanoparticles, viz., heavy metals such as cadmium, chromium, nickel, and copper, without fouling. Electrospun membranes are therefore utilised previous to reverse osmosis or ultrafiltration. More importantly, a membrane's properties, like mechanical constancy, permeability, pore size, hydrophilicity, and also charge density, are effortlessly adjustable, making it exceptionally good and extensively appropriate for polluted water remediation.

Nanomaterials can also be combined with inorganic or polymeric membranes to fulfil a variety of tasks. (Table 14.2). To achieve this goal, Arancibia et al. (2016) used a range of metal oxide nanoparticles showing hydrophilic nature, viz., zeolite, TiO_2, and Al_2O_3,; bactericidal nanoparticles, viz., nano-Ag and carbon nanotubes; and also photocatalytic nanomaterials, viz., bi-metallic nanoparticles, TiO_2 (Gehrke et al., 2015). The usage of hydrophilic-natured metal-oxide nanoparticles decreases fouling by growing the membrane's hydrophilicity. Likewise, adding TiO_2, zeolite, or alumina to polymeric ultrafiltration membranes boosted the membranes' fouling resistance, surface hydrophilicity, and water penetrability. Nanomaterials with bactericidal properties, such as carbon and nano-Ag nanotubes, have also been found to reduce membrane biofouling.

TABLE 14.2
List of Nanomaterials Used for Wastewater Remediation

Sr. No.	Nanoparticle		Application	References
1	Zero-Valent Metal Nanoparticles	Silver Nanoparticles.	Antimicrobial properties against a varied range of microorganisms such as fungi, viruses, and bacteria.	Kalhapure et al., 2015; Krishnaraj et al., 2012; Borrego et al., 2016;
		Iron Nanoparticles	Adsorption, reduction, precipitation, and oxidation (in the existence of DO) are utilised to eliminate a varied range of pollutants, including phenols, nitroaromatic compounds, halogenated organic compounds, heavy metals, organic dyes, inorganic anions such as metalloids and nitrates, radio elements, and phosphates.	Hoag et al., 2009; Arancibia et al., 2016; Markov´a et al., 2013; Ling et al., 2015; Ling & Zhang, 2015; Liang et al., 2014
		Zinc Nanoparticles	Halogenated organic compound degradation, especially CCl_4,	Tratnyek et al., 2010
2.	Metal Oxides Nanoparticles	TiO_2 Nanoparticles	acts as a photocatalyst for the destruction of a wide variety of contaminants, including phenols, polycyclic aromatic hydrocarbons, chlorinated organic compounds, pesticides, dyes, heavy metals, and cyanide. The photocatalytic activities of TiO_2 NPs can destroy a wide-ranging variety of microbes, including bacteria, algae, fungi, viruses and protozoa.	Chen et al., 2016; Guo et al., 2015; Nguyen et al., 2016; Alalm et al., 2015; Ohsaka et al., 2008; Kim et al., 2016; Foster et al., 2011;
		ZnO Nanoparticles	Photocatalytic activity Linking ZnO NPs with other semiconductors such as TiO_2, CeO_2, CdO, SnO_2, and reduced graphene oxide (GO), reduced graphene oxide (RGO), is a promising technique for enhancing photodegradation efficiency.	Behnajady et al., 2006; Samadi et al., 2014; Liu, Yang et al., 2014; Uddin et al., 2012; Pant et al., 2012; Dai et al., 2014; Zhou et al., 2012
		Iron Oxides Nanoparticles	Sorbent materials used in the heavy-metals removal from the wastewater	Lei et al., 2014; Ngomsik et al., 2012; Tan et al., 2014
3.	Carbon Nanotubes		have outstanding adsorption capacities and high adsorption efficiencies for a larger range of contaminants, including dyes, dichlorobenzene, Pb^{2+}, Zn^{2+}, Cu^{2+}, Cd^{2+}, and ethyl benzene.	Peng et al., 2003; Lu et al., 2008; Cho et al., 2010; Li et al., 2002; Madrakian et al., 2011
4.	Nanocomposites		Integrating TiO_2 NPs in TFN (Thin Film Nanocomposite) nanofiltration membranes were developed.	Peyravi et al., 2014; Tesh & Scott, 2014

14.4 RESTRICTION IN USAGE OF NANOPARTICLES WASTEWATER REMEDIATION

All of the nanomaterials investigated have great physiochemical features, viz., shape, size, and surface with volume ratio, making them a better choice for replacing conventional equivalents, and have been a hot issue in the field of study for a variety of applications. The nanoparticle has developed as one of the greatest progressive wastewater-remediation techniques based on these features. Although there are a number of obstacles and shortcomings with using nanoparticles in wastewater remediation, these systems are unlikely to be commercialised on a significant scale. Some of the chief apprehensions are dispersal, retaining and retrieval, functionality damage over time, accurate condition effectiveness, rate balance, and unknown downhill harmfulness of nanoparticles.

Based on the above all properties, the nanoparticle has evolved into one of the greatest complex wastewater-remediation methods. Despite the significant problems and disadvantages of using nanoparticles in waste water remediation, these methods are unlikely to be widely adopted (Wiesner et al., 2006). Membrane filtration and immobilisation procedures are employed to retain nanoparticles, which restricts the active dosage of applied nanoparticles with the existing surface area in the reactor, limiting total disinfection efficacy. The use of nanoparticles has also been limited due to activity loss over time, and also, a decrease in the number of efficiency cycles. Although Ag nanoparticles have high bactericidal properties, their activity lessens with time due to the Ag^+ ions loss (Chaturvedi et al., 2012). Though doping prevents e-/h+ recombination, which alleviates some of the issues associated with nanoparticle inactivation, the system's efficiency degrades with time. Aside from that, the vast majority of waste-degradation efficiency experiments are conducted in a laboratory setting under replicated settings. In reality, unusable water is significantly more intricate, comprising a diverse variety of complexing ions and chemicals.

Most of the nanoparticle systems perform better in replicated conditions, but when applied to an actual wastewater system, their efficacy decreases. With the exception of some metal oxide nanomaterials, the characterisation and manufacturing of nanoparticles is prohibitively costly. Retrieval and reusability are crucial for the nanoparticles. The nanoparticles released into bodies of water may have a harmful influence on living beings as well as the aquatic environments. The harmfulness of these nanoparticles is unclear. Even though exposure to some nanoparticles appears to be non-toxic in the short term, it may have long-term harmful implications. Nanoparticles have the potential to produce chronic inflammation and stress (Usmani et al., 2017). The disintegration of nanomaterials from water may result in the release of metal ions, which might be harmful. As a result, a full understanding of these limits, as well as minimising potential repercussions and dangers associated with these nanoparticles, is crucial for increased usage and adoption of nanotechnology in wastewater remediation.

REFERENCES

Adhikari, M. D., Mukherjee, S., Saikia, J., Das, G., & Ramesh, A. (2014). Magnetic nanoparticles for selective capture and purification of an antimicrobial peptide secreted by food-grade lactic acid bacteria. *Journal of Materials Chemistry B, 2*(10), 1432–1438.

Ahmad, A., Mohd-Setapar, S. H., Chuong, C. S., Khatoon, A., Wani, W. A., Kumar, R., & Rafatullah, M. (2015). Recent advances in new generation dye removal technologies: Novel search for approaches to reprocess wastewater. *RSC Advances, 5*, 30801–30818.

Alalm, M. G., Tawfik, A., & Ookawara, S. (2015). Comparison of solar TiO_2 photocatalysis and solar photo-Fenton for treatment of pesticides industry wastewater: Operational conditions, kinetics, and costs. *Journal of Water Process Engineering, 8*, 55–63.

Albrecht, M. A., Evans, C. W., & Raston, C. L. (2006). Green chemistry and the health implications of nanoparticles. *Green Chemistry*, *8*, 417–432.

Anjum, M., Miandad, R., Waqas, M., Gehany, F., & Barakat, M. A. (2016). Remediation of wastewater using various nano-materials. *Arabian Journal of Chemistry, 12*(8), 4897–4919.

Arancibia-Miranda, N., Baltazar, S. E., García, A., Muñoz-Lira, D., Sepúlveda, P., Rubio, M. A., & Altbir, D. (2016). Nanoscale zero valent supported by Zeolite and Montmorillonite: Template effect of the removal of lead ion from an aqueous solution. *Journal of Hazardous Materials, 301*, 371–380.

Asahi, R., Morikawa, T., Ohwaki, T., Aoki, K., & Taga, Y. (2001). Visible-light photocatalysis in nitrogen-doped titanium oxides. *Science, 293*(5528), 269–271.

Banerjee, A., Choudhury, M., Chakravorty, A., Raghavan, V., Biswas, B., Sana, S. S., . . . Ramakrishna, S. (2021). Microbiological degradation of organic pollutants from industrial wastewater. In *Nanobiotechnology for green environment* (pp. 83–114). CRC Press.

Bayoumi, T. A., & Saleh, H. M. (2018). Characterization of biological waste stabilized by cement during immersion in aqueous media to develop disposal strategies for phytomediated radioactive waste. *Progress in Nuclear Energy*, 107, 83–89.

Behnajady, M. A., Modirshahla, N., & Hamzavi, R. (2006). Kinetic study on photocatalytic degradation of C.I. Acid Yellow 23 by ZnO photocatalyst. *Journal of Hazardous Materials, 133*(1–3), 226–232.

Bhardwaj, L. K., Rath, P., & Choudhury, M. (2022). A comprehensive review on the classification, uses, sources of nanoparticles (NPs) and their toxicity on health. *Aerosol Science* and *Engineering, 7*, 426. https://doi.org/10.1007/s41810-022-00163-4

Borrego, B., Lorenzo, G., Mota-Morales, J. D., et al. (2016). Potential application of silver nanoparticles to control the infectivity of Rift Valley fever virus in vitro and in vivo. *Nanomedicine: Nanotechnology, Biology and Medicine, 12*(5), 1185–1192.

Briggs, A. M., Cross, M. J., Hoy, D. G., Blyth, F. H., Woolf, A. D., & March, L. (2016). Musculoskeletal health conditions represent a global threat to healthy aging: A report for the 2015 world health organization world report on ageing and health. *Gerontologist, 56*, 243–255.

Chaturvedi, S., Dave, P. N., & Shah, N. K. (2012). Applications of nano-catalyst in new era. *Journal* of *Saudi Chemical Society*, *16*(3), 307–325.

Chen, Z. P., Li, Y., Guo, M., et al. (2016). One-pot synthesis of Mn-doped TiO2 grown on graphene and the mechanism for removal of Cr(VI) and Cr(III). *Journal of Hazardous Materials, 310*, 188–198.

Cho, H.-H., Wepasnick, K., Smith, B. A., Bangash, F. K., Fairbrother, D. H., & Ball, W. P. (2010). Sorption of aqueous Zn[II] and Cd[II] by multiwall carbon nanotubes: The relative roles of oxygen-containing functional groups and graphenic carbon. *Langmuir, 26*(2), 967–981.

Daer, S., Kharraz, J., Giwa, A., & Hasan, S. W. (2015). Recent applications of nanomaterials in water desalination: A critical review and future opportunities. *Desalination, 367*, 37–48.

Dai, K., Lu, L., Liang, C., Dai, J., Zhu, G., Liu, Z., Liu, Q., & Zhang Y. X. (2014). Graphene oxide modied ZnO nanorods hybrid with high reusable photocatalytic activity under UV-LED irradiation. *Materials Chemistry and Physics, 143*(3), 1410–1416.

Danilczuk, M., Lund, A., Sadlo, J., Yamada, H., & Michalik, J. (2006). Conduction electron spin resonance of small silver particles. *Spectrochimica Acta—Part A: Molecular and Biomolecular Spectroscopy, 63*(1), 189–191.

Das, R., Vecitis, C. D., Schulze, A., Cao, B., Ismail, A. F., Lu, X., Chene, J., & Ramakrishna, S. (2017). Recent advances in nanomaterials for water protection and monitoring. *Chemical Society Reviews, 46*, 6946–7020.

Foster, H. A., Ditta, I. B., Varghese, S., & Steele, A. (2011). Photocatalytic disinfection using titanium dioxide: Spectrum and mechanism of antimicrobial activity. *Applied Microbiology and Biotechnology, 90*(6), 1847–1868.

Fujishima, A., Zhang, X., & Tryk, D. A. (2008). TiO2 photocatalysis and related surface phenomena. *Surface Science Reports, 63*(12), 515–582.

Gazit, E. (2007). Self-assembled peptide nanostructures: The design of molecular building blocks and their technological utilization. *Chemical Society Reviews, 36*(8), 1263–1269.

Gehrke, I., Geiser, A., & Somborn-Schulz, A. (2015). Innovations in nanotechnology for water treatment. *Nanotechnology*, *Science* and *Applications, 8*, 1.

Gitis, V., & Hankins, N. (2018). Water treatment chemicals: Trends and challenges. *Journal of Water Process Engineering, 25*, 34–38.

Gogoi, N., & Choudhury, M. (2022). Green materials synthesis for wastewater treatment. In C. Baskar, S. Ramakrishna, & A. Daniela La Rosa (Eds.), *Encyclopedia of green materials*. Springer. https://doi.org/10.1007/978-981-16-4921-9_31-1

Guo, M., Song, W., Wang, T., Li, Y., Wang, X., & Du, X. (2015). Phenyl functionalization of titanium dioxide-nanosheets coating fabricated on a titanium wire for selective solid-phase microextraction of polycyclic aromatic hydrocarbons from environment water samples. *Talanta, 144*, 998–1006.

Gupta, V. K., Tyagi, I., Sadegh, H., Shahryari-Ghoshekandi, R., Makhlouf, A. S. H., & Maazinejad, B. (2015). Nanoparticles as adsorbent; a positive approach for removal of noxious metal ions: A review. *Science*, *Technology* and *Development, 34*, 195–214.

Han, X., Kuang, Q., Jin, M., Xie, Z., & Zheng, L. (2009). Synthesis of titania nanosheets with a high percentage of exposed (001) facets and related photocatalytic properties. *Journal of the American Chemical Society, 131*(9), 3152–3153.

Hattori, A., Yamamoto, M., Tada, H., & Ito, S. (1998). A promoting effect of NH4F addition on the photocatalytic activity of sol-gel TiO_2 films. *Chemistry Letters, 27*(8), 707–708.

Hayat, K., Gondal, M. A., Khaled, M. M., Ahmed, S., & Shemsi, A. M. (2011). Nano ZnO synthesis by modified sol gel method and its application in heterogeneous photocatalytic removal of phenol from water. *Applied Catalysis A, 393*(1), 122–129.

Hoag, G. E., Collins, J. B., Holcomb, J. L., Hoag, J. R., Nadagouda, M. N., & Varma, R. S. (2009). Degradation of bromothymol blue by 'greener' nano-scale zero-valent iron synthesized using tea polyphenols. *Journal of Materials Chemistry, 19*(45), 8671–8677.

Hodges, B. C., Cates, E. L., & Kim, J. (2018). Challenges and prospects of advanced oxidation water treatment processes using catalytic nanomaterial. *Nature Nanotechnology, 13*, 642–650.

Huang, Z., Zheng, X., Yan, D., Yin, G., Liao, X., Kang, Y., Huang, D., & Hao, B. (2008). Toxicological effect of ZnO nanoparticles based on bacteria. *Langmuir, 24*(8), 4140–4144.

Istratie, R., Stoia, M., Păcurariu, C., & Locovei, C. (2016). Single and simultaneous adsorption of methyl orange and phenol onto magnetic iron oxide/carbon nanocomposites. *Arabian Journal of Chemistry, 12*(8), 3704–3722.

Jagadevan, S., Jayamurthy, M., Dobson, P., & Thompson, I. P. (2012). A novel hybrid nano zerovalent iron initiated oxidation–Biological degradation approach for remediation of recalcitrant waste metalworking fluids. *Water Research, 46*(7), 2395–2404.

Jones, N., Ray, B., Ranjit, K. T., & Manna, A. C. (2008). Antibacterial activity of ZnO nanoparticle suspensions on a broad spectrum of microorganisms. *FEMS Microbiology Letters, 279*(1), 71–76.

Kalfa, O. M., Yalçınkaya, Ö., & Türker, A. R. (2009). Synthesis of nano B_2O_3/TiO_2 composite material as a new solid phase extractor and its application to preconcentration and separation of cadmium. *Journal of Hazardous Materials, 166*(1), 455–461.

Kalhapure, R. S., Sonawane, S. J., Sikwal, D. R., et al. (2015). Solid lipid nanoparticles of clotrimazole silver complex: An efficient nano antibacterial against Staphylococcus aureus and MRSA. *Colloids and Surfaces B: Biointerfaces, 136*, 651–658.

Kandasamy, S., & Prema, R. S. (2015). Methods of synthesis of nano particles and its applications. *Journal of Chemical and Pharmaceutical Research, 7*(3), 278–285.

Khalil, A., Gondal, M. A., & Dastageer, M. A. (2011). Augmented photocatalytic activity of palladium incorporated ZnO nanoparticles in the disinfection of *Escherichia coli* microorganism from water. *Applied Catalysis A: General, 402*(1), 162–167.

Kikuchi, Y., Sunada, K., Iyoda, T., Hashimoto, K., & Fujishima, A. (1997). Photocatalytic bactericidal effect of TiO 2 thin films: Dynamic view of the active oxygen species responsible for the effect. *Journal of Photochemistry* and *Photobiology A: Chemistry, 106*(1), 51–56.

Kim, J. S., Kuk, E., Yu, K. N., Kim, J,-H., Park, S. J., Lee, H. J., & Hwang, C.-Y. (2007). Antimicrobial effects of silver nanoparticles. *Nanomedicine, 3*(1), 95–101.

Kim, S. H., Lee, S. W., Lee, G. M., Lee, B.-T., Yun, S.-T., & Kim, S.-O. (2016). Monitoring of TiO2-catalytic UV-LED photo-oxidation of cyanide contained in mine wastewater and leachate. *Chemosphere, 143*, 106–114.

Krishnaraj, C., Ramachandran, R., Mohan, K., & Kalaichelvan, P. T. (2012). Optimization for rapid synthesis of silver nanoparticles and its effect on phytopathogenic fungi. *Spectrochimica Acta—Part A: Molecular and Biomolecular Spectroscopy, 93*, 95–99.

Kuwahara, Y., Kamegawa, T., Mori, K., & Yamashita, H. (2010). Design of new functional Titanium oxide-based photocatalysts for degradation of organics diluted in water and air. *Current Organic Chemistry, 14*(7), 616–629.

Lei, Y., Chen, F., Luo, Y., & Zhang, L. (2014). Three-dimensional magnetic graphene oxide foam/Fe3O4 nanocomposite as an efficient absorbent for Cr(VI) removal. *Journal of Materials Science, 49*(12), 4236–4245.

Li, L., Fan, M., Brown, R. C., Van, Leeuwen, J., Wang, J., Wang, W., & Zhang, P. (2006). Synthesis, properties, and environmental applications of nanoscale iron-based materials: A review. *Critical Reviews in Environmental Science and Technology, 36*(5), 405–431.

Li N, Xia T, Nel AE. (2008). The role of oxidative stress in ambient particulate matter-induced lung diseases and its implications in the toxicity of engineered nanoparticles. *Free Radical Biology and Medicine, 44*(9), 1689–1699.

Li, Y.-H., Ding, J., Luan, Z., Di, Z., Zhu, Y., Xu, C., Wu, D., & Wei, B. (2003). Competitive adsorption of Pb2+, Cu2+ and Cd2+ ions from aqueous solutions by multiwalled carbon nanotubes. *Carbon, 41*(14), 2787–2792.

Li, Y.-H., Wang, S., Wei, J., et al. (2002). Lead adsorption on carbon nanotubes. *Chemical Physics Letters, 357*(3–4), 263–266.

Liang, D.-W., Yang, Y.-H., Xu, W.-W., Peng, S.-K., Lu, S.-F., & Xiang, Y. 2014. Nonionic surfactant greatly enhances the reductive debromination of polybrominated diphenyl ethers by nanoscale zero-valent iron: Mechanism and kinetics. *Journal of Hazardous Materials, 278*, 592–596.

Liau, S. Y., Read, D. C., Pugh, W. J., Furr, J. R., & Russell, A. D. (1997). Interaction of silver nitrate with readily identifiable groups: Relationship to the antibacterial action of silver ions. *Letters in Applied Microbiology, 25*(4), 279–283.

Ling, L., & Zhang, W.-X. (2015). Enrichment and encapsulation of uranium with iron nanoparticle. *Journal of the American Chemical Society, 137*(8), 2788–2791.

Ling, L., Pan, B., & Zhang, W.-X. (2015). Removal of selenium from water with nanoscale zero-valent iron: Mechanisms of intraparticle reduction of Se(IV). *Water Research, 71*, 274–281.

Liu, C., Hong, T., Li, H., Wang, L. (2018). From club convergence of per capita industrial pollutant emissions to industrial transfer ects: An empirical study across 285 cities in China. *Energy Policy*, 121, 300–313.

Liu, F., Yang, J., Zuo, J., Ma, D., Gan, L., Xie, B., Wang, P., & Yang, B. (2014, August 1). Graphene-supported nanoscale zero-valent iron: Removal of phosphorus from aqueous solution and mechanistic study. *Journal of Environmental Sciences (China), 26*(8), 1751–1762. https://doi.org/10.1016/j.jes.2014.06.016. Epub 2014 July 16. PMID: 25108732.

Longwane, G. H., Sekoai, P. T., Meyyappan, M., & Moothi, K. (2019). Review simultaneous removal of pollutants from water using nanoparticles: A shift from single pollutant control to multiple pollutant control. *Science* of the *Total Environment, 656*, 808–833.

Lu, C., Su, F. S., & Hu, S. (2008). Surface modification of carbon nanotubes for enhancing BTEX adsorption from aqueous solutions. *Applied Surface Science, 254*(21), 7035–7041.

Lyon, D. Y., Brown, D. A., & Alvarez, P. J. J. (2008). Implications and potential applications of bactericidal fullerene water suspensions: Effect of nC60 concentration, exposure conditions and shelf life. *Water Science and Technology, 57*(10), 1533–1538.

Ma, H., Guo, Y., Qin, Y., & Li, Y. Y. (2018). Review nutrient recovery technologies integrated with energy recovery by waste biomass anaerobic digestion. *Bioresource Technology, 269*, 520–531.

Macak, J. M., Zlamal, M., Krysa, J., & Schmuki, P. (2007). Self-organized TiO_2 nanotube layers as highly efficient photocatalysts. *Small, 3*(2), 300–304.

Madrakian, T., Afkhami, A., Ahmadi, M., & Bagheri, H. (2011). Removal of some cationic dyes from aqueous solutions using magnetic-modified multi-walled carbon nanotubes. *Journal of Hazardous Materials, 196*, 109–114.

Marková Z, Šišková KM, Filip J, Čuda J, Kolář M, Šafářová K, Medřík I, Zbořil R. Air stable magnetic bimetallic Fe-Ag nanoparticles for advanced antimicrobial treatment and phosphorus removal. *Environ Sci Technol. 2013 May 21;47*(10):5285–93. doi: 10.1021/es304693g. Epub 2013 May 6. PMID: 23590411.

Mir, N. A., Haque, M. M., Khan, A., Umar, K., Muneer, M., & Vijayalakshmi, S. (2012). Semiconductor mediated photocatalysed reaction of two selected organic compounds in aqueous suspensions of Titanium dioxide. *Journal of Advanced Oxidation Technologies, 15*, 380–391.

Morones, J. R., Elechiguerra, J. L., Camacho, A., Holt, K., Kouri, J. B., Ramírez, J. T., & Yacaman, M. J. (2005). The bactericidal effect of silver nanoparticles. *Nanotechnology, 16*(10), 2346.

Ngomsik, A.-F., Bee, A., Talbot, D., & Cote, G. (2012). Magnetic solid-liquid extraction of Eu(III), La(III), Ni(II) and Co(II) with maghemite nanoparticles. *Separation and Purification Technology, 86*, 1–8.

Nguyen, A. T., Hsieh, C.-T., & Juang, R.-S. (2016). Substituent effects on photodegradation of phenols in binary mixtures by hybrid H2O2 and TiO_2 suspensions under UV irradiation. *Journal of the Taiwan Institute of Chemical Engineers, 62*, 68–75.

Ni, M., Leung, M. K. H., Leung, D. Y. C., & Sumathy, K. (2007). A review and recent developments in photocatalytic water-splitting using TiO_2 for hydrogen production. *Renewable and Sustainable Energy, 11*(3), 401–425.

Ohno, T., Akiyoshi, M., Umebayashi, T., Asai, K., Mitsui, T., & Matsumura, M. (2004). Preparation of S-doped TiO_2 photocatalysts and their photocatalytic activities under visible light. *Applied Catalysis A: General, 265*(1), 115–121.

Ohsaka, T., Shinozaki, K., Tsuruta, K., & Hirano, K. (2008). Photoelectrochemical degradation of some chlorinated organic compounds on n-TiO2 electrode. *Chemosphere, 73*(8), 1279–1283.

Pal, S., Tak, Y. K., & Song, J. M. (2007). Does the antibacterial activity of silver nanoparticles depend on the shape of the nanoparticle? A study of the gram-negative bacterium *Escherichia coli*. *Applied and Environmental Microbiology, 73*(6), 1712–1720.

Palimi, M. J., Rostami, M., Mahdavian, M., & Ramezanzadeh, B. (2014). Surface modification of Fe2O3 nanoparticles with 3-aminopropyltrimethoxysilane (APTMS): An attempt to investigate surface treatment on surface chemistry and mechanical properties of polyurethane/Fe_2O_3 nanocomposites. *Applied Surface Science, 320*, 60–72.

Pant, H. R., Park, C. H., Pant, B., Tijing, L. D., Kim, H. Y., & Kim, C. S. (2012). Synthesis, characterization, and photocatalytic properties of ZnO nano-flower containing TiO_2 NPs. *Ceramics International, 38*(4), 2943–2950.

Peng, X., Li, Y., Luan, Z., Di, Z., Wang, H.,Tian, B., & Jia, Zhiping (2003). Adsorption of 1,2-dichlorobenzene from water to carbon nanotubes. *Chemical Physics Letters, 376*(1–2), 154–158.

Peyravi, M., Jahanshahi, M., Rahimpour, A., Javadi, A., & Hajavi, S. (2014). Novel thin film nanocomposite membranes incorporated with functionalized TiO_2 nanoparticles for organic solvent nanofiltration. *Chemical Engineering Journal, 241*, 155–166.

Rajpal, A., Ali, M., Choudhury, M., Almohana, A. I., Alali, A. F., Munshi, F. M. A., Khursheed, A., & Kazmi, A. A. (2022). Abattoir wastewater treatment plants in India: Understanding and performance evaluation. *Frontiers in Environmental Science, 10*, 881623. https://doi.org/10.3389/fenvs.2022.881623

Qi, L., Xu, Z., Jiang, X., Hu, C., & Zou, X. (2004). Preparation and antibacterial activity of chitosan nanoparticles. *Carbohydrate Research, 339*(16), 2693–2700.

Qu, X., Alvarez, P. J. J., & Li, Q. (2013). Applications of nanotechnology in water and wastewater treatment. *Water Research, 47*(12), 3931–3946.

Rajasulochana, P., & Preethy, V. (2016). Comparison on efficiency of various techniques in treatment of waste and sewage water—A comprehensive review. *Technology Resources, 4*, 175–184.

Rajput, N. (2015). Methods of preparation of nanoparticles—A review. *International Journal of Advanced Engineering Technology*, *7*(4), 1806–1811.

Ramakrishna, S., Fujihara, K., Teo, W.-E., Yong, T., Ma, Z., & Ramaseshan, R. (2006). Electrospun nanofibers: Solving global issues. *Materials Today, 9*(3), 40–50.

Richter, K. E., & Ayers, J. M. (2018). An approach to predicting sediment microbial fuel cell performance in shallow and deep water. *Applied Sciences*, *8*(12), 2628. https://doi.org/10.3390/app8122628

Saha, B., Das, S., Saikia, J., & Das, G. (2011). Preferential and enhanced adsorption of different dyes on iron oxide nanoparticles: A comparative study. *Journal of Physical Chemistry C, 115*(16), 8024–8033.

Saikia, J., Saha, B., & Das, G. (2011). Efficient removal of chromate and arsenate from individual and mixed system by malachite nanoparticles. *Journal of Hazardous Materials, 186*(1), 575–582.

Saikia, J., Sikdar, Y., Saha, B., & Das, G. (2013). Malachite nanoparticle: A potent surface for the adsorption of xanthene dyes. *Journal of Environmental Chemical Engineering, 1*(4), 1166–1173.

Samadi, M., Pourjavadi, A., & Moshfegh, A. Z. (2014). Role of CdO addition on the growth and photocatalytic activity of electrospun ZnO nanofibers: UV vs. visible light. *Applied Surface Science, 298*, 147–154.

Sawai, J. (2003). Quantitative evaluation of antibacterial activities of metallic oxide powders (ZnO, MgO and CaO) by conductimetric assay. *Journal of Microbiological Methods, 54*(2), 177–182.

Seery, M. K., George, R., Floris, P., & Pillai, S. C. (2007). Silver doped titanium dioxide nanomaterials for enhanced visible light photocatalysis. *Journal of Photochemistry and Photobiology A: Chemistry, 189*(2–3), 258–263.

Sekoai, P. T., Ouma, C. N. M., Du Preez, S. P., Modisha, P., Engelbrecht, N., Bessarabov, D. G., & Ghimire, A. (2019). Application of nanoparticles in biofuels: An overview. *Fuel, 237*, 380–397.

Shi, Q., Liang, H., Feng, D., Wang, J., & Stucky, G. D. (2008). Porous carbon and carbon/metal oxide microfibers with well-controlled pore structure and interface. *Journal* of the *American Chemical Society, 130*(15), 5034–5035.

Sizmur, T., Fresno, T., Akgül, G., Frost, H., & Moreno-Jiménez, E. (2017). Biochar modification to enhance sorption of inorganics from water. *Bioresource Technology, 246*, 34–47. https://doi.org/10.1016/j.biortech.2017.07.082. Epub 2017 July 17. PMID: 28781204.

Sökmen, M., Candan, F., & Sümer, Z. (2001). Disinfection of E. coli by the Ag-TiO_2/UV system: Lipid peroxidation. *Journal of Photochemistry* and *Photobiology A: Chemistry, 143*(2), 241–244.

Sung-Suh, H. M., Choi, J. R., Hah, H. J., Koo, S. M., & Bae, Y. C. (2004). Comparison of Ag deposition effects on the photocatalytic activity of nanoparticulate TiO_2 under visible and UV light irradiation. *Journal of Photochemistry and Photobiology A: Chemistry, 163*(1), 37–44.

Tan, L., Xu, J., Xue, X., Lou, Z., Zhu, J., Baig, S. A., & Xu, X. (2014). Multifunctional nanocomposite Fe_3O_4@SiO_2–mPD/SP for selective removal of Pb(ii) and Cr(vi) from aqueous solutions. *RSC advances*, 4(86), 45920–45929.

Tang, W.-W., Zeng, G.-M., Gong, J.-L., et al. (2014). Impact of humic/fulvic acid on the removal of heavy metals from aqueous solutions using nanomaterials: A review. *Science of the Total Environment, 468–469*, 1014–1027.

Tesh, S. J., & Scott, T. B. (2014). Nano-composites for water remediation: A review. *Advanced Materials, 26*(35), 6056–6068.

Torres, G. R., Lindgren, T., Lu, J., Granqvist, C.-G., & Lindquist, S. -E. (2004). Photoelectrochemical study of nitrogen-doped titanium dioxide for water oxidation. *Journal of Physical Chemistry B, 108*(19), 5995–6003.

Tratnyek, P. G., Salter, A. J., Nurmi, J. T., & Sarathy, V. (2010). Environmental applications of zerovalent metals: Iron vs. zinc. In *Nanoscale materials in chemistry: Environmental applications,* Vol. 1045 of *ACS Symposium Series* (Chapter 9, pp. 165–178). American Chemical Society.

Uddin, M. T., Nicolas, Y., Olivier, C., Toupance, T., Servant, L., Muller, M. M., Kleebe, H.J., Ziegler, J., & Jaegermann, W. (2012). Nanostructured SnO2–ZnO heterojunction photocatalysts showing enhanced photocatalytic activity for the degradation of organic dyes. *Inorganic chemistry, 51*(14), 7764–7773. doi: 10.1021/ic300794j. Epub 2012 Jun 26. PMID: 22734686.

Umar, K. (2018). Water contamination by organic-pollutants: TiO2 photocatalysis, in: Modern Age Environmental Problems and their Remediation, Springer, 2018, pp. 95–109.

Umar, K., Haque, M. M., Mir, N. A., & Muneer, M. (2013). Titanium dioxide-Mediated photocatalyzed mineralization of two selected organic pollutants in aqueous suspensions. *Journal of Advanced Oxidation Technologies, 16*, 252–260.

Umar, K., Ibrahim, M. N. M., Ahmad, A., & Rafatullah, M. (2019). Synthesis of Mn-Doped TiO2 by novel route and photocatalytic mineralization/intermediate studies of organic pollutants. *Research on Chemical Intermediates, 45*, 2927–2945.

Usmani, M. A., Khan, I., Bhat, A. H., Pillai, R. S., Ahmad, N., Mohamad Haafiz, M. K., & Oves, M. (2017). Current trend in the application of nanoparticles for waste water treatment and purification: A review. *Current Organic Synthesis*, *2*, 206–226.

Vecitis, C. D., Zodrow, K. R., Kang, S., & Elimelech, M. (2010). Electronic-structure-dependent bacterial cytotoxicity of single-walled carbon nanotubes. *ACS Nano, 4*(9), 5471–5479.

Vinodgopal, K., Wynkoop, D. E., & Kamat, P. V. (1996). Environmental photochemistry on semiconductor surfaces: Photosensitized degradation of a textile azo dye, acid orange on TiO2 particles using visible light. *Environmental Science & Technology, 30*(5), 1660–1666.

Wang, J, Wang, Z., Carolina, L. Z. V., Wolfson, J. M., & Pingtian, G. (2019). Review on the treatment of organic pollutants in water by ultrasonic technology. *Ultrasonics Sonochemistry*, *55*, 273–278.

Wegmann, M., Michen, B., & Graule, T. (2008). Nanostructured surface modification of microporous ceramics for efficient virus filtration. *Journal of the European Ceramic Society, 28*(8), 1603–1612.

Wiesner, M. R., Lowry, G. V., Alvarez, P., Dionysiou, D., & Biswas, P. (2006). Assessing the risks of manufactured nanomaterials. *Environmental Science & Technology, 40*(14), 4336–4345.

Wu, Y., Pang, H., Liu, Y., Wang, X., Yu, S., Fu, D., Chen, J., & Wang, X. (2019). Environmental remediation of heavy metal ions by novel-Nanomaterials: A review. *Environmental Pollution, 246*, 608–620.

Xiu, Z.-M, Zhang, Q.-B, Puppala, H. L., Colvin, V. L., & Alvarez, P. J. J. (2012). Negligible particle-specific antibacterial activity of silver nanoparticles. *Nano Letters, 12*(8), 4271–4275.

Yan, J., Han, L., Gao, W., Xue, S., & Chen, M. (2015). Biochar supported nanoscale zerovalent iron composite used as persulfate activator for removing trichloroethylene. *Bioresource Technology, 175*, 269–274.

Yang, K., & Xing, B. (2010). Adsorption of organic compounds by carbon nanomaterials in aqueous phase: Polanyi theory and its application. *Chemical Reviews, 110*(10), 5989–6008.

Yang, K., Wu, W., Jing, Q., & Zhu, L. (2008). Aqueous adsorption of aniline, phenol, and their substitutes by multi-walled carbon nanotubes. *Environmental Science & Technology, 42*(21), 7931–7936.

Yaqoob, A. A., & Ibrahim, M. N. M. (2019). A review article of nanoparticles; synthetic approaches and wastewater treatment methods. *International Research Journal of Engineering and Technology, 6*, 1–7.

Zhang, Y., Wu, B., Xu, H., Liu, H., Wang, M., He, Y., & Pan, B. (2016). Nanomaterials-enabled water and wastewater treatment. *Nano Impact, 3*, 22–39.

Zhang, Z., Wang, C.-C., Zakaria, R., & Ying, J. Y. (1998). Role of particle size in nanocrystalline TiO_2-based photocatalysts. *Journal of Physical Chemistry B, 102*, 10871–10878.

Zhou, X., Shi, T., & Zhou, H. (2012). Hydrothermal preparation of ZnO-reduced graphene oxide hybrid with high performance in photocatalytic degradation. *Applied Surface Science, 258*(17), 6204–6211.

15 Waste Management
Learning and Challenges from a Case Study of a University Model in India

Lijo Thomas, Joseph Varghese Kureethara, and Martin Klein

15.1 INTRODUCTION

Waste management is one of the major problems for almost every urban and rural city in India (Kumar et al., 2017). The situation is getting worse day by day because of a greater number of people moving to urban cities. Universities, especially those from the urban cities, are like small towns by themselves because of the population and other complex systems and activities. The universities are the key sites to create and disseminate knowledge and best practices for societal transformation. It has got far-reaching potential to catalyze the transition of society (Stephens et al., 2008). Waste management is probably the greatest test that urban India faces. It is not simply India but the entire world that faces the hazard of piles of waste. The best solutions for waste management cannot be implemented by governments or officers who are miles away from our homes. It should be started within one's own home. The important points to remember while disposing of waste is its segregation. Successful segregation of waste implies that less waste goes to the landfill, which makes it less expensive and better for individuals and the environment (Mondal et al., 2023). It is, additionally, critical to segregate for public health. Specifically, hazardous waste can cause long-term medical conditions, so it is vital that they are discarded effectively and securely and not mixed with ordinary waste emerging from your home or office. While waste management can begin at our home, there ought to be sufficient awareness and instructive projects that the government should direct to make individuals mindful of the potential danger of non-degradable materials. As the problem is gaining recognition, more and more organizations are trying to combat the problem (Dutta et al., 2022).

Waste management involves multidisciplinary activities, such as engineering principles, economic, urban and regional planning, management techniques and social sciences, to minimize the overall waste activity of the system under consideration. A systematic approach to waste management, encompassing waste of all kinds of resources at all stages, should be adopted. However, the material constitutes the major fraction of the total production cost, while materials wasted are of critical importance. Educational institutions are the best platforms, where

DOI: 10.1201/9781003258377-15

waste management can be taught, practiced and implemented (de Vega et al., 2008). The influence of such programmes on students and their impact later on society is immeasurable (Zhang et al., 2011). Several countries have insisted that waste management is part of the curriculum at various levels of education. In several countries, various student movements take care of waste management programmes (Meneses, 2015). On the whole, there have been tremendous changes in the population about the need for proper waste management. It involves segregation, reuse and disposal of waste. In India, several universities and institutions have taken individual and independent steps in the last couple of decades to promote environment-friendly activities. In this chapter, we take the case of CHRIST (Deemed to be University) of India that has taken steps to clean and green the campus through various programmes and practices.

15.2 CHRIST (DEEMED TO BE UNIVERSITY) WASTE MANAGEMENT MODEL

Social responsibility is an important core value of CHRIST (Deemed to be University), India. The university has twin visions, viz., 'Excellence and Service'. Social responsibility is a way of fulfilling its vision 'Service.' Focusing on its vision and commitment to social development, the University established the Centre for Social Action (CSA) to oversee all the service learning and social outreach programmes of the University. The centre has been serving more than a hundred urban slums in India through its diverse woman and child development programmes and other sustainable, economic development programmes.

The University's waste management system is scientifically managed by the wing *Parivarthana*, which means regeneration, under CSA, since 2008. The centre, in collaboration with other student organizations, such as Student Welfare Office, NCC and Student Council, educates and implements a zero-waste model on campus. The faculty, staff and students actively get involved in the cause (Figure 15.1-A).

As a university, a vast amount of different types of waste is generated on the campus. For a long time, it was managed differently, without much thought. Nevertheless, there was the possibility of adopting comprehensive, scientific and environment-friendly approaches in waste management through decentralized solid waste collection and management practices. As expected, thus the disposal of waste ended up in incineration. This caused environmental pollution, leading to health issues for the population inside and around the campus. However, the Centre for Social Action (CSA) of the university took consultation from the erstwhile CHF International (now Global Communities) for a zero-waste campaign at the University campus. This paved way for many changes in the university, which we discuss in detail in this chapter.

15.3 THE WORKS OF *PARIVARTHANA*

Parivarthana is the visible face of recycling activities of the university. It involves in-house recycling and transferring the recyclables to other professional recyclers. The University promotes both the activities of *Parivarthana* by cash and kind

(Figure 15.1-B). The women who work with *Parivarthana* take the proceeds from the sale of recycled items and recyclables.

The University generates about a tonne of waste daily. The waste includes both wet and dry waste. The dry waste is about 80%, and what remains is wet waste. Due to the awareness programmes conducted in the university, the waste segregated at source is primarily segregated by students. Students are motivated by the campaign for a clean and green environment. This eventually leads the university to be a Zero-Waste Zone. The collected, segregated waste is then categorized by the segregation unit of *Parivarthana* for the making of handmade sheets, pen, files, greeting cards, visiting cards, bags, vanity bags, etc.

15.3.1 Handmade Paper Unit

The waste papers are used mainly for constructing handmade recycled papers. After the categorization of the waste, pure paper wastes are water-treated using a Hydro-Beater Pulper along with fibre cotton and water. It is then converted to pulp. The pulp is then pressed and dried to form rectangular sheets. Each unit of production is about 30 sheets. These thinned sheets are polished. From the polished rectangular sheets, files and folders, carry bags (small and big), writing/scribbling pads, books (small, medium and big), photo frames, photo albums (small and big), gift boxes, greetings/message cards and diaries are made by the skilled workers who are part of the *Parivarthana* Handmade Paper Unit. On-campus and off-campus orders are honoured simultaneously. This unit of *Parivarthana* thus makes economically viable, technically feasible, commercially profitable and environmentally suitable paper products.

15.3.2 Sorting and Compost Unit

A lot of waste is generated on the 20-thousand-student campus around the clock. To keep the University clean as well as accomplishing the zero-waste campus motto, a dedicated team collects those large quantities of waste products, such as food, fruits and vegetables, which get batched and chopped in special containers. The wet waste is then collected from the campus composting unit, an interesting standalone experiment to convert wet waste into manure or compost. As the first step the food, fruit and vegetable waste gets shredded and collected in a container. The dry leaves line the compost tank as the first layer and then wet waste covers it, culled from the campus and spread evenly in the composting tank. On the entire bed of leaves and food waste, just sprinkle a spoonful of moist bio-inoculam. Leach and drain out the water from the tank as and when it is needed. After the leached water is added to the tank and the content gets mixed, the waste starts composting after 25 days. It is then dissected further into fine powder by the compost shredder for gardening. The fertile manure is then scattered around the campus for mulching.

(A)

Sl. No	Type of Waste	Quantity (Kgs)
1	Dry Waste	82,023
2	Wet Waste (Food Waste)	149,920
3	Wet Waste (Compost)	20,701
4	Rejected Waste	13,931 (5.22%)
	Total	266,575

(B)

S. No	Items	Quantity (Kgs)	Amount INR
1	Sale of Waste—Drv	77.414	586,177
2	Fee Collection	From 18 stalls in Main Campus	386,390
3	Compost	2,466	49,530
4	Bio-Gas per Year	6,265	469,281
	Total		1,022,097

(C)

SL No.	Items	Quantity (Kgs) '	Sl. No.	Items	Quantity (Kgs) '	SL No.	Items	Quantity (Kgs)
1	**Colour papers**	20,154	6	**Newspaper**	4,040	11	**Aluminum**	842
2	**Carton box**	10,630	7	**Note books**	6,018	12	**Milk cover**	572
3	**White paper**	13,198	8	**Tetra packs**	1,932	13	**Polythene cover**	305
4	**Water bottle**	5,282	9	**Kadak**	1,467	14	**Others**	5,841
5	**Solid plastic items**	3,923	10	**Magazines**	3,210		**Total**	77,414

FIGURE 15.1 (A)—Particulars of waste collected (2019–2020). (B)—Revenue generated from *Parivarthana* activities (2019–2020). (C)—Details of recycled items sold through *Parivarthana* (2019–2020).

15.3.3 Biogas Model (Floating Dome ARTI and Biotech Model)

The live-demo water-jacket model of the Biotech and ARTI Biogas plants at the campus—they are like information and education centres for all diverse stakeholders of the university. Just local raw materials can erect the system, with skeletal skills to collate and assemble the materials. But, while they are functioning efficiently, note that the exceptional biogas plant has brought about change. A number of stakeholders and representatives from the community are dropping in to understand a number of its features. The CSA has erected an upgraded form of the Floating Dome ARTI and Biotech Model, located at the paper recycling unit. Its focus is on energy from food waste, which has brought about a mini revolution in the biogas world. It is designed to showcase a small, household-sized model. It uses a daily plant supply of about 2 to 3 kg, which generate gas to make tea for the unit's workers. The focus is to not just supply gas to the plant workers but also to showcase a reasonably priced, functional model for stakeholders.

15.3.4 Energy from Food Waste

The biogas plant, which has been recently revived, is central to demonstrate the production of energy from food waste and its disposal, utilization and management. The unit is important as a demonstration for stakeholders related to different kinds of efficient and impactful models in the University. The unit offers a clear contrast between diverse and efficient technologies and inputs for multiple stakeholders. The plant produces 500 kgs of food waste and 50 cubic feet of methane gas that is like a daily generation of 25 kgs of LPG. The Biotech digester has another interesting aspect to offer—multiple years of implementation research, demonstrating a working model that uses a daily supply of 500 kgs of food waste to offer significant gas to the canteen. A floating device with a dome-like roof has space for collecting biogas manufactured from slurry within. Another output tank is crafted such that it can immediately convert the waste into manure.

15.3.5 Sewage Treatment Plant (STP)

The main focus of water treatment at the University is to allow waste to be discharged so that it poses no threat to either health or the environment. Wastewater can be used optimally for irrigation, which is an effective way of disposing as well as recycling water. It is used in the University for vegetation as well as gardening, which is an important and unique treatment practice (30 KLD). The process allows wastewater to be drawn through pipelines and collected in a sump, once it is screened through bar-screen chambers, so that the floating and sinkable solids are sieved. The garbage is passed through a diffuser in a sump and then collected in an activated sludge process (ASP) system to minimize the pollutants. An air blower is linked to the grid piping, and the aeration tank is exposed to air. The wastewater sludge, too, is pulled into another settling tank or tube settler. This is the spot in which the solid sewage sinks right down to the bottom and the clearer supernatant overflows into an alternative tank. The superfluous sewage is directed to the ASP tank through pumps. The clarified water tank is disinfected with hypochlorite solution. Meanwhile, the secondary treated sewage is sorted through a tertiary treatment plant connected to pressure sand filters and activated carbon filters acting as polishers. Sludge is drawn out from the settling tank and passed on to drying beds.

15.4 WATER MANAGEMENT

Most of our water crises are brought about by negligent and careless management practices and our neglect of water and water bodies. In order to conserve water, CHRIST (Deemed to be University) has flagged off a few projects. Firstly, the university across campuses step up rain water harvesting and sewage-water-treatment systems, in which water can be treated and reused in restrooms and to irrigate plants and wash vehicles. Every day, 400 thousand litres of water are used up by over 21 thousand students, faculty and non-teaching staff. It is gathered after rainwater harvesting in large tanks. From this, 3 lakh litres of water is preserved and recycled for washing, flushing and watering the plants on the campus. The University's model of water conservation can be adopted by any other institution. Other steps to preserve and create awareness by the public about the water bodies in the city includes adoption of lakes, such as Arekere, Agara and Hulimavu lakes by the University's CSA and different academic departments.

Many programmes have been adopted to revive the lakes and optimize ecological responsibility. The drive included lake clean-up drives, door-to-door awareness drives, street plays and surveys, documentaries and even competitions to sensitize the public. The University partners with residents to go on campaigns to ensure that the government will be able to maintain the lakes. Moreover, the University partners with a number of NGOs and voluntary forums in order to flag off campaigns for community cooperation and water conservation. Also, joining hands with national movements such as #RallyForRivers helps to bring the University into public attention for ecological preservation. By partnering with these organizations, a number or awareness-creating competitions such as painting, stall making, meme making, street plays, photography and social media drives within and outside the University can encourage the youth. Rallies and drives like the 2017 National Cadet Corps 'Virtual Water' campaign helped to make people more aware about the amount of water that is wasted in making consumer products. There are collaborations with public works, ports and inland Water Transport Department—officially called the PWD. The attempt is to create awareness among citizens for conservation of water though digital and social media campaigns. They can be pioneered through contests, classroom assignments and sensitization drives. Even street plays and flash mobs at public places such as thoroughfares can draw attention; and other seminars, open debates and public talks by campaigners can drive the movement forward. During the Social Responsibility Week, the university organizes inter-deanery and intra-deanery competitions. Through these competitions, participating students and audiences are more routed in the need for water harvesting and recycling. We list here some of the items.

15.4.1 Stall

Each deanery was to set up a stall that elucidates the theme of 'Water Conservation' in creative and innovative ways, so as to spread the message of the Social Responsibility Week. The presentation could involve activities and games of various kinds. Only seven participants per deanery were allowed to participate in the competition. The judging criteria included innovative efforts, crowd attraction, comprehensibility and cohesion.

15.4.2 Slam Poetry

Participants were required to perform their poems based on the theme of 'Water Conservation.' The participants could be between 1 and 4 people. No plagiarism or vulgarity/profanity was allowed, nor was the usage of instruments or props. The judging criteria included originality, relevance to the theme, expressions and audibility.

15.4.3 Advertising Through Memes

The students are required to make memes on behalf of the Public Works Department, Ports and Inland Water Transport Department (the water regulating authority in Bengaluru) for sensitizing the citizens about the 'Need to conserve water.' Participants divided themselves into groups of three and designed at least two memes on behalf of the PWD to educate the public about the need for water conservation in the city. The memes were required to be submitted in soft copy in JPEG format. Popular culture references and dark humour were allowed to be used. No profanity/blasphemy/vulgarity/innuendoes were allowed. The judging criteria included humour, content, relevance to the theme and Likes on social media (which would be shared on the Centre for Social Action Page).

15.4.4 Mad Ads

This was a competition to sell the idea of water conservation through advertising a product that helps to conserve water. The students were to come up with an advertisement with jingles and taglines to market the product. Four members per team were allowed to participate. The total time provided was 25 minutes (20 minutes for research and preparation + 5 minutes to perform). The teams strictly needed to adhere to the time limit and were required to plot the ad around the water-conservation theme. The judging criteria included product chosen should minimize the usage of water, relevance to the theme, acting, catchiness of jingles and taglines that describe the product.

15.4.5 Photo Story

Participants were required to create a story using images and a paragraph describing each picture, which is part of the sequence that forms the story. The story must revolve around the theme: water conservation. The participants would have to do this in two days. Each team would have four participants and the story can have a maximum of 15 pictures. The pictures must be clicked by the participants. The story must have a title and must be relevant to the theme.

15.4.6 Obituary

Participants must personify water as a human being and creatively write and design an obituary on MS Word. The time given was one hour. Each team would consist of two members. The format would be the one which appears in The New York Times. The date of expiry can also be given. The judging criteria included creativity, format and content and design.

15.4.7 Painting

The participants have to describe their ideas and thoughts about water conservation through art. The judging criteria was creativity and aesthetics.

15.4.8 Observing Images

Participants need to analyze an image being flashed and present their thoughts. Each participant is given 40 minutes to write a paragraph of 350–500 words about their perceptions about the images. The still-image photos will be relevant to the theme: water conservation. The judging criteria included clarity of message, creativity, originality, impact and presentation.

15.4.9 Report on Water Conservation

The Deanery of Sciences students were required to work in teams to analyze the water consumption of Christ University or any one block of the University and write a report on it. The report must include the water consumption level, water consumption patterns and techniques to conserve water. Each team can have three members. No plagiarism was allowed. The word limit was 1,500 words. The students must submit the report in four days. The judging criteria included factual accuracy and the practicality of the techniques.

15.4.10 Improvise a Device

Students from the Deanery of Sciences were invited to take an existing device and redesign it so that it becomes efficient and conserves water. They will need to describe how the existing device or product is and how they intend to improvise it. They would also need to mention the functioning of the device once improvised. Each team can have three participants per team; the blueprint (a concise and brief report with a sketch) of the device was to be submitted at the end of the two days. No plagiarism was allowed. The criteria for judgment were on the innovative idea, conservation of water and the concise blueprint.

15.4.11 Debate

The students of the School of Law debated in two teams of five members on the topic: Is water conservation cost effective?

15.4.12 CSR Campaign Strategy

The management students participated in a competition in which they constructed a creative campaign strategy for conserving water to fulfil an imaginary company's CSR. Each team would have up to six participants.

15.4.13 Poster Designing (Online and Handmade)

Students were asked to participate in teams of 2–3 members and create both online and handmade posters. The theme for the competition was water conservation. Judging criteria included innovation and catchiness, reflection of the theme, originality and Instagram polls.

15.5 EDUCATION AND AWARENESS/RESEARCH AND PUBLICATION

The Centre for Social Action of the university organizes several innovative competitions with regard to social awareness regarding water conservation and usage by joining with various partner organizations, such as NGOs, Volunteering Forums and Social Work Organizations.

In the education system, awareness on waste segregation practices and levels of waste segregation are to be taught. In the university, apart from the above, using the waste as recycled paper, compost and biogas plant as an energy source and sewage treatment plant for water recycling are available for students to see and learn. An information centre on the foregoing is also available in the university. A total of 21 thousand students and faculty members are made aware of solid waste management practices in public and in domestic spaces. Beginning from their academic departments, students and faculty members are exposed to solid waste management systems practiced by the university. This is a training for life for them. Segregation at source is taught with the labeling of waste bins for particular types of wastes, viz., dry, wet, glass and paper. Students are encouraged to use these methods at home. Primary segregation is done with a large waste bin with multiple openings and multiple small waste bins. The primary segregation was successful, as it helped the university to segregate up to 70% of the waste by the faculty and students themselves. The people who are working with *Parivarthana* transfer the segregated and unsegregated waste for secondary segregation. The self-help group (SHG) members of the nearby Janakiram Layout collect the tetra-packs for further processing. All the papers and recyclable materials are processed in the *Parivarthana* itself. When papers are recycled into files, greeting cards, etc., tetra packs are used to manufacture aesthetically appealing vanity bags. The thin lamination on the tetra packs gives glazing looks to the bags and increases the durability of them. Student volunteers of the CSA narrate the entire process as stories using theatre, flash-mobs, dances, etc.

15.5.1 Information Centre

The University has a very important project named 'From Food Waste to Good Energy.' The information centre plays a key role in propagating the expectations of the project. It provides assistance and consultancies to people who are interested in knowing this sustainable energy project. Another important activity of the information centre is advocacy of the efficient use of water. Recycled water can be used for sanitation, gardening and cleaning purposes. The information centre arranges workshops, seminars and symposiums. The information centre is the publicity office for the university's waste management initiatives. Handbills, posters, pamphlets, brochures, short-videos, training modules, etc., are also supplied by the information centre. The centre also is keenly involved in supplying campaign materials in vernacular.

15.5.2 *Anuspandhana* Business Unit

A minor venture into social entrepreneurship is the *Anuspandhana* Business Unit. It has its primary motif in empowering women. Initially, the CSA imparted systematic training to women from various backgrounds to be the stakeholders in this business unit. They were trained various aspects of a business such as marketing, human resources management, accounts and business development skills. The skilling, reskilling and upskilling programmes were implemented through SHGs and other volunteer programmes. CSA started *Anuspandhana* Business Unit in March 2010. Initially, 20 women were trained to make high-quality products from used and discarded tetra packs and embroidered cement bags. The *Anuspandhana* is a registered company. The proceeds from the company go to the women who work with the business unit. A part of income is used for training the women in strong business and life skills as well as ensuring access to better health care and education for their children.

15.5.3 Centre for Social Research

The Centre for Social Research (CSR) at the University was established in 1999 to promote social researches, evaluation and impact studies. The CSR gives more focus on the evaluation and impact of the University level interventions implemented on and off the campus. The research team at the Centre closely observes, collects both qualitative and quantitative data from all the stakeholders of the waste management projects and interventions and provides feedback to the concerned centres for their improvements. The CSR also takes initiatives to organize capacity-building workshops for introducing more research-based models to student volunteers. It also inculcates research culture among the student volunteers and helps them to connect with more scientific and better practices of waste management from across the globe.

15.5.4 Centre for Publication and Centre for Concept Design

Though the Centre for Publication was established to promote research publications for the students and faculties at the University, the centre also supports the dissemination of waste management, and other students lead social and behavioral improvement projects on campus and to the public. The support from the Centre for

Publication is a huge avenue for the students to express, promote and publicize the need for waste management and communicate to the entire university's stakeholders about the significance of everyone's involvement in continuing to keep the campus zero waste through the university's magazines and newsletters.

The Centre for Concept Design (CCD) was established at the University to incorporate digital media as a pedagogical tool. The CCD also welcomes the students to use their Green View Studio to develop a few high standard promotional videos for educating the University's stakeholders to keep the campus clean and zero waste. The media professionals at the Centre also help/train the student volunteers to create their own memes and educational videos on waste management.

15.6 SUSTAINABILITY MODEL

The entire waste management programme is implemented at the university through a systematic model for its sustainability. The model has evolved through more than two decades of experimentation and evaluations. The university identified the significance of student ownership and peer tutoring/mentoring for effective transferring and sustainability of this zero-waste philosophy. Initially, instructions were given through university leadership; later, it was taken up by the Center for Social Action (CAS) through its diverse projects and awareness programmes. Today, the entire implementation of the zero-waste programme is completely planned, organized, implemented and evaluated by the student volunteers with the support of mainly five centres in the university, which are under the supervision of the Office of Student Services. Though all of these five centres are managed and led by students, each of these are utilizing different strategies and pedagogies to implement, educate and disseminate a zero-waste philosophy among students and faculty on campus.

15.6.1 Centre for Social Action (CSA)

The CSA comes up with diverse annual and capstone projects under the leadership of the student volunteers. They come up with specific projects like Clean My Village, Clean Up Meet, Lake Rejuvenation, River Care, Neighborhood Care, etc. They adopt villages, lakes and roads for a year and then hand them over to their juniors. They conduct diverse awareness programmes. They also do some competitions. The CSA's *Parivarthana* (recycling wing) is fully set aside as an office with trained professionals to train the students, implement and supervise the projects on campus and off campus.

15.6.2 Student Welfare Office (SWO)

Student Welfare Office was established to identify and develop the student's artistic and cultural talents. The centre also gets involved with zero-waste awareness programme utilizing performing arts like music, theater, dance, etc. They also organize art and literary competitions based on this topic. SWO volunteer body is the true ambassadors and role models carry through this vision by showcasing themselves as role models on campus. The SWO media team creates a lot of memes to promote and educate through this zero-waste campaign.

15.6.3 Sustainable Development Goals Cell (SDGC)

SDG cell on campus organizes conferences, workshops and seminars on all the 17 SDGs. Since Zero-Waste philosophy is so rooted on the university campus, they focus more on SDGs 6, 11 and 12 (Figure 15.2).

15.6.4 National Cadent Corps (NCC)

At the University, NCC cadets are trained inside college during regular parades and are both mentally and physically prepared to become sincere and efficient civilians to protect and promote organizations and the country's core values. With this intention, the cadets are strictly trained, and they are oriented to conduct various social-awareness campaigns. Cadets are sent to different locations at various intervals to clean the wastes and spread awareness of waste management. They also do many waste management campaigns and events inside and outside of the campus.

They join *Swatch Bharath Abhiyan*, India's waste management initiative, so-called to spread awareness to people about keeping India clean. The movement encourages people to take strict care of their surroundings and to demotivate anyone trying to

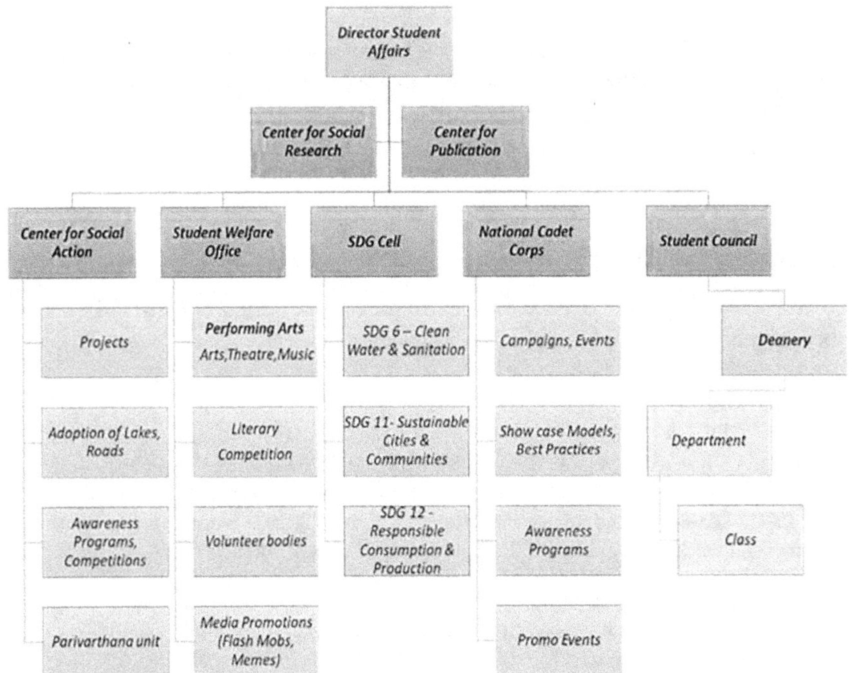

FIGURE 15.2 University Sustainability Model.

dispose of waste in the road and other public places. The cadets put their full efforts into creating posters, role plays and slogans which are used during such events. The cadets identify waste-dump locations in the neighbourhood, and they clean the area. They also plant some trees there and erect posters to remind and educate the public to reduce the unethical waste disposal method of people.

15.6.5 Student Council

Student Council is the governing student-leadership body, hierarchically organized from the university representatives to deanery level to departmental level to the classroom. This established and systematically organized student leadership is an effective mode of communicating, regulating and establishing any policies of the university. It has got its legitimate and peer-derived power to influence classmates towards its vision (Figure 15.1-C).

15.7 CONCLUSION

Educational institutions have a significant role in helping to protect the environment. Through the curriculum and design of the teaching-learning activities, waste management skills have to be inculcated to the students so that a generation of well-informed citizens will be formed. CHRIST (Deemed to be University) in India believes in the holistic education of its students. Hence, as part of its hidden curriculum, the University educates its students to love and protect the environment.

In 2019–20 alone, a total of 166,575 kilograms of waste were generated on the campus (see Table 15.1-A). Through an efficient waste management procedure, 14 items were segregated (see Table 15.1-C) and are either sold or recycled. Around a million INR had been procured from the waste sold in the campus. These facts and the various processes at the university present a unique and sustainable model of waste management. The model is replicable in other Indian universities.

ACKNOWLEDGEMENT

The authors would like to thank the Centre for Social Action of CHRIST (Deemed to be University) for supplying the necessary data.

REFERENCES

de Vega, C. A., Benítez, S. O., & Barreto, M. E. R. (2008). Solid waste characterization and recycling potential for a university campus. *Waste Management, 28*, S21–S26.

Dutta, J., Kerketta, A., Choudhury, M., Ranga, M. M. (2022). Women warriors of waste management. In C. Baskar, S. Ramakrishna, S. Baskar, R. Sharma, A. Chinnappan, & R. Sehrawat (Eds.), *Handbook of solid waste management*. Springer. https://doi.org/10.1007/978-981-16-4230-2_104

Kumar, S., et al. (2017). *Challenges and opportunities associated with waste management in India*. Royal Societies.

Meneses, G. L. (2015). Electronic waste (e-waste) management at selected colleges of the Bulacan State University: Perspectives for program development. *Open Access Library Journal, 2*(04), 1.

Mondal, T., Choudhury, M., Kundu, D., Dutta, D., & Samanta, P. (2023). Landfill: An eclectic review on structure, reactions and remediation approach. *Waste Management*, *164*, 127–142. https://doi.org/10.1016/j.wasman.2023.03.034

Stephens, J. C., Hernandez, M. E., Roman, M., Graham, A. C., Scholz, R. W. (2008). Higher education as a change agent for sustainability in different cultures and contexts. *International Journal of Sustainability in Higher Education*, *9*(3), 317–338.

Zhang, N., Williams, I. D., Kemp, S., & Smith, N. F. (2011). Greening academia: Developing sustainable waste management at Higher Education Institutions. *Waste Management*, *31*(7), 1606–1616.

16 Biochar Production from Crop Residues

M. Vigneshwaran, A. Varalakshmi, G. Sunil Kumar, M. Karan, V. Santhosh Kumar, G. Sajith Kumar, Einstein Mariya David, Saiyyeda Ferdous, and Theivasigamani Parthasarathi

16.1 INTRODUCTION

Organic carbon in the soil is an essential soil health quality indicator; it has a crucial function in various biogeochemical activities in the mud (Doetterl et al., 2015). Soil acts as a source and sinks of carbon (Solomon et al., 2007; Stavi & Lal, 2013), and a huge amount of carbon (1500–2000 Pg) has accumulated in the shape of organic carbon in soil (Wei et al., 2015). About two-thirds of the carbon in the terrestrial ecosystem is made up of soil organic carbon (SOC) that is actively exchanged with atmospheric constituents (Genxing et al., 2003). In the past 20 years, emissions of carbon dioxide to the atmosphere have increased tremendously due to the anthropogenic activity field (Woolf et al., 2010), thereby triggering an alarming situation for the earth by increasing global temperature, extreme weather events like flood and drought, glacier melting, rise in sea level, etc. (Hansen et al., 2017). Therefore, a small change in soil organic carbon sink or source greatly affects the global carbon cycle.

The concentration of carbon dioxide has increased from 300 to 408 ppm in the last 50 years, and it is projected that, by the end of this decade, it will soar up to 600–700 ppm; thereby, the surface temperature will be increased up to 4.5 to 5 °C (Anwar et al., 2018). The fundamental concept of carbon sequestration is to trap the atmospheric carbon into a soil organic carbon pool and stabilize them for a long duration of time so that the concentration of carbon dioxide in the atmosphere can be reduced. With particular discerption of the source and sink as well as a measurement and monitoring technique (Olson et al., 2014), "The process of moving CO2 from the environment into the soil of a land unit through unit vegetation, vegetation wastes, and other organic solids that are stored or kept in the unit as a component of the soil organic matter," according to the definition (humus). Short-period (not instantaneously released back into the environment) to long-period (millennia) retention of sequestered carbon in the ground (terrestrial pool) is possible. During and after research, the sequestered SOC process should enhance net SOC accumulation over the earlier pre-treatment threshold.

The rise in population, rapid urbanization, and industrialization, along with the rising living standard due to modernization, have influenced the huge production of solid wastes. These wastes are generated from agro-industrial wastes, municipal solid-trash, marine waste, forestry waste, etc.; only a little portion of agroforestry

DOI: 10.1201/9781003258377-16

garbage is used for internal and farming purposes like cooking, feedstock, and composting; in addition, the rest of the waste materials are disposed of through burning or dumping into the field, which pollutes the environment. It has been predicted that, globally, almost by 2025, 19 billion tonnes of solid debris would have been produced yearly (Yoshizawa et al., 2004).

Decomposition of organic waste through composting is a time-intensive process and laborious job, and during decomposition, it releases some greenhouse gases like CH_4, CO_2, and N_2O (Sánchez et al., 2015); therefore, composting is not a good alternative and beneficial strategy for massive waste management, whereas biochar production is an easy, rapid, eco-friendly as well as cost-effective process (Gayathri et al., 2021; Hemavathy et al., 2020), and it holds the ability to increase carbon retention in soil, including curtailing the effect of greenhouse emissions (Woolf et al., 2010). Biochar is a long-lasting substance that is a carbonaceous, permeable component produced under a low level or absence of oxygen through the technique of pyrolysis (exothermic disintegration) with organic substances (Atkinson et al., 2010), and it has high aromaticity, high thermal stability, and strong absorbability. Furthermore, it can sequester carbon to mitigate the effects of climate change, decrease carbon dioxide emissions to the atmosphere by reducing the return of photo-synthetically fixed carbon (Enders et al., 2012), restore degraded land, purify water by eliminating contaminants (Inyang & Dickenson, 2015), and improve soil fertility (Randolph et al., 2017). Municipal solid waste generation increases day by day throughout the world, and it became challenging to manage the waste efficiently. There is a positive correlation between economic growth, population growth, urbanization, and industrialization (Rana et al., 2015). Rapid urbanization and industrialization create an extra load on society and the ecosystem as a result of the unrestricted use of environmental assets (Gerdes & Gunsilius, 2010). Figure 16.1 shows the wastes generated from the different sources, including municipal solid waste, construction and demolition, industrial, commercial, reservoir, effluent treatment, garbage management and land restoration, and energy production units (UNEP, 2015). Almost 33% of the municipal garbage output is not handled in an environmentally safe way on a yearly basis (S. Wang et al., 2020). The open scrapping of municipal solid-garbage is usual in economically developing nations; it discharges several types of contaminants which directly affect the environment (soil, water, and air). This waste discharges greenhouse gases, particularly methane gas, volatile organic compounds, potentially toxic elements, leachate, etc. (Sridevi et al., n.d.). Open dumping of waste creates two major problems: the first is the greenhouse gas emissions (particularly CO_2 and CH_4); and the second is the penetration of leachates into the soil profile and groundwater (Zhao et al., 2016). Methane gas is more potent to global warming than carbon dioxide. In 2010, it was estimated that the annual global emission of methane gas was 31MT from the solid waste landfills—i.e., 11% of the total global emission (GMI, 2013). They also release odour because of organic mixtures such as ethylbenzene, toluene, xylene, benzene (Kumarathilaka et al., 2016), and hazardous substances (Zhang & Sun, 2018) and create bad sights. Toxic substances (including heavy metals) released from the open dump are hazardous to humans, plants, animals, soil, and water bodies. Therefore, efficient management of solid waste to mitigate the soil, air, and water pollution is very important, and it can be performed through the conversion of trash into a renewable energy source. Solid waste can be managed with various methods—specifically, incineration, anaerobic digestion, vermicomposting, pyrolysis, and composting. The majority of these management methods contribute less to soil C storage because of the fast decomposition rate of organic carbon; as a result, carbon dioxide emission is high and less efficient in sustaining the C in soil (Lehmann, 2007; Paustian et al., 2016). Biochar fabrication through pyrolysis is getting more attention in the past two decades, owing to its multiple benefits like management of waste, mitigating climate change, energy generation, soil amendments, etc. (Awad et al., 2018; Usman et al., 2016). This approach will benefit the environment with the detrimental consequence of solid waste in the near future.

Land degradation and declining soil fertility are extensively documented as a global problem linked with reduction of organic-carbon content, major and micronutrients depletion, desertification, salinization, soil erosion, acidification, soil sealing, soil contamination, waterlogging, loss of biodiversity, etc. This deterioration affects a large number of people in the world. Land degradation plays a crucial role in climatic variations because carbon dioxide and nitrous oxide emission into the atmosphere is greater in degraded land. Land degradation is directly interlinked with the soil quality indicators (pore size distribution, aggregate stability, bulk density, saturated hydraulic conductivity, surface crust, electrical conductivity, cation exchange capacity, pH, soil organic matter soil, microbial biomass carbon, and microbial respiration); it has been reported by several researchers that soil health quality deterioration is greater in the area where land degradation is high. As per the FAO, desertification is classified as a deterioration in soil health that causes the environment's ability to support products or services to its users to be reduced (FAO, 2010). A shrinking in organic matter content in the soil leads to a decrease in soil fertility.

16.2 BIOCHAR—PRODUCTION AND BENEFITS

Pyrolysis is a process of thermochemical disintegration of biomass when oxygen is scarce or absent. It occurs at a high temperature, ranging between 300 and 700 °C. During this process, long-chain hydrocarbon is broken down into simpler molecules, and a new, solid product is formed, known as biochar (solid), along with this bio-oil—a condensable liquid—and non-condensable gases (syngas); namely, carbon mono oxide, carbon dioxide, hydrogen, and methane are formed as the co-products (Tan et al., 2020; Ye et al., 2020). The pyrolysis procedure can be categorized into two subgroups; namely, rapid pyrolysis and slow pyrolysis. The basic disparities between slow pyrolysis and rapid pyrolysis are atmospheric conditions, continuance, and heating rate. In terms of physical characteristics, the biochar is black in colour (like charcoal), porous, lightweight, and it has a fine texture and substantial surface area. Biochar contains 70% of carbon (C), and the remaining percentage contains nitrogen (N), hydrogen (H), and oxygen (O), in terms of other elements. The class of biomass and the methods used to produce biochar influence the physical and chemical properties of biochar. Biochar is different from charcoal. The methods used to make charcoal and biochar are different in terms of heat; charcoal is made at 400 °C whereas biochar is made from 600 to 1000 °C (Cairns et al., 2021). Charcoal can harm the environment, whereas biochar will not (Cairns et al., 2021).

Biochar as a soil amendment has numerous advantages (Josiah, 2010) like: increases yield, lowers acidity of the soil, promotes seed germination, increases soil fertility, decreases nutrient runoff, improves water retention, improves availability of carbon in soil, makes the plants stronger, decreases contamination of soil, and improves soil health.

Biochar can improve the growth of plants in poor soil. Most of the plants require a pH of 6.5 to 7 for good growth and development, but most of the soil in the world is acidic and has a pH of 4 to 4.5. In acidic soils, the plants can't take up nutrients that are available in the soil. When biochar is blended with soil, it increases the soil pH, which ensures the uptake of nutrients by the plants.

Biochar is attracted to chemicals of all kinds. Gently stick clean carbon biochar to the soil, and it is covered with scales of magnesium, iron, phosphorus, potassium, calcium, and sulfur after six months, which leads to the formation of a mineral ball. When a combination of biochar and fertilizer is applied to the soil, that promotes a slow release of nutrients and remains for a longer period when compared to the application of fertilizer alone. Biochar can easily adsorb metals like cadmium, lead, and mercury when plants can't take them up.

16.3 BIOCHAR—THE KEY INGREDIENT

16.3.1 Plant and Animal Biomass

P-rich biomass residues: sewage sludge from the municipality, biogas fibre, manure from cattle, and excrement from chickens.

C-rich biomass residues: wood chippings, wheat straw, rice, bamboo, and corn/maize are excellent biomass products.

16.3.2 Food-Processing Residues

Food waste like meat, pasta, rice, bread, noodles, vegetable peelings, etc., is used as biomass to produce biochar. The biochar that is created from lawn scraps speeds up the deterioration and mineralization rates of food waste material into compost. 350 and 450 °C are the two different temperature ranges that were applied at the rate of 10 and 15% (w/w). Valuation of food waste composting is done by overall food waste in-vessel compost bioreactor (Waqas et al., 2018).

16.3.3 Forestry Cutting

Forestry cutting products like woody plant material is used to make biochar in the process of pyrolysis; in this method, wood materials are heated for hundreds of degrees without burning. In pyrolysis, gases are formed that are captured as a source of energy and used for turbine or internal combustion engines for heating. Yearly, 41 million tonnes of charcoal are used for industrial purposes worldwide, but the disorganized production method leads to massive deforestation, emission of GHGs, and other pollutants. GHG emissions arising from land every year lead to "slash-and-burn" agriculture with a "slash-and-char" system (Laghari et al., 2021).

16.4 BIOCHAR PROPERTIES

16.4.1 Nutrient Retention of the Soil

Loss of nutrients during the application of biochar mainly depends on the soil's nutrient retention properties. In non-calcareous soils, the retention of the P is based on the soil nature but not on the feedstock (biomass) nature. As a result, applying biochar produced from the same biomass type at the same rate may have varied effects, depending on the specific soil conditions. For instance, when an equal quantity of biochar is added to clay soil and sandy soil, the sandy soil starts to emit P more quickly compared to the clay soil. The temperature during which biochar is formed might have no influence on the P-emitting characteristics of the soil. As a result, biochar made using complex or basic processes would most likely function similarly, depending on the soil type.

16.4.2 Development of Soil Aggregates and Stabilizing of Soil Organic Matter

Biochar's impact on soil aggregation and organic matter stability varies according to the soil condition.

1. While biochar amendment has a significant impact and improves aggregate stability in fine-textured soil, it had no effect on coarse-textured soil.
2. Biochar significantly improves the C absorption in macro aggregates of finely textured soil and increases the physical shield of SOM (soil organic matter stabilization) in the soil.

16.4.3 Soil Microbial Properties

The biochar's influence on soil microbiological properties includes soil respiration, nitrogen mineralization rates, enzyme activities of bacteria-fungi ratio, and soil-borne pathogens.

16.4.4 Soil Physical Properties

- Bulk density is reduced from 3–31%.
- 14–64% porosity is increased.
- Penetration resistance has no effect, and it is limited.
- 226% increases the instability of wet aggregate.
- Consistency of the soil is improved.
- The stability of dry soil aggregate has mixed effects
- 130% increase in the availability of water.
- Saturated hydraulic conductivity is increased in soil that is fine-textured and decreases in soil that is coarse-textured.
- Particle density and tensile strength are reduced.
- Infiltration of water is altered.
- Moderates soil thermal properties.

- The effect on water repellence of soil is minimal.
- Overall, it enhances the physical environment of the soil (Blanco-Canqui, 2017).

16.4.5 Soil Sequestration of Biochar and Amelioration of Climate Change

The application of biochar to agricultural lands is widely seen as a soil-based greenhouse-alleviation approach for long-term environmental sustainability. Using biochar as a source of nutrition C sequestration should be increased, both in biomass and in the soil. The biochar with lower N content was proven to be more appropriate for minimizing N_2O excretions from the soil, and those generated at extremely high temperatures may have an enormous capability for C sequestration. Biochar's C resists decomposition, which is one of its most essential characteristics—i.e., can hold C in soils for 100–1000 years, as demonstrated by the *terra preta* soil of the Amazonian region of Northern Brazil. Only a tiny fraction of biochar C (3%) is bioavailable, according to a concept of disintegration and priming impact on biochar stability in soil, while the rest contributes to long-term soil stability. Furthermore, converting forestry and agricultural waste into biochar might lower CO_2 and CH_4 greenhouse emissions from feedstocks during the breakdown of organic matter (Singh et al., 2022; Nair et al., 2017). Biochar properties in soil permeability include: minimizes bulk density, progressive soil aggregation, engages with soil mineral particles, soil packing reduction, and increases soil porosity by reducing particle density.

16.5 METHODS OF BIOCHAR PREPARATION

16.5.1 Pyrolysis

It's a thermal disintegration process of biological material in an anaerobic condition within the range of temperatures (250–900 °C) named pyrolysis. Under this condition, depolymerization occurs, which breaks the linkage between lignocellulose. When the temperature increases, the syngas production also increases, which further decreases the production of biochar. Mainly, pyrolysis is categorized into two processes: rapid and slow pyrolysis based on the increased thermal reading, temperature, continuance, and pressure.

16.5.1.1 Fast or Rapid Pyrolysis

Rapid pyrolysis is popular to produce bio-oil. The heating rate of feedstock is very fast about 10–10000 °C min^{-1}, heated at an extremely high temperature (600–1000 °C) and a short residence time (< 2 seconds) to boost the oil production (Choi et al., 2017; D. Wang et al., 2020). Fast pyrolysis produces 75% bio-oil, 12% biochar, and 13% syngas (Tomczyk et al., 2020). In this method, rapid disintegration of feedstock occurs and produces vapour and biochar. After condensation of vapours, liquid that is dark brown in color (bio-oil) is collected. Due to high temperature in fast pyrolysis, the production of biochar is comparatively less, as compared to slow pyrolysis, because feedstock is quickly heated and releases vapour. These pyrolysis vapours have less continuance in the extreme heat territory; they leave less carbon behind (D. Wang et al., 2020).

16.5.1.2 Slow Pyrolysis

Slow pyrolysis, also known as carbonization, emerges as a result of a lengthy procedure—i.e., reduced warming rate and lengthy residence time (D. Wang et al., 2020). The primary goal of slow pyrolysis is to maximize the production of biochar. It is commonly operated at low to moderate temperature—i.e., 300–700 °C with a heating rate of 5–7 °C min^{-1} and a long continuance (min to an hour) (Pandey et al., 2020). At this temperature, a complex organic compound like lignin has very little thermal degradability; therefore, a large portion of the carbon in feedstocks is preserved as a biochar (Tan et al., 2020). Slow pyrolysis produces 35–45% biochar, along with 25–35% bio-oil and 20–30% syngas (Lai et al., 2013; Méndez et al., 2015).

16.5.2 Hydrothermal Carbonization

It is considered an economically feasible technique for making biochar because it can be done at a low temperature of about (180–250 °C). During the process, the biomass along with water combined and is housed in a controlled reactor. To increase the stability, we need to raise the temperature. The products that are produced at different temperatures are as follows: CO, CH_4, CO_2, and H_2 are generated at a temperature above 400 °C, mentioned as hydrothermal carbonization.

16.5.3 Gasification

Gasification is a thermochemical decomposition of biomass into a gaseous product known as syngas. It comprises carbon monoxide, CO_2, methane, H_2O, and hydrocarbon in the presence of gasifying agents like air, O_2, steam, etc., and high temperature (> 800 °C). Gasification is an effective thermochemical process by which the syngas is produced as a key product, whereas biochar is produced as a by-product. Gasification is a two-step process: step one is dying of biomass feedstock; and the second step is oxidation and combustion of feedstock. During drying, moisture is being processed that is present in biomass, which is evaporated without any sort of energy recovery. The oxidizing and burning interactions of gasification agents are the primary energy sources for the process, which react with the dry feedstock available in the gasifier and release pyrolytic volatiles (Yaashikaa et al., 2020). The disintegration rate of biomass differs with temperature, and it depends upon the complexity of the organic compound. The overall decomposition of biomass ends at nearly 400–500 °C, where biochar is produced as a key product and further elevation of temperature leads to secondary decomposition to produce syngas (Mohan et al., 2006).

16.5.4 Torrefaction and Flash Carbonization

Torrefaction is also known as mild pyrolysis; the feedstock is heated in this procedure between 200–300 °C in the existence of an inert environment at a modest rate of heating and long continuance (20–120 min) (L. Wang et al., 2017). The inert atmosphere-air is used to remove the moisture, oxygen, and carbon dioxide present in biomass feedstock (Yu et al., 2017). Decarboxylation and dehydration are the key reactions in the degradation of feedstock in the torrefaction process (Pala et al., 2014). The process of torrefaction could be categorized into sets of three, viz., oxidative torrefaction, wet torrefaction, and steam torrefaction (Yaashikaa et al., 2020).

16.5.4.1 Steam Torrefaction

Steam torrefaction is usually performed at a temperature not exceeding 260 °C with a continuance of about 10 min.

16.5.4.2 Wet Torrefaction

In this process, biomass is introduced to the condition of a temperature of 180–260°C under a period of 5–240 min in the presence of water.

16.5.4.3 Oxidative Torrefaction

The oxidative torrefaction procedure is performed in presence of oxidizing agents (gases). About 40% of biomass is transformed into solid products, and this process of torrefaction declines with an increase in pressure.

Table 16.1 shows slow pyrolysis produces more biochar due to its slow heating rates and long continuance. In contrast to fast pyrolysis, the heating rate is high and short residence time leads to high production of bio-oil. Torrefaction is the most efficient technique to produce the maximum amount of biochar, whereas gasification can produce the maximum amount of bio-oil.

16.5.5 Thermochemical Degradation Process

TABLE 16.1
Thermochemical Degradation Process

Method of degradation	Temp. °C	Residence time	Heating rate °C min-1	Pressure	Aeration	Product: Biochar (%)	Product: Bio-oil (%)	Product: Syngas
Slow pyrolysis	300–700	Min to hr.	0.1–10	Vaccum atmospheric	Oxygen free or limited	35	30	35
Fast pyrolysis	600-1000	Seconds	10–10000	Atmospheric	Oxygen free	12	75	13
Mild pyrolysis (steam torrefaction)	290	10–60 min	<= 50	Atmospheric	Limited or no oxygen	80	0	20
Hydrothermal carbonization (wet torrefaction)	180–260	5–240 min	—	—	—	50–80	5–20	2–5
Oxidative torrefaction	260		—	—	—	—	—	—
Gasification	> 800	10–20 second	—	Elevated atmospheric	—	5	85	10

16.6 BIOCHAR ON ABIOTIC STRESS ALLEVIATION

Biochar is an organic compound prepared from the combustion of biomass in limited- or zero-oxygen environments, which is carbon-rich and can amend the properties of soil. This property of the biochar enables it to show some mitigation on drought and salinity. Salinity in the soil causes many adhere effects on plant growth such as plant height, yield, osmoregulation, and nutrient uptake. Drought stress, combined with salinity, emerges to be the most frequently occurring abiotic stress across the world. Advanced irrigation technology, along with soil amendment by the application of biochar, further enhances water efficiency, promotes nutrient retention, and improves overall soil health. The application of biochar shows some positive effects on plant growth and yield, even under stressed conditions. It also significantly improves nutrient uptake and photosynthesis. The chlorophyll content of the leaf increases, along with which leaf nitrogen content decreases. The soil's water retention capacity improves dramatically under the impact of biochar. Whereas, in plants, there is a decreased amount of Na^+ uptake and an increase in K^+ uptake (since the soil's water-holding capacity increases, it also increases the soil moisture content, which causes a dilution effect in plants), it also helps in the regulation of stomatal movement and phytohormones. In addition, biochar also contains calcium, magnesium, and other nutrients which are essential for plant growth. Biochar could immobilize the soil N and immobilize and decrease the N availability due to its high C:N ratio when applied in an improper amount, even though there is a considerable increase in the yield in plants. Biochar also absorbs the excess Na^+ available in the soil, which directly reduces the salinity of the soil.

16.7 BIOCHAR EFFECTS ON PLANTS

Biochar enhances soil fertility, thereby promoting plant growth parameters, alleviating aluminium toxicity in the soil and decreasing the soil acidity. Pyrolysis method results in highly alkaline biochar. As the alkalinity of biochar increases, acidity decreases. Applying biochar to low-nutrient soil has been shown to increase the availability of nutrients, increase internal porosity, and increase plant biomass and the presence of non-polar and polar surface sites on biochar, all of which play an important role in metallic uptake and the marginal effect. Biochar integration into acidic soil reduces soil Al toxicity through increasing soil exchangeable base cations, which improves the fertility of the soil (Shetty & Prakash, 2020) (Figure 16.1).

16.8 POSITIVE AND NEGATIVE EFFECTS OF BIOCHAR

Positive effects of biochar include reduced leaching of glyphosate into the soil, reduces nitrogen loss from the soil, used as a soil improver to lessen possible environmental concerns, and high plant productivity.

Negative effects of biochar include erosion results in land loss, compaction of the soil during the application, contamination risk, removal of crop residues, and reduction in worm life rates.

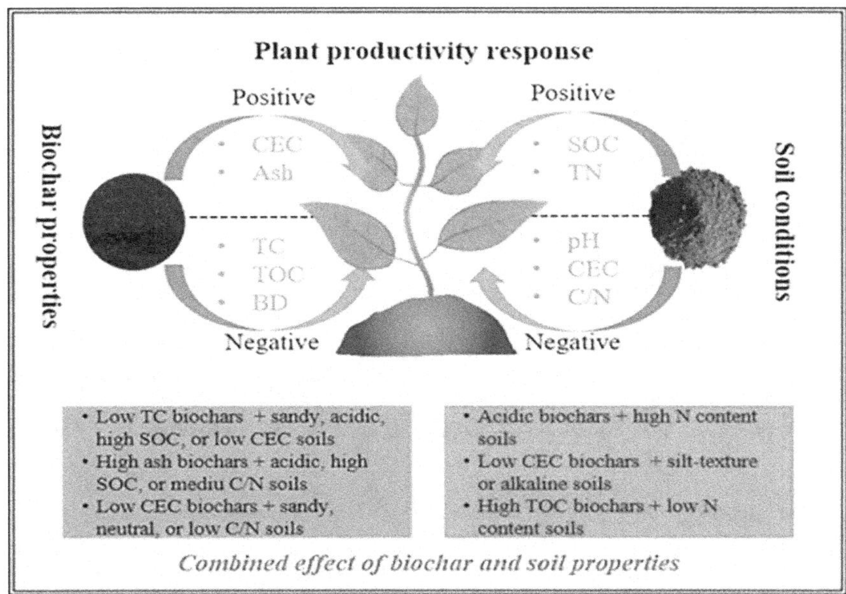

Abbreviation: CEC: cation exchange capacity; TC: total carbon content; TOC: total organic carbon content; BD: bulk density; TN: total N content; SOC: soil organic carbon content

FIGURE 16.1 Effect of biochar and soil properties (Dai et al., 2020).

16.9 CONCLUSION AND FUTURE PROSPECTS

The application of biochar can improve food security and increase cropland diversity by enhancing the soil nutrient availability, soil pH, soil porosity, and water retention, resulting in a higher yield of the crops. Biochar acts as an alternative tool to fertilize the soil in the areas with low rainfall and low-quality soil. Further, it improves soil aggregation and soil microbes, increase soil respiration, enhances carbon stability, reduces N_2O emission, and promotes plant growth. In this chapter, we discussed the importance of biochar in plant growth under poor soil as well as discussed the characteristics of biochar. Biochar is readily fabricated from biomass waste materials; after pyrolysis, the C-rich material can be used as soil organic amendment. In many studies, it is used as a soil amendment to fertilize poor soil and increase crop production. Biochar mitigates abiotic stresses by regulating ion homeostasis, improving photosynthetic processes and reducing oxidative damage and heavy metal uptake in tomato, maize, and other crops. Limited application of biochar significantly reduces plant disease, it was observed. The supplementation of biochar in sandy-loam soil with reduced irrigation results in increased yield and fruit quality. The combination of biochar and the type of soil significantly affects crop production, but it was not well defined. In the future, the ratio of biochar and soil will be recommended for the farmers to use biochar technology.

REFERENCES

Anwar, M. N., Fayyaz, A., Sohail, N. F., Khokhar, M. F., Baqar, M., Khan, W. D., Rasool, K., Rehan, M., & Nizami, A. S. (2018). CO2 capture and storage: A way forward for sustainable environment. *Journal of Environmental Management*, *226*, 131–144. https://doi.org/10.1016/J.JENVMAN.2018.08.009

Atkinson, C. J., Fitzgerald, J. D., & Hipps, N. A. (2010). Potential mechanisms for achieving agricultural benefits from biochar application to temperate soils: A review. *Plant and Soil*, *337*(1), 1–18. https://doi.org/10.1007/S11104-010-0464-5

Awad, Y. M., Lee, S. S., Kim, K. H., Ok, Y. S., & Kuzyakov, Y. (2018). Carbon and nitrogen mineralization and enzyme activities in soil aggregate-size classes: Effects of biochar, oyster shells, and polymers. *Chemosphere*, *198*, 40–48. https://doi.org/10.1016/j.chemosphere.2018.01.034

Blanco-Canqui, H. (2017). Biochar and Soil Physical Properties. *Soil Science Society of America Journal*, *81*(4), 687–711. https://doi.org/10.2136/SSSAJ2017.01.0017

Cairns, S., Chaudhuri, S., Sigmund, G., Robertson, I., Hawkins, N., Dunlop, T., & Hofmann, T. (2021). Wood ash amended biochar for the removal of lead, copper, zinc and cadmium from aqueous solution. *Environmental Technology and Innovation*, *24*, 101961. https://doi.org/10.1016/j.eti.2021.101961

Cha, J. S., Park, S. H., Jung, S. C., Ryu, C., Jeon, J. K., Shin, M. C., & Park, Y. K. (2016). Production and utilization of biochar: A review. *Journal of Industrial and Engineering Chemistry*, *40*, 1–15. https://doi.org/10.1016/J.JIEC.2016.06.002

Choi, J. H., Kim, S. S., Ly, H. V., Kim, J., & Woo, H. C. (2017). Effects of water-washing Saccharina japonica on fast pyrolysis in a bubbling fluidized-bed reactor. *Biomass and Bioenergy*, *98*, 112–123. https://doi.org/10.1016/j.biombioe.2017.01.006

Dai, Y., Zheng, H., Jiang, Z., & Xing, B. (2020). Combined effects of biochar properties and soil conditions on plant growth: A meta-analysis. *Science of The Total Environment*, *713*, 136635. https://doi.org/10.1016/J.SCITOTENV.2020.136635

Doetterl, S., Stevens, A., Six, J., Merckx, R., van Oost, K., Casanova Pinto, M., Casanova-Katny, A., Muñoz, C., Boudin, M., Zagal Venegas, E., Boeckx, P., Doetterl, S., Stevens, A., Six, J., Merckx, R., van Oost, K., Casanova Pinto, M., Casanova-Katny, A., Muñoz, C., . . . Boeckx, P. (2015). Soil carbon storage controlled by interactions between geochemistry and climate. *NatGe*, *8*(10), 780–783. https://doi.org/10.1038/NGEO2516

Enders, A., Hanley, K., Whitman, T., Joseph, S., & Lehmann, J. (2012). Characterization of biochars to evaluate recalcitrance and agronomic performance. *Bioresource Technology*, *114*, 644–653. https://doi.org/10.1016/J.BIORTECH.2012.03.022

FAO. (2010). Land degradation assessment in drylands. *Seventh Session of the Conference of the Parties, Item 10 of the Provisional Agenda*, 4684–4684. www.google.com/search?q=FAO+2014+Land+Degradation+Assessment+in+Dryland.+Retrieved+&rlz=1C1CHBF_enIN887IN887&ei=E8ZfYrq8CfKgseMPtNSPwAc&ved=0ahUKEwj6kYehnqL3AhVyUGwGHTTqA3gQ4dUDCA4&uact=5&oq=FAO+2014+Land+Degradation+Assessment+in+Dryland.+Retriev

Gabhane, J. W., Bhange, V. P., Patil, P. D., Bankar, S. T., & Kumar, S. (2020). Recent trends in biochar production methods and its application as a soil health conditioner: A review. In *SN applied sciences* (Vol. 2, Issue 7, pp. 1–21). Springer. https://doi.org/10.1007/s42452-020-3121-5

Gayathri, R., Gopinath, K. P., & Kumar, P. S. (2021). Adsorptive separation of toxic metals from aquatic environment using agro waste biochar: Application in electroplating industrial wastewater. *Chemosphere*, *262*. https://doi.org/10.1016/J.CHEMOSPHERE.2020.128031

Genxing, P., Lianqing, L., Xuhui, Z., Jingyu, D., Yunchao, Z., & Pingjiu, Z. (2003). Soil organic carbon storage of china and the sequestration dynamics in agricultural lands. *Advances in Earth Science*, *18*(4), 609. https://doi.org/10.11867/J.ISSN.1001-8166.2003.04.0609

Gerdes, P., & Gunsilius, E. (2010). The waste experts : Enabling conditions for informal sector integration in solid waste management. In *Management*. GTZ. www.gtz.de/recycling-partnerships

GMI. (2013, September). Successful applications of anaerobic digestion from across the world. *Global Methane Initiative (GMI)*, 1–24. www.globalmethane.org/documents/GMI Benefits Report.pdf

Hansen, J., Sato, M., Kharecha, P., Von Schuckmann, K., Beerling, D. J., Cao, J., Marcott, S., Masson-Delmotte, V., Prather, M. J., Rohling, E. J., Shakun, J., Smith, P., Lacis, A., Russell, G., & Ruedy, R. (2017). Young people's burden: Requirement of negative CO2 emissions. *Earth System Dynamics*, *8*(3), 577–616. https://doi.org/10.5194/ESD-8-577-2017

Hemavathy, R. V., Kumar, P. S., Kanmani, K., & Jahnavi, N. (2020). Adsorptive separation of Cu(II) ions from aqueous medium using thermally/chemically treated Cassia fistula based biochar. *Journal of Cleaner Production*, *249*. https://doi.org/10.1016/J.JCLEPRO.2019.119390

Inyang, M., & Dickenson, E. (2015). The potential role of biochar in the removal of organic and microbial contaminants from potable and reuse water: A review. *Chemosphere*, *134*, 232–240. https://doi.org/10.1016/J.CHEMOSPHERE.2015.03.072

Kumarathilaka, P., Jayawardhana, Y., Basnayake, B. F. A., Mowjood, M. I. M., Nagamori, M., Saito, T., Kawamoto, K., & Vithanage, M. (2016). Characterizing volatile organic compounds in leachate from Gohagoda municipal solid waste dumpsite, Sri Lanka. *Groundwater for Sustainable Development*, *2–3*, 1–6. https://doi.org/10.1016/j.gsd.2016.04.001

Laghari, M., Müller-Stöver, D. S., Puig-Arnavat, M., Thomsen, T. P., & Henriksen, U. B. (2021). Evaluation of biochar post-process treatments to produce soil enhancers and phosphorus fertilizers at a single plant. *Waste and Biomass Valorization*, *12*(10), 5517–5532. https://doi.org/10.1007/S12649-021-01358-5

Lai, W. Y., Lai, C. M., Ke, G. R., Chung, R. S., Chen, C. T., Cheng, C. H., Pai, C. W., Chen, S. Y., & Chen, C. C. (2013). The effects of woodchip biochar application on crop yield, carbon sequestration and greenhouse gas emissions from soils planted with rice or leaf beet. *Journal of the Taiwan Institute of Chemical Engineers*, *44*(6), 1039–1044. https://doi.org/10.1016/j.jtice.2013.06.028

Lehmann, J. (2007). A handful of carbon. *Nature*, *447*(7141), 143–144. https://doi.org/10.1038/447143a

Lehmann, J., & Stephen, J. (2015). Biochar for environmental management: Science, technology and implementation. In *Science and technology* (Vol. 1). https://books.google.co.in/books?hl=en&lr=&id=gWDABgAAQBAJ&oi=fnd&pg=PP1&dq=Biochar+for+Environmental+Management:+Science,+Technology+and+Implementation+(2nd+ed.).+Routledge.+&ots=tZXlARC-oV&sig=M-wughaZUFkJ4ZHvLIr5lu5Jwng

Méndez, A., Paz-Ferreiro, J., Gil, E., & Gascó, G. (2015). The effect of paper sludge and biochar addition on brown peat and coir based growing media properties. *Scientia Horticulturae*, *193*, 225–230. https://doi.org/10.1016/j.scienta.2015.07.032

Mohan, D., Pittman, C. U., & Steele, P. H. (2006). Pyrolysis of wood/biomass for bio-oil: A critical review. *Energy and Fuels*, *20*(3), 848–889. https://doi.org/10.1021/EF0502397

Nair, V. D., Nair, P. K. R., Dari, B., Freitas, A. M., Chatterjee, N., & Pinheiro, F. M. (2017). Biochar in the agroecosystem-climate-change-sustainability Nexus. *Frontiers in Plant Science*, *8*. https://doi.org/10.3389/FPLS.2017.02051

Olson, K. R., Al-Kaisi, M. M., Lal, R., & Lowery, B. (2014). Experimental consideration, treatments, and methods in determining soil organic carbon sequestration rates. *Soil Science Society of America Journal*, *78*(2), 348–360. https://doi.org/10.2136/SSSAJ2013.09.0412

Pala, M., Kantarli, I. C., Buyukisik, H. B., & Yanik, J. (2014). Hydrothermal carbonization and torrefaction of grape pomace: A comparative evaluation. *Bioresource Technology*, *161*, 255–262. https://doi.org/10.1016/J.BIORTECH.2014.03.052

Pandey, D., Daverey, A., & Arunachalam, K. (2020). Biochar: Production, properties and emerging role as a support for enzyme immobilization. *Journal of Cleaner Production, 255*. https://doi.org/10.1016/J.JCLEPRO.2020.120267

Paustian, K., Lehmann, J., Ogle, S., Reay, D., Robertson, G. P., & Smith, P. (2016). Climate-smart soils. *Nature, 532*(7597), 49–57. https://doi.org/10.1038/nature17174

Rana, R., Ganguly, R., & Gupta, A. K. (2015). An assessment of solid waste management system in Chandigarh City, India. *Electronic Journal of Geotechnical Engineering, 20*(6), 1547–1572. www.researchgate.net/publication/276275442_An_Assessment_of_Solid_Waste_Management_System_in_Chandigarh_City_India

Randolph, P., Bansode, R. R., Hassan, O. A., Rehrah, D., Ravella, R., Reddy, M. R., Watts, D. W., Novak, J. M., & Ahmedna, M. (2017). Effect of biochars produced from solid organic municipal waste on soil quality parameters. *Journal of Environmental Management, 192*, 271–280. https://doi.org/10.1016/J.JENVMAN.2017.01.061

Sánchez, A., Artola, A., Font, X., Gea, T., Barrena, R., Gabriel, D., Sánchez-Monedero, M. Á., Roig, A., Cayuela, M. L., & Mondini, C. (2015). Greenhouse gas emissions from organic waste composting. *Environmental Chemistry Letters, 13*(3), 223–238. https://doi.org/10.1007/S10311-015-0507-5

Shetty, R., & Prakash, N. B. (2020). Effect of different biochars on acid soil and growth parameters of rice plants under aluminium toxicity. *Scientific Reports, 10*(1), 1–10. https://doi.org/10.1038/s41598-020-69262-x

Singh, S., Singh, P., Sharma, A., & Choudhury, M. (Eds.). (2022). *Agriculture waste management and bioresource: The circular economy perspective*. John Wiley & Sons.

Solomon, S., Qin, D., Manning, M., Chen, Z., Marquis, M., Averyt, K. B., Tignor, M., Miller, H. L. (2007). IPCC (2007) *"Climate change 2007: The physical science basis", Contribution of Working Group I to the Fourth Assessment Report of the Intergovermental Panel on Climate Change*. Cambridge University Press. www.researchgate.net/publication/260164000_IPCC_Intergovernmental_Panel_on_Climate_Change_Climate_Change_2007_The_Physical_Science_Basis_Contribution_of_Working_Group_I_to_the_Fourth_Assessment_Report_of_the_IPCC_Cambridge_United_Kingdom_Cambr

Sridevi, V., Sridevi, V., Modi, M., Ch, M. V. V., Lakshmi, A., & Kesavarao, L. (n.d.). *A review on integrated solid waste management*. Retrieved April 18, 2022, from http://citeseerx.ist.psu.edu/viewdoc/summary?doi=10.1.1.300.9119

Stavi, I., & Lal, R. (2013). Agroforestry and biochar to offset climate change: A review. *Agronomy for Sustainable Development, 33*(1), 81–96. https://doi.org/10.1007/S13593-012-0081-1/FIGURES/5

Tan, Z., Yuan, S., Hong, M., Zhang, L., & Huang, Q. (2020). Mechanism of negative surface charge formation on biochar and its effect on the fixation of soil Cd. *Journal of Hazardous Materials, 384*, 121370. https://doi.org/10.1016/j.jhazmat.2019.121370

Tomczyk, A., Sokołowska, Z., & Boguta, P. (2020). Biochar physicochemical properties: Pyrolysis temperature and feedstock kind effects. In *Reviews in environmental science and biotechnology* (Vol. 19, Issue 1, pp. 191–215). Springer. https://doi.org/10.1007/s11157-020-09523-3

Tripathi, M., Sahu, J. N., & Ganesan, P. (2016). Effect of process parameters on production of biochar from biomass waste through pyrolysis: A review. *Renewable and Sustainable Energy Reviews, 55*, 467–481. https://doi.org/10.1016/J.RSER.2015.10.122

Usman, A. R. A., Al-Wabel, M. I., Ok, Y. S., Al-Harbi, A., Wahb-Allah, M., El-Naggar, A. H., Ahmad, M., Al-Faraj, A., & Al-Omran, A. (2016). Conocarpus Biochar Induces Changes in Soil Nutrient Availability and Tomato Growth Under Saline Irrigation. *Pedosphere, 26*(1), 27–38. https://doi.org/10.1016/S1002-0160(15)60019-4

Vijayaraghavan, K. (2020). The importance of mineral ingredients in biochar production, properties and applications. *Critical Reviews in Environmental Science and Technology, 51*(2), 113–139. https://doi.org/10.1080/10643389.2020.1716654

Wang, D., Jiang, P., Zhang, H., & Yuan, W. (2020). Biochar production and applications in agro and forestry systems: A review. *Science of The Total Environment*, *723*, 137775. https://doi.org/10.1016/J.SCITOTENV.2020.137775

Wang, L., Barta-Rajnai, E., Skreiberg, O., Khalil, R., Czégény, Z., Jakab, E., Barta, Z., & Grønli, M. (2017). Impact of torrefaction on woody biomass properties. *Energy Procedia*, *105*, 1149–1154. https://doi.org/10.1016/J.EGYPRO.2017.03.486

Wang, S., Yan, W., & Zhao, F. (2020). Recovery of solid waste as functional heterogeneous catalysts for organic pollutant removal and biodiesel production. *Chemical Engineering Journal, 401*, 126104. Elsevier. https://doi.org/10.1016/j.cej.2020.126104

Waqas, M., Nizami, A. S., Aburiazaiza, A. S., Barakat, M. A., Ismail, I. M. I., & Rashid, M. I. (2018). Optimization of food waste compost with the use of biochar. *Journal of Environmental Management*, *216*, 70–81. https://doi.org/10.1016/J.JENVMAN.2017.06.015

Wei, Y., Córdoba, P., Caramanna, G., Maroto-Valer, M., Nathaniel, P., & Steven, M. D. (2015). Influence of a CO2 long term exposure on the mobilisation and speciation of metals in soils. *Geochemistry*, *75*(4), 475–482. https://doi.org/10.1016/J.CHEMER.2015.10.003

Woolf, D., Amonette, J. E., Street-Perrott, F. A., Lehmann, J., & Joseph, S. (2010). Sustainable biochar to mitigate global climate change. *Nature Communications*, *1*(1), 1–9. https://doi.org/10.1038/ncomms1053

Yaashikaa, P. R., Kumar, P. S., Varjani, S., & Saravanan, A. (2020). A critical review on the biochar production techniques, characterization, stability and applications for circular bioeconomy. In *Biotechnology reports* (Vol. 28, p. e00570). Elsevier. https://doi.org/10.1016/j.btre.2020.e00570

Ye, Z., Liu, L., Tan, Z., Zhang, L., & Huang, Q. (2020). Effects of pyrolysis conditions on migration and distribution of biochar nitrogen in the soil-plant-atmosphere system. *Science of the Total Environment*, *723*, 138006. https://doi.org/10.1016/j.scitotenv.2020.138006

Yoshizawa, S., Tanaka, M., & Shekdar, A. (2004). Global trends in waste generation, In: Recycling, Waste Treatment and Clean Technology. In *TMS mineral*. www.scirp.org/(S(i43dyn45teexjx455qlt3d2q))/reference/ReferencesPapers.aspx?ReferenceID=2069723

Yu, K. L., Lau, B. F., Show, P. L., Ong, H. C., Ling, T. C., Chen, W. H., Ng, E. P., & Chang, J. S. (2017). Recent developments on algal biochar production and characterization. *Bioresource Technology*, *246*, 2–11. https://doi.org/10.1016/J.BIORTECH.2017.08.009

Zhang, L., & Sun, X. (2018). Influence of sugar beet pulp and paper waste as bulking agents on physical, chemical, and microbial properties during green waste composting. *Bioresource Technology*, *267*, 182–191. https://doi.org/10.1016/J.BIORTECH.2018.07.040

Zhao, X.-G., Jiang, G.-W., Li, A., & Wang, L. (2016). Economic analysis of waste-to-energy industry in China. *Waste Management (New York, N.Y.)*, *48*, 604–618. https://doi.org/10.1016/J.WASMAN.2015.10.014

17 Sustainable Waste Management and Women's Empowerment

Sheeja Karalam, Theresa Nithila Vincent, and Abraham Francis

17.1 URBAN COMMUNITY AND SLUMS

A predictable link exists between an individual and his or her surroundings, which causes a desire to feel a sense of belonging, which drives the individual's quest for uniqueness in the greater community. Slums in cities are areas where residents do not have even the basic necessities for healthy and safe lives. As defined by the United Nations Human Settlements Programme (UN-Habitat): "A slum is a group of people who live together in a community and lack one or more of the following conditions: access to improved water, access to improved sanitation, adequate living space, housing durability, and secure tenure" (UN-Habitat, 2003). Slum formation and expansion in cities are typically due to large-scale migration into cities from rural areasl (Kenya, UN-Habitat, April 2007). High population-densities, poor living-conditions, low vaccination-rates, limited health-related data, and inadequate health-service have all been found linked to a greater rate of disease transmission in slums. Due to restricted space in the houses, overcrowding causes diseases to spread faster and further. Slum residents are also more susceptible to some ailments due to their poor living conditions. For example, poor water-quality is a leading cause of many deadly diseases, such as malaria, diarrhoea, and trachoma. Improving living conditions, such as increased cleanliness and access to basic amenities, can aid in mitigating the effects of diseases. Slums frequently feature extremely tight alleyways that prevent vehicles (including emergency vehicles) from passing through. Garbage accumulates in large amounts as a result of lack of services, such as regular garbage collection.2 The lack of infrastructure is attributable to the informal nature of settlements and the failure of government authorities to plan for the poor. The most vital aspect of any community cooperative initiative is people's engagement. Ambitious community development programmes in India have largely failed due to a lack of meaningful participation by the general public. Getting people to participate and keeping them engaged frequently needs unique insights and years of experience working with people. In India, a slum is defined as "those houses in need of repair, stability, natural light, and air, a dump system, water supply, drainage, and sanitary conveniences, storage facilities, food preparation, and cooking facilities, and wastewater disposal, the building deemed to be unfit if it is so for defective in one or more of the same

DOI: 10.1201/9781003258377-17

matter and not found reasonably suitable for occupation in that condition," according to Chapter II, Section 3 of the Slum Area Act, 1956.

17.2 SUSTAINABLE WASTE MANAGEMENT

Waste management is a worldwide problem faced by major cities. Humans produce a great deal of trash, much of which now has an impact on the environment affecting air, water, and land. According to the United Nations, over 11.2 billion tonnes of solid garbage is collected globally, with humans accounting for nearly all of it. As a result, we must not only manage this garbage but also devise plans for long-term waste management. It is critical to consider sustainability in this regard, so that all the trash can be managed effectively, rather than getting dumped in landfills (Mondal et al., 2023). The collection, transportation, valuation, and disposal of diverse types of trash in a manner that does not damage the environment, human health, or future generations are referred to as sustainable waste management. Sustainable waste management aims to limit the number of natural resources consumed by reusing, as much as possible, the materials obtained from nature and producing as little garbage as possible (Paul et al., 2019). To be sustainable, a waste management approach must incorporate environmental, economic and social factors (Morrissey & Browne, 2004).

17.3 SUSTAINABLE WASTE MANAGEMENT IN URBAN SLUM COMMUNITIES

The uncontrollable solid waste problem is a grave concern for all countries, particularly developing ones. Solid waste management is one of the world's most pressing environmental health issues and continues to overwhelm local and national governments as urban populations grow and consumption patterns shift (Mukama et al., 2016). Household waste-collection models frequently do not extend to low-income and slum areas, resulting in more visible waste in these communities, which leads to increased flooding, health issues, traffic problems, air pollution, the spread of diseases, and other issues (Mytty, 2015). This poses several environmental and health risks, such as contamination of the water we drink, ecosystem degradation and soil pollution, as well as greenhouse gas emissions from anaerobic waste decomposition. The problems are significant and deeper in slums of developing countries. These are the nations in which solid waste management systems are not sufficient. The involvement of communities as well as their awareness, attitudes, and practices has a direct impact on effective solid waste management (Mukama et al., 2016). Increased municipal solid waste (MSW) generation and inadequate waste management infrastructure and routine non-compliance with waste management rules are all contributing to India's urban waste management crisis. It is believed that more than 90% of garbage in India is dumped in public places, rather than being transferred to properly constructed disposal sites (Kumar et al., 2017). The main problems observed in the practice of solid waste disposal by urban slum families are a lack of dustbins, irregular visits by municipal vans for household-waste collection and a lack of knowledge about the importance of waste segregation (Nirgude et al., 2014).

Waste accumulation and slum formation are overlapping problems that arise as a result of a lack of vision for sustainability. Conjointly, they have long-term effects on any country's economic, environmental, and social frameworks. The traditional approaches to slum development follow a top-down approach in communicating decisions as well as utilising resources, in which the governing body plays a major role. However, to be effective, the slum improvement process should involve the slum-dwellers themselves, by making them active participants in their community and using their capacities and social relationships as the most important resource for viable implementation. Thus, a bottom-up approach is considered to be more effective. This would encourage slum dwellers to develop a sense of identity and ownership over their surroundings, and they would undoubtedly work to improve their living conditions, resulting in a more sustainable solution to their slum problem (Elgizawy et al., 2016). Individuals in the community are capable of dealing with their own problems. This means that, while community members may face events that make them feel jaded and hopeless, they can still develop attitudes and abilities that allow them to work to shape their community to fulfil their needs. Active participation is required in the formulation and implementation of an effective communication process that allows for the identification of common goals and the implementation of collective action.

17.4 SOCIO-ECONOMIC STATUS OF WOMEN IN URBAN SLUM COMMUNITIES

Women have also historically been a marginalised demographic in India, as they have traditionally lacked the same level of education as men, and it is still difficult for a young woman from a slum to rise above it. Most of the poor women residing in slums happen to be first-generation immigrants. They shifted to the city with their husbands, driven by financial requirements and aspirations. Their meagre financial resources are being depleted, and their role in the official sector is limited to that of consumers rather than producers (www.careindia.org/wp-content/uploads/2017/05/Poor-Women-in-Urban-in-India.pdf). Nevertheless, these women are not docile, but rather, they are outspoken and very street smart. Many of these ladies are observed to be huddled together in close-knit social groups, sharing their interests and constantly looking for opportunities to supplement the incomes of their husbands (Rambarran, 2014).

17.5 COMMUNITY-BASED PARTICIPATORY RESEARCH

Community-based participatory research (CBPR) is a type of research that involves a collective, reflective, and systematic inquiry, in which researchers and community stakeholders participate as equal partners in all stages of the research process, to educate, improve practice, or effect social change (Tremblay et al., 2018). CBPR aims to make studies more valuable for both researchers and the communities being investigated (Butterfoss et al., 1993; Schulz et al., 1998). CBPR fosters mutual trust, which improves the quantity and quality of data collected.

> **Community-based participatory research** is a collaborative research approach that is designed to ensure and establish structures for participation by communities affected by the issue being studied, representatives of organisations, and researchers in all aspects of the research process to improve health and well-being through taking action, including social change.
>
> **Viswanathan et al., 2004**

In CBPR, researchers must collaborate with the community to identify and justify the most effective research methodologies, while balancing research, rigour, and community responsiveness. Community people can comprehend complicated research challenges when they are presented clearly and deliberately. Capacity building in the community becomes considerably easier to achieve in a CBPR model. Because of its potential to mobilise community action, CBPR is well-positioned to be an effective model to turn around pressing problems of the community with a long-term, viable solution promotion (Viswanathan et al., 2004). CBPR focusses on research aimed at improving situations and practices by questioning the power relationships prevalent in the community. CBPR also uses collaborative framing procedures strategically to create a representation of a social problem (cause), mobilise behind the cause, and define a collective-action strategy that leads to system reforms to address the problem (Tremblay et al., 2018). Pre-existing trust, relationships, shared values, and research agendas enable the smooth flow of CBPR. It fosters a positive form of dissemination, and ultimately, the implementation of the recommendations (Donnelly et al., 2018).

CBPR approach is similar to the Design Thinking Approach, which is iterative. Through a series of discussions, brainstorming sessions, and real-world experiments, the issue is identified. CBPR approaches the problem with an open mind and does not focus on the end results, but rather, on the process to realise the intended outcome, which is the transformation of the individual and society. If the research process is to be a means of supporting change, CBPR emphasises the necessity of incorporating members of a study population as active and equal participants in all phases of the research endeavour (Holkup et al., 2004).

17.6 CASE STUDY AT LR NAGAR URBAN SLUM IN BENGALURU, INDIA

The city of Bengaluru has a population of over 10 million, making it the third most populous metropolis and the fifth most populated urban agglomeration.3 Bustling with traffic, Bengaluru classifies as a diverse city, where various cultures and religions coexist. Also called the Garden City and the City of Lakes because it has more than 1,000 parks and a large number of lakes, it has skyscrapers, sprawling apartments, and also badly inhabited slums. Bengaluru's population is bursting at the seams. Traffic problems, bad roads, cleanliness, and sanitation are just a few of the issues facing the city. In the light of COVID-19, the need for proper waste management has become a matter of great importance for the good health and well-being of the community members, especially the women. The government sector operates the waste management system for urban dwellers. A new system was introduced in the slum area by the Greater Bangalore metropolitan authorities: the *Bruhat Bengaluru Mahanagara Palike* (*BBMP*) for the wet and dry waste collection during the COVID-19 first lockdown period, at the main road near the LR Nagar slum. Predominantly, only women members of the household are involved in waste management in slums. To sustain the new waste management system in the community, a collective initiation of women members in the urban slum community became vital. Keeping the aforementioned factors in mind, the current study about waste management practices among urban slum-dwellers and their implications on women empowerment evolved. This study is based on a community-based participatory method, which includes partnership building, regular interchange among partners, and experience sharing among researchers (Tremblay et al., 2018). Conclusions were drawn out of the debate and negotiations of interpretations between the researchers and community members were reached.

17.6.1 Methods

17.6.1.1 Aim

The study aimed to investigate the development of waste management practices by women in urban slums during the COVID-19 pandemic period as well as the potentials and challenges of sustainable waste management and women's empowerment.

17.6.2 Objectives

- To study the profile of the women participants.
- To examine the role of women at home and in sustainable waste management processes in the community.
- To understand the existing waste management processes in the urban slums
- To develop the immediate, long-term actions to be taken for sustainable waste management in the community through women's empowerment.

17.6.3 Design

The Community Based Project Research approach (CBPR) was applied for this study, wherein the main women leaders in the urban slum community were co-researchers looking for a viable solution to the problem. The process of CBPR depends on the data or information found in the course of the study, which keeps adding newer insights as the investigation progresses.

17.6.3.1 Sampling

Ten women members of the slum community were selected to lead the research process for the study, applying purposive sampling.

17.6.3.2 Tools for Data Collection

- **Profile of the women and their role at home**: The body-mapping exercise and semi-structured interview guides were developed in consultation with the women participants, related to their social lives in the slum community. The body-mapping exercise with the women members portrays their personality characteristics, health issues, skills, desires, and a semi-structured interview-guide to assess their roles. Two women members from the community and the community-based organisation facilitated the whole research process with all the participants.

- The existing waste management process in the urban slums and immediate, long-term actions to be taken for sustainable waste management through women's empowerment were assessed through in-depth interviews and FGD. The engagement of the women participants of the community in a dialogue through regular meetings, in-depth interviews, focus group discussions, and question-answer sessions supported the researchers to understand the actual situation of waste management in the slum community, perceiving the role of women in sustainable waste management through their empowerment. The data comprehends and examines the role of women's sustainable waste management in a slum community.

17.6.3.3 Data Analysis

The data obtained from the women participants were analyzed qualitatively, with supporting literature, related to social life in the slum, the prevalent process of waste management, and sustainable waste management through women's empowerment.

17.6.4 Ethical Considerations

- Co-researchers of this study had taken consent from the government's system on waste management in urban dwellers to support a field study.
- The protection of the participants from unwarranted stress or deprivation from local political interferences on their participation in the field study.
- Consent from the participants on the use of electronic technology appropriately during the field study.

17.6.5 Limitations

The government system on waste management in slums and local political interferences were the limitations of the study.

17.6.6 Results

The co-researchers were the 10 women participants in LR Nagar, an urban slum in Bengaluru, India. The qualitative data were analysed for the following themes and the possibilities of application of various tools for the CBPR in the urban slum communities.

17.6.6.1 Profile of the Women and Their Role at Home

The age group of the participants falls in the 25–50 years category. Three out of the 10 women participants work outside their homes in tasks such as bookbinding, housekeeping, and daily wage work. On the other hand, seven of the 10 women participants were homemakers. "*We are spending the major time of our days in cooking meals, taking care of the children, and completing other household chores.*"

Three of the participants who go out to work cook meals beforehand and finish their household chores before leaving for their workplaces. "*Our husbands never get involved in any work we do at home and only expend their attention into their work; and they just wake up, freshen up, and go to work, then share a meal in the night with the family and go to sleep.*" It was also found that all the participants and their spouses spend their leisure time watching television. Hence, it can be noted that the division of labor is unequal at home.

All 10 participants had taken part in the body mapping data collection methods after gathering at the community centre. They made their drawings with personality characteristics, health issues, skills equipped, and their desires.

Physical Features: Participants mentioned their physical features such as sensitive skin and curly hair. This could directly show the cause and effect of maintaining their physical features while managing their finances, working in their workspace, and taking care of their families (Figure 17.1A).

Personality Characteristics: The participants mentioned their personality characteristics such as optimism, restlessness, fear of taking risks, and other traits. Two of the participants stated that they were hardworking. This may directly correlate with their initiative to manage waste in their homes and surroundings. Two of the participants also stated that they were responsible citizens, also implying that they would take the initiative and sustain waste management in their communities (Figure 17.1B).

Health Issues: The participants mentioned the varied health issues they were suffering from—ranging from thyroid to joint aches. Four of the participants mentioned that they had leg pain, three of them stated that they had recurring headaches, and two of them explained that they had body pain, backache, and thyroid issues. The other health issues that were discussed included obesity and joint pains. These issues could be a contributing factor to workspace stress and poor lifestyle choices (Figure 17.1C).

Skills: The participants penned down skills such as cooking, finance management, teaching, leadership, and singing. Three of the participants mentioned that they had sublime management skills, which might be acquired through their lifestyle and conditioning to their surroundings and workspace. Skills such as leadership, finance management, teaching, and management might be put to effective use in building an efficient and sustainable waste-management system in urban slums (Figure 17.1D).

Desires: The participants expressed their desires that emphasise their need for engaging in leisure activities, pursuing their hobbies, and spending more time with their families. Five of the participants mentioned that they need more time and resources to watch television. Two of the participants want to invest more time in using their mobile phones. Four of the

(A)

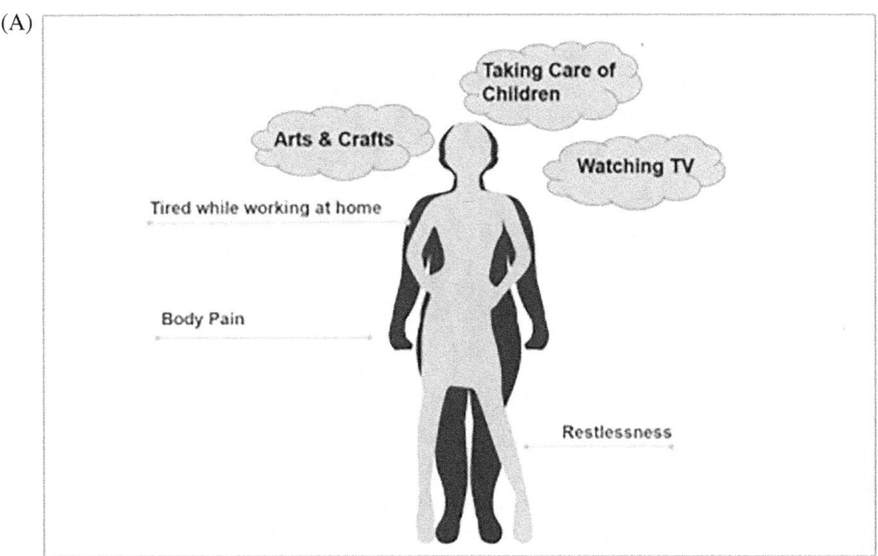

(B)

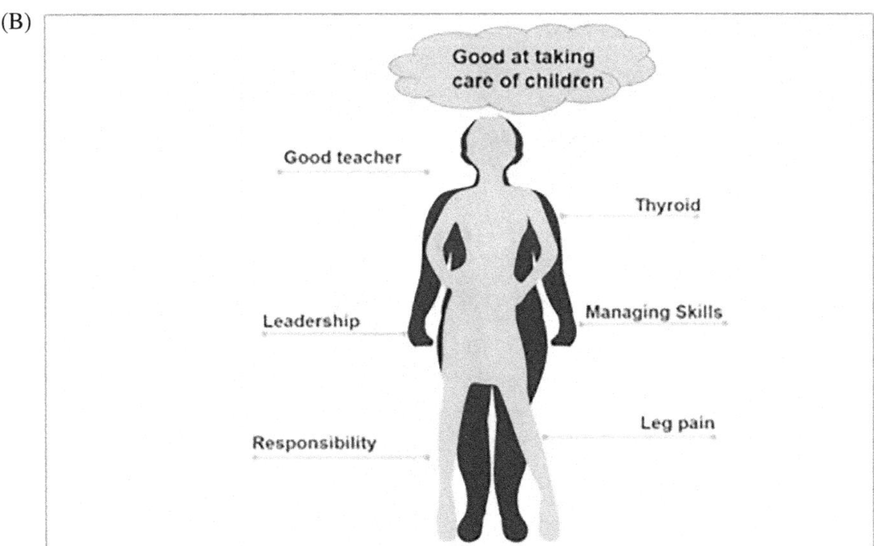

FIGURE 17.1 A–F (A) Ms SS; (B) Ms AM; (C) Ms VI; (D) Ms MI; (E) Ms NI; (F) Ms AJ.

(C)

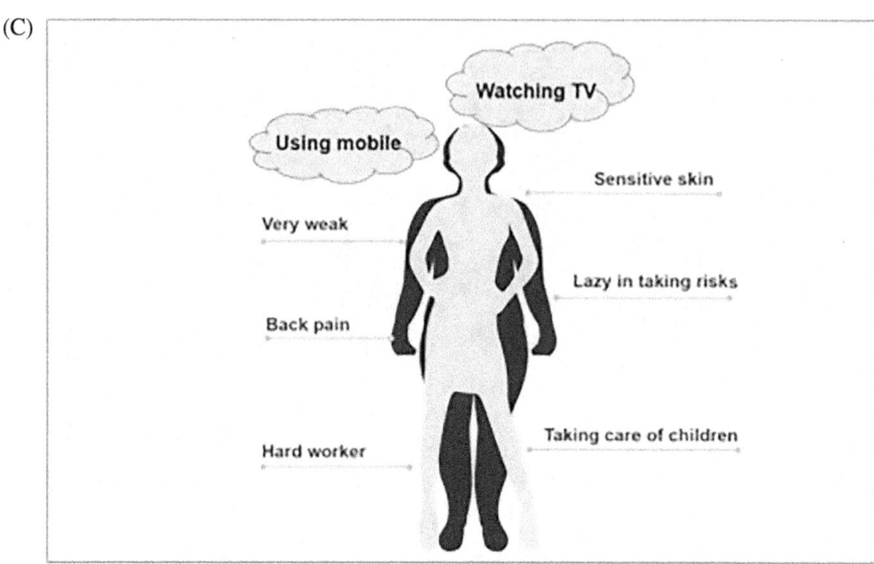

(D)

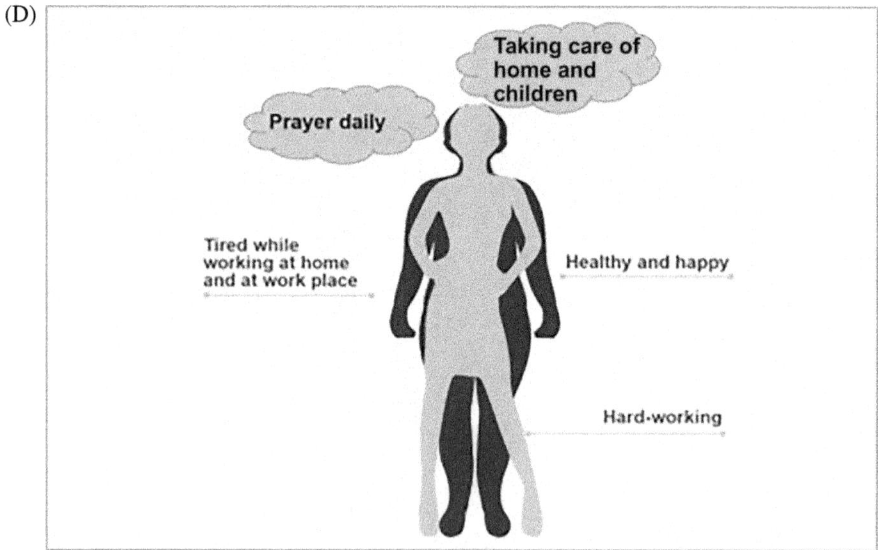

FIGURE 17.1 (Continued)

(E)

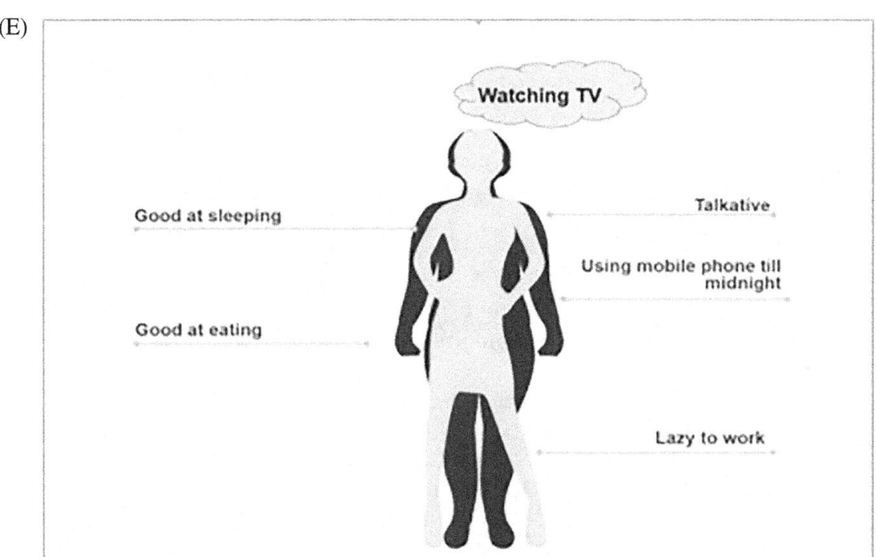

(F)

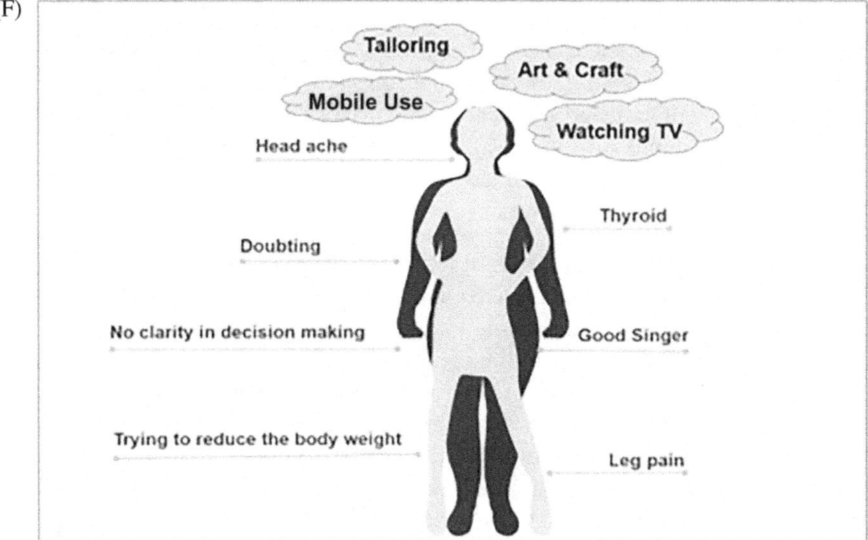

FIGURE 17.1 (Continued)

participants wanted to learn and spend time on tailoring. Three of them wanted to engage in arts and crafts. Four participants wanted to take care of their children well, while two of them wanted to work further. One participant wanted to spend more time in a religious activity such as prayer. This would improve their mental well-being and would also help the participants be more content and satisfied (Figure 17.1E) (Figure 17.1F).

17.6.6.2 The Existing Waste Management Process in the Urban Slums and Immediate, Long-Term Actions to be Taken for Sustainable Development Through Women's Empowerment

17.6.6.2.1 The Existing Waste Management Process in the Urban Slum

All women participants narrated their experiences and agreed to the observations of the co-researchers in the slum community.

> We are upset by the improper waste collection and clearing system presently in this community. There are two wet waste bins, two dry waste bins, and one e-waste bin. These bins are located near the main road; hence, they are not easily accessible to us—the residents in this slum." "The Government vehicle comes only twice or thrice a month to collect the waste from the bins. This affects us and our surroundings and also leads to irregular cleaning of the bins, garbage overflow and open scattering of garbage on the roads. It puts our health at risk.

Only a small garbage vehicle comes every day, near the main road. The driver does not wait for people to throw the waste. If the people pay Rs. 10 or 20, then the vehicle will wait for a few minutes. Due to this, the people dispose of the waste in the open drains. The small garbage vehicle collects waste and throws them into the main bin. By the time the government vehicle arrives, all the waste is spread on the road, and people dump much more waste, as they don't have any more places to dispose of it. Due to the above issues, people and children are affected by allergies and mosquito bites that lead to illnesses.

17.6.7.2.1.1 Plans for the Waste Management

The regular awareness programs were conducted by co-researchers on SDGs related to women's empowerment and waste management. All the stakeholders in the community supported the women to draw up plans for sustainable waste management. During the follow-up FGDs, co-researchers discussed their awareness of waste management and its impact on their involvement in waste management and sustainability. It was evident from the responses and observations that their learning about the segregation of the waste disposal enabled them to demand extra garbage management in their locality. They emphasized that they were independent and confident when involved in this community-based participatory research.

They said, "*We got the motivation and will to work towards sustainable waste management in the slums.*" Nine participants mouthed similar lines on the awareness they obtained, which encouraged them to work efficiently for waste management issues and emphasised that peer interactions were more engaging and enjoyable during the data collection processes.

They said: "At home, we need to look after our kids, cook for them and look into all their needs. Now, after joining these research activities, we feel that the waste management is also our responsibility for the community, and when we all come together, we discuss our points with unity, in a happy manner. We don't fight—we interact well with each other." They added: "We are interested in developing a sustainable waste management system in our surroundings and area. We are going to take permission from the leaders of the area on whether they want to implement any action regarding waste management. We are very happy because we learned about cleanliness and segregation of waste due to awareness of waste management and we have taken the initiative to spread awareness of waste management in our surroundings and neighborhood. Our children also learned about waste disposal from us. We had a good experience being part of this research."

17.6.6.3 Actions for Sustainable Waste Management by the Women Leaders in the Slum Through CBPR

Immediate plans: Eight out of 10 participants stated that they would clean their surroundings by themselves and maintain health and hygiene. Three of the 10 participants also found it necessary to teach their neighbors how they should manage their waste. Hence, in the short term, the participants and the individuals living in the urban slums would focus on maintaining their surroundings on their own. Concurrently, the participants would work at educating their neighbors on waste management to ensure that a large part of the society they live in is sanitary.

Long Term: The availability of garbage bins is minimal. Apart from this, the waste-collecting government vehicles do not regularly collect wastes from slums. Hence, in the long term, two out of the 10 participants would want to request extra bins, which can help them to store the waste inside their homes, while ensuring that their surroundings are healthy and hygienic. Two out of 10 participants want to learn more about the process of segregating waste into biodegradable and non-biodegradable wastes, which could aid in easy recycling. Six of the 10 participants state that, in the long term, they would ensure that they would teach their neighbours the importance of waste management and the process to be followed to ensure that the waste is segregated meticulously. This ensures that the peer-to-peer learning method is followed, which also clarifies that the community would be taking initiatives towards waste management, and hence, would make it sustainable (Figure 17.2).

- The immediate action to be taken is to resolve the sewage problem faced by the slums of Bengaluru, with the support of government authorities.
- There is a need to get extra waste bins and regulate the collection of waste by the waste-collecting vehicles.
- What needs to be taken up immediately is elimination of the inconvenience of transport of waste due to improper roads, which needs to be rectified.
- Extra garbage bins to store the waste in a hygienic form.
- Including resolving sewage issues, regulating waste-collecting vehicles, repairing improper roads, and procuring extra bins are some of the actions that need to be taken immediately, with efficient execution, to ensure that the sustainable waste management in the urban slum is well-managed.

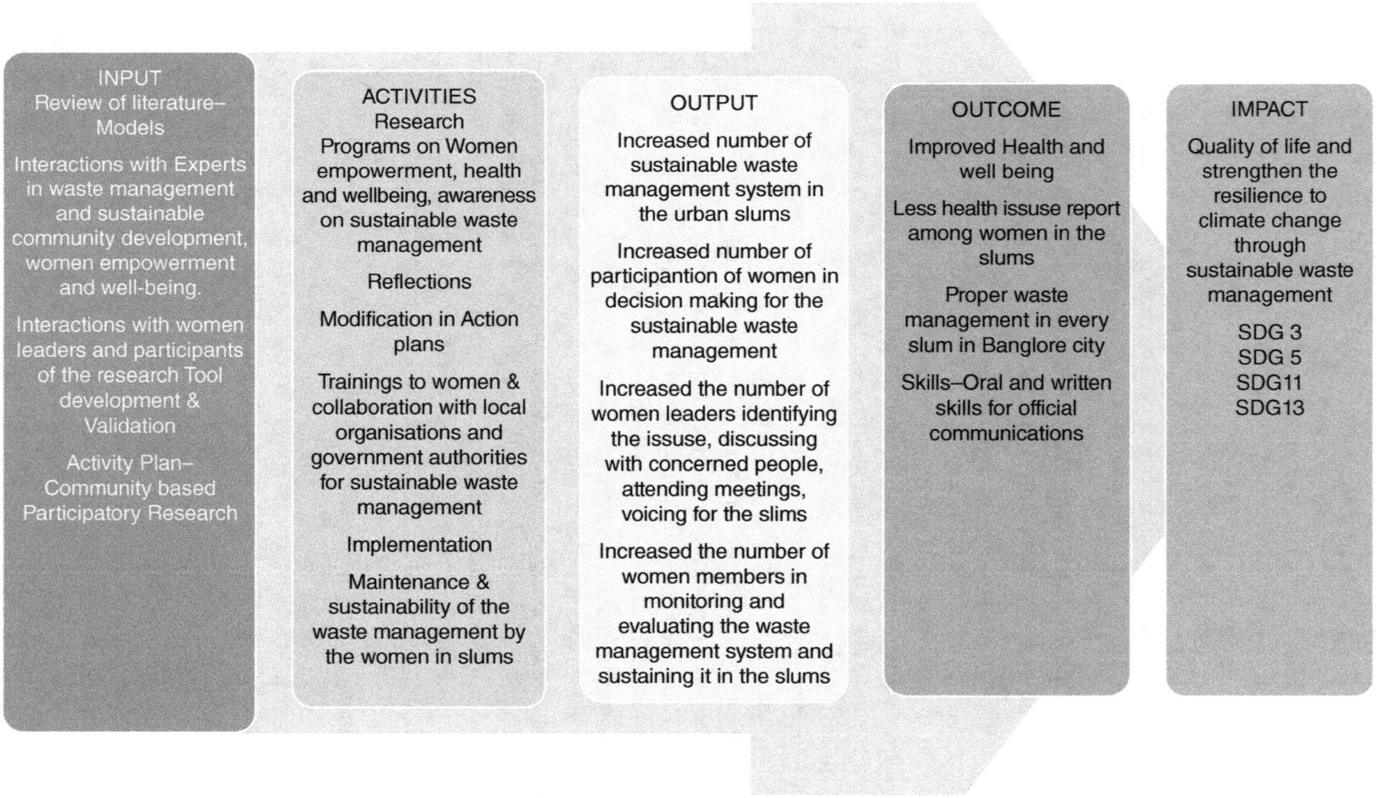

INPUT
Review of literature–Models
Interactions with Experts in waste management and sustainable community development, women empowerment and well-being.
Interactions with women leaders and participants of the research Tool development & Validation
Activity Plan–Community based Participatory Research

ACTIVITIES
Research
Programs on Women empowerment, health and wellbeing, awareness on sustainable waste management
Reflections
Modification in Action plans
Trainings to women & collaboration with local organisations and government authorities for sustainable waste management
Implementation
Maintenance & sustainability of the waste management by the women in slums

OUTPUT
Increased number of sustainable waste management system in the urban slums
Increased number of participantion of women in decision making for the sustainable waste management
Increased the number of women leaders identifying the issuse, discussing with concerned people, attending meetings, voicing for the slims
Increased the number of women members in monitoring and evaluating the waste management system and sustaining it in the slums

OUTCOME
Improved Health and well being
Less health issuse report among women in the slums
Proper waste management in every slum in Banglore city
Skills–Oral and written skills for official communications

IMPACT
Quality of life and strengthen the resilience to climate change through sustainable waste management
SDG 3
SDG 5
SDG11
SDG13

FIGURE 17.2 Community-based participatory research process of the study.

This model can be replicated in other slums in India. The results of the study will be evidence on work during the post-COVID-19 pandemic situation and the possibilities of application of various project tools the urban-slum communities. Students, academicians, and individuals working within the urban communities can be part of the project to learn the process to achieve long term and short-term outcomes.

17.7 DISCUSSIONS AND CONCLUSIONS

The co-researchers emphasized how involvement in the research and the awareness creation about sustainable practices of waste management impending towards a goal to achieve Sustainable Development Goals created the will in them to consciously work towards creating a better environment in the slum community. The rapid urbanisation in the country, with its physical, climatic, geographical, ecological, social, cultural, and linguistic diversities has made it even more challenging to manage the waste generated (Bhalla et al., 2013). Waste management is an important aspect of sustainability. Wastes are potential resources, and effective waste management with resource extraction is fundamental to effective waste management strategies. The lack of accountability of the slum dwellers on their residential area and insufficient coordination capacity are affecting the participation. Women's empowerment is a significant aspect of SDG 3 as well, wherein SDG 5 focuses on gender equality. The involvement of women's empowerment in such projects is vital. This would help women who are unable to take up other work to engage in activities that yield financial benefits to their families and empower the women (Ortiz Rodríguez et al., 2016). The women members accentuated their involvement in this community-based participatory research as co-researchers. They mentioned how being part of waste management helped them to also be consciously involved in waste segregation and management and contribute towards the creation of a better environment for the future. The improper disposal of waste led to the pollution of water bodies, soil, and even air. The awareness creation about sustainable practices of waste management, leading to achieving the SDGs, created a desire in them to consciously work towards creating a better environment. The improper utilisation and overconsumption of natural resources have led to significant climate change. Though waste can be recycled, the impact that excessive waste generation has on the environment and health is significant. The wastes that are generated can be classified into solid, liquid, and demolition waste, each type with different processes for management (O'Leary et al., 1988.

This community-based participatory research, supported to attain the SDG 11 of Sustainable Cities and Urban slum Communities, has contributed to the sustainable development of cities, especially in Bengaluru. The women members in the urban-slum community and the stakeholders of waste management intend to support the SDG 13 of climate action, wherein the protection of environment is taken care of through waste segregation and support to recycling and composting.

The unheard voices of women members give evidence to local leaders to modify their strategies to function in urban slums. The empowered woman will be an asset not only to her home but also to her community. Thus, their strength can be extended to other women to sustain the evolved waste management system. The CBPR promoted collaboration between researchers' academic institutions and women leaders of the urban-slum community as partners in the research process for the co-creation of knowledge. One of the primary goals of CBPR was to improve a community's ability to address the potential and challenges of sustainable waste management and women's empowerment in urban slums by developing effective interventions that could be sustained over time. The immediate change among co-researchers in co-creating knowledge includes: the increased number of activities for sustainable waste management system in the urban slum, increased participation of women in decision-making for sustainable waste management in the slums, increased number of women leaders identified in the issues, discussions with concerned people, attending meetings, voicing for the slums, and increased women members in monitoring and evaluating the waste management system and sustaining it in the slums (Dutta et al., 2022). In the light of COVID-19, the need for proper waste management is a matter of great importance for good health and well-being of the community members. A collective initiation of women leaders in the slum communities are vital now.

ACKNOWLEDGEMENT

The authors would like to thank the instructors of the Mentor Training Program on Community-based Participatory Research, jointly conducted by PRIA, New Delhi and the University of Victoria, Canada, and CHRIST (Deemed to be University), Bangalore, India for the financial support of this work.

NOTES

1 www.smartcitiesdive.com/ex/sustainablecitiescollective/emergence-urban-slums/119186/.
2 https://en.wikipedia.org/wiki/Slum
3 www.bangalorecitytour.com/about-bangalore

REFERENCES

Bhalla, B., Saini, M. S., & Jha, M. K. (2013). Effect of age and seasonal variations on leachate characteristics of municipal solid waste. *International Journal of Research in Engineering and Technology*, 2, 223–232. https://doi.org/10.15623/ijret.2013.0208037.

Butterfoss, F. D., Goodman, R. M., & Wandersman, A. (1993). Community coalitions for prevention and health promotion. *Health Education Research*, *8*(3), 315–330.

Donnelly, S., Ní Raghallaigh, M., & Foreman, M. (2018). Reflections on the use of community based participatory research to affect social and political change: Examples from research with refugees and older people in Ireland. *European Journal of Social Work*, *22*, 1–14. https://doi.org/10.1080/13691457.2018.1540406

Dutta, J., Kerketta, A., Choudhury, M., Ranga, M. M. (2022). Women warriors of waste management. In C. Baskar, S. Ramakrishna, S. Baskar, R. Sharma, A. Chinnappan, & R. Sehrawat (Eds.), *Handbook of solid waste management*. Springer. https://doi.org/10.1007/978-981-16-4230-2_104

Elgizawy, S., Elhaggar, S., & Nassar. K. (2016, January). *Approaching sustainability of construction and demolition waste using zero waste concept*. Submitted for publication to the Journal of Natural Science.

Holkup, P. A., Tripp-Reimer, T., Salois, E. M., & Weinert, C. (2004). Community-based Participatory research: An approach to intervention research with a Native American Community. *ANS. Advances in Nursing Science*, *27*(3), 162.

Kumar, S., Smith, S. R., Fowler, G., Velis, C., Rena, Kumar, R., & Cheeseman, C. (2017). Challenges and opportunities associated with waste management in India. *Royal Society Open Science (R. Soc. Open Sci), 4*, 160764.

Mondal, T., Choudhury, M., Kundu, D., Dutta, D., & Samanta, P. (2023). Landfill: An eclectic review on structure, reactions and remediation approach. *Waste Management*, *164*, 127–142. https://doi.org/10.1016/j.wasman.2023.03.034

Morrissey, A. J., & Browne, J. (2004). Waste management models and their application to Sustainable waste management. *Waste Management*, *24*(3), 297–308.

Mukama, T., Ndejjo, R., Musoke, D., Musinguzi, G., Halage, A. A., Carpenter, D. O., & Ssempebwa, J. C. (2016). Practices, concerns, and willingness to participate in solid waste management in two urban slums in central Uganda. *Journal of Environmental and Public Health*, *2016*, 1–7. https://doi.org/10.1155/2016/6830163

Mytty, K. M. (2015). *The role of actors and incentives in municipal solid waste Management: a case study on Muzaffarnagar, India* [DEGREE OF MASTER IN CITY PLANNING, MASSACHUSETTS INSTITUTE OF TECHNOLOGY]. https://dspace.mit.edu/bitstream/handle/1721.1/98941/922070118-MIT.pdf?sequence=1&isAllowed=y

Nirgude, A., Naik, P., Prasad, V., & Nagaraj, K. (2014). Solid waste disposal practices in an urban slum area of South India. *Indian Journal of Applied Research*, *4*(1).

O'Leary, P. R., Walsh, P. W., & Ham, R. K. (1988). Managing solid waste. *Scientific American*, *259*(6), 36–45. http://www.jstor.org/stable/24989301

Ortiz Rodríguez, J., Pillai, V. K., & Ribeiro Ferreira, M. (2016). The impact of women's agency and autonomy on their decision-making capacity in Nuevo Leon, Mexico. *Acta Universitaria*, *26*(5), 70–78. https://doi.org/10.15174/au.2016.976

Paul, S., Choudhury, M., Deb, U., Pegu, R., Das, S., & Bhattacharya, S. S. (2019). Assessing the ecological impacts of ageing on hazard potential of solid waste landfills: A green approach through vermitechnology. *Journal of Cleaner Production*, *236*, 117643. https://doi.org/10.1016/j.jclepro.2019.117643

Rambarran, R. (2014). *The socio-economic status of women in the urban slums of India*. MPRA Paper 62736, University Library of Munich.

Schulz, A. J., Parker, E. A., Israel, B. A., Becker, A. B., Maciak, B. J., & Hollis, R. (1998). Conducting a participatory community-based survey for a community health Intervention on Detroit's east side. *Journal of Public Health Management and Practice*, *4*, 10–24.

Tremblay, M. C., Martin, D. H., McComber, A. M., McGregor, A., & Macaulay, A. C. (2018). Understanding community-based participatory research through a social movement Framework: A case study of the Kahnawake Schools Diabetes Prevention Project. *BMC Public Health*, *18*(1), 1–17.

United Nations Centre Human Settlements Programme (UN-Habitat). (2007). *Situation analysis of informal settlements in Kampala: Cities without slums: Sub-regional programme for eastern and southern Kivulu (Kagugube) and Kinawataka (Mbuya 1)* UN-Habitat.

United Nations Human Settlements Programme Staff, & United Nations Human Settlements Programme. (2003). *The challenge of slums: Global report on human settlements, 2003*. UN-HABITAT.

Viswanathan, M., Ammerman, A., Eng, E., Garlehner, G., Lohr, K. N., Griffith, D., & Whitener, L. (2004). *Community-based participatory research: Assessing the evidence: Summary*. AHRQ Evidence Report Summaries.

www.careindia.org/wp-content/uploads/2017/05/Poor-Women-in-Urban-in-India.pdf.

18 Remote-Sensing Techniques in Solid Waste Management

Swati Singh, K. V. Suresh Babu, and Shivani Singh

18.1 INTRODUCTION

There are substantial problems of solid waste issues in spatio-temporal landscapes on a global scale (Clapp, 2002). Waste management refers to the system of waste collection including transportation, reduction, and recycling (Demirbas, 2011). The main reason for regulating waste treatment and disposal is to reduce negative environmental impacts, thereby creating a healthier future. It is attributed to human activities, which must be handled to avoid adverse effects on human and environmental health. The integrity of ecosystem processes is greatly affected by waste; this constrains its aesthetic sensibilities and well-thought-out economics (Bera & Sadowska, 2020). Therefore, we require proper waste management to control the spread of pollution for environmental protection like soil, water, and air pollution, etc. The initial methods of waste management are based on various parameters, which may differ between developed countries, rural and urban residential areas, or industrial environments (Shekdar, 2009). From a global scientific perspective, there is an urgent requirement to focus on solid waste management (SWM) for a sustainable environment. The great harbinger of technological development that has given rise to innovations in the SWM sector is undoubtedly the development of Remote-Sensing and GIS systems (Ziaeian et al., 2009; Gautam et al., 2020). In this review, a brief description of RS and GIS application for SWM is presented.

18.2 SUSTAINABLE SOLID WASTE MANAGEMENT (SWM): APPROACH AND CHALLENGES

SWM is an essential segment of environmental management (Kaza et al., 2018). The main objective of SWM is to eliminate solid waste products to the maximum possible extent and ensure a sustainable environment. The Solid Waste (Management and Handling) Rules, 2000, were promoted for waste management, and accountability has been allocated to urban local bodies for municipal waste management (Lal Patel et al., 2011). The current ad-hoc method to manage waste collection and transportation results in effective resource utilization. Several systems have been developed for management, including a variety of policies and technologies such as recycling, incineration, waste prevention at the production point, and sanitary landfilling (Abdel-Shafy & Mansour, 2018). Managing a waste management program is

 DOI: 10.1201/9781003258377-18

not an easy task; its traditional methods are quite cost-intensive and challenging. These challenges are allied with the major characteristics of the waste sector—e.g., collection of waste, transportation, treatment, and disposal methods (Figure.18.1-A). The sustainable SWM program has been shifted from traditional dumping methods,

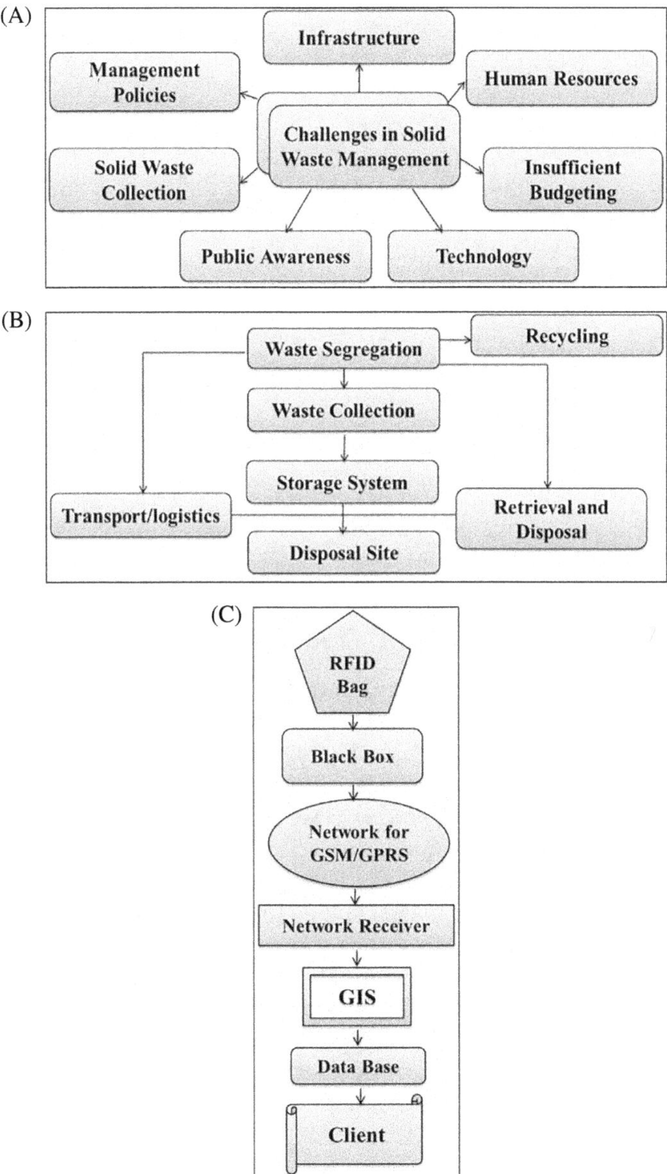

FIGURE 18.1 (A)—Major challenges in solid waste management. (B)—Elements of solid waste management. (C)—SWM model using GIS and GPS. (D)—Architecture of GIS functionality.

(D)

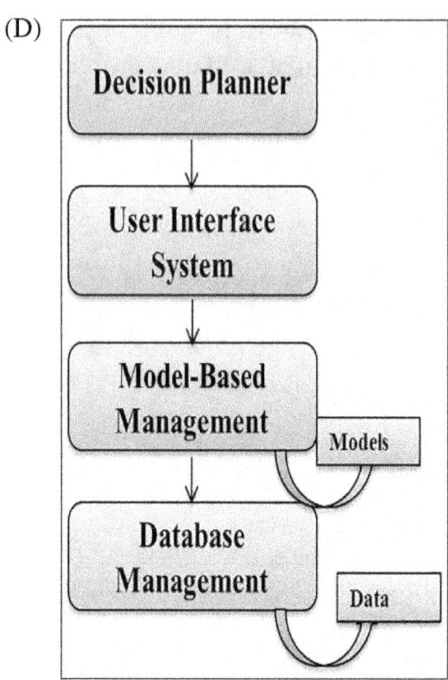

FIGURE 18.1 (Continued)

which are very cost-intensive and harmful, to the economy and environment (Paul et al., 2019; Das et al., 2019).

18.3 IMPACT ON ENVIRONMENT AND ECONOMY

Solid waste production, often neglected during economic and industrial development, harms the environment (Kaza et al., 2018). Exposed dumping sites are a major concern, where the wastes are dumped with poor covering practices (Nkwachukwu et al., 2010). These dumpsites are extensively used in developing countries due to low-cost involvement and modest process (Das et al., 2019). There are many open dumpsites around the world today, although it is extremely difficult to update their exact numbers (Binion & Gutberlet, 2012). It is reported that dumping sites led to soil and vegetation degradation and produce highly toxic pollutants, ultimately hindering the plant physiology and soil ecosystem (Choudhury et al., 2022; Amusan et al., 2005; Ali et al., 2014). It is also recognized that the discharge of metal contaminants from effluents can pollute ground/surface water (Ashraf et al., 2013; Abdalla & Khalil, 2018). Many environmentalists suggested that, primarily, environmental pollution increases with the per-capita gross domestic product (GDP); at a certain point, emission rate and GDP become dissociated (Smulders, 2004; Huang et al., 2008). Then simultaneous increases in GDP can be allied with the decrease in environmental pollution as advanced productivity and treatment methods expand with national incomes (Sun et al., 2020). In the case of SWM, technological developments could increase with growing incomes as it will be associated with green design and recycling (Yu et al., 2021).

18.4 METHODS OF SOLID WASTE MANAGEMENT

Today, the complexities associated with global SWM are becoming more complex due to its various sources, increasing amounts, and different compositions of waste. Mostly, managed efforts are made to reuse the available resources in waste management. Methods of waste management and recycling are not newly conceptualized activities. Solid waste management includes many activities, from its initial production to its final disposal process (Demirbas, 2011). The involved waste management activities can be grouped under six functional elements (Figure. 18.1-B). These are: collection, segregation, transportation, storage systems, and disposal. Management of waste, primarily in the rural and metropolitan environment, is the responsibility of the municipality, while the waste produced by industrial organizations has its responsibility, which is self-management (Dutta & Goel, 2017).

18.5 APPLICATION OF REMOTE SENSING (RS) AND GIS

RS and GIS system is widely used in various fields such as agriculture (Acharya et al., 2018; Singh 2022a), pollution monitoring (Yang et al., 2010), health services (Nykiforuk & Flaman, 2011), forest-fire mapping (Singh & Suresh Babu, 2021; Singh, S. 2022b), and solid waste management (Pandey et al., 2012), etc. The integration of RS, GIS, and the GPS (Global Positioning System) helps in recording spatial data that can be used for object-specific cartographic representation and analysis (Milla et al., 2005). Remote sensing has revolutionized waste management globally and is now widely used in solid waste management (Table.18.1). It can efficiently manage spatial characteristics, planning, and other operating resources. It deals with time management at all stages and allows accurate assessment with various substitutes (Taye, 2018).

18.6 SOLID WASTE MANAGEMENT USING REMOTE SENSING AND GIS

RS and GIS technology has proved to be very useful in providing an efficient solution for monitoring and management of solid waste. The most applied GIS application is used in collection, dumping facilities, transport optimization, and waste rate estimation in terms of socio-economic and demographic dynamics (Nishanth et al., 2010). The GIS solution is developed in two phases: monitoring and computation of the collected information to optimize the implementations. This allows consultants to improve critical decision-making efficiency. The detailed application of RS and GIS in SWM are deliberated as follows:

18.6.1 Solid Waste Generation

Solid waste generation is affected by various physical, geographical, and socio-cultural factors (Adzawla et al., 2019). In recent decades, waste generation has been witnessing a steady increase with the population around the world. Globally 2.01 billion tons of solid waste was generated in the year 2016, a total of 0.74

TABLE 18.1
Application of RS and GIS in SWM

Sl. No	Study area	Method	Applications	Reference
1	Niger Republic	RS and GIS	Selection of dumping site	Twumasi & Merem, 2007
2	Ethiopia	GIS, and multi-criteria analysis	Suitable dumping sites	Ebistu & Minale, 2013
3	India	ArcGIS	Bin location	Nithya et al., 2012
4	Turkey	GIS and AHP	Disposal-site selection	Şener et al., 2011
5	Malaysia	ArcGIS	Routes optimization	Malakahmad et al., 2014
6	Iran	ArcGIS	Assessment of landfill site	Davami et al., 2014
7	India	Spatial analyst tool and AHP	Landfill-site selection	Kumar & Hassan, 2013
8	Kenya	ArcGIS	Decision-support system	Muriithi & Mutua, 2012
9	India	Multi-objective decision	Waste-collection routes	Sudhir et al., 1996
10	Sri Lanka	ArcGIS	Disposal site selection	Balasooriya et al., 2014
11	Nigeria	RS and GIS	Solid waste management	Oyinloye, 2013
12	India	Weighted mean center	settlement map	Senthil et al., 2012
13	India	ArcGIS	Waste estimation	Shoba & Rasappan, 2013
14	Nigeria	Satellite imagery GPS, and ArcGIS	Waste management	Sule et al., 2014
15	India	ArcGIS	Suitable landfill site selection	Reddy, 2005
16	India	RS and GIS	Modeling framework for waste management municipal solid waste	Singh, 2019a
17	China	Machine learning	Municipal solid waste management (MSWM)	Xia et al., 2021
18	Malaysia	GIS, machine learning, wireless-sensor network	Solid waste management	Akram et al., 2021
19	India	GIS, machine learning and fuzzy AHP	Sanitary landfill site selection	Mohsin et al., 2021
20	UAE	GIS and machine learning	Site suitability mapping	Al-Ruzouq et al., 2019

kilograms/day/person (World Bank; Richard, 2021). By 2050, global waste generation is estimated to increase by approx. 70% to 3.4 billion metric tons (Kaza et al., 2018). This is mainly due to the increasing population and the dynamics of economic development. Generally, the ground-based survey is conducted to assess the relationship between the region-specific waste composition and estimation (volume and weight). This classification is very important for solid waste handling and recycling.

In recent years, the GIS system has shown to be a boon for managing the vast amount of solid waste. Collected data samples of the area-specific waste amount and population distribution are prepared as different layers in Arc GIS software for consequent analysis (Dutta & Goel, 2017). These superimposed or stacked layers are used for classification to define categories for the targeted study area. Therefore, RS and GIS techniques can be significantly used for the specific area based on solid-waste identification and classification.

18.6.2 Storage System

Solid waste storage is an essential part of an effective SWM plan. Having suitable bins for garbage collection is very important for storage and also makes transportation easier. GIS techniques are used to locate the dustbins with effective circulation. As per Indian guidelines, the maximum space between dustbins should not exceed 500 meters and should be at a distance of 250 meters from the waste generation and collection point for households (Paul et al., 2017). The ESRI (Environmental Systems Research Institute) Arc/Info platforms can be used significantly for such investigations. Complete data is essential for the management of the system and contains the following essential aspects: thematic maps, road networks, and altitude survey data for appropriate management. Attribute data and thematic maps are encompassed in a GIS-based 'Arc Macro Language' based model. GIS helps in optimizing the storage bin locations based on the road network and landscape survey. Further computation drivers finalize the bins' number and location in the specific area. Different types of settlement maps can be prepared using ArcGIS software, and then, total generated waste from that particular region can be estimated. WMC (Weighted Mean Center) is implemented to find alternative bin sites with the help of different static software such as SPSS, geographic coordinates, and settlement records (Dutta & Goel, 2017).

18.6.3 Waste Collection and Routing Competence

Solid waste collection and transportation involves 80 to 90% of the whole management budget in low-income countries, whereas collection competencies stay below 50%. In comparison to high-income countries, collection competencies are more than 90%, and collection competencies are below 10% of the overall cost (Hoornweg & Bhada-Tata, 2012). Several difficulties that occurred during waste collection are: waste production rate, transport, manpower expenditures, and absence of suitable planning. GIS techniques are used to define optimum roads for assembling solid waste by using network investigation. Primarily, the assortment of all ancillary data prepared from several sources (Kumar et al., 2009). Then the NETWORK segment of ArcGIS was used to deliver a spatial analysis of the network which comprises directions, adjacent facility center, and area investigation. ArcGIS System supports managers for controlling speed limits, turning, and transport services in a day at different periods. Network Analyst Users can similarly find out the effective traffic routes, decide the nearby vehicle convenience, travel guidelines, and find an adjacent location. Distance and time are the two optimum criteria used for waste collection. The GIS-based analysis helped to understand very fast in comparison to manual-based investigations. A model developed based on the integration of GPS, GIS, Radio Frequency Identification (RFID) for SWM is schematically presented in Figure 18.1-C.

In this model, the RFID system is placed in the waste-collection vehicle, which helps in locating bins, collection date and time, conditions, and obtained weight (Dutta & Goel, 2017). This innovative technology offers information concerning the real-time status of bins, collection, and location, and other factors that are suitable for solid wastes monitoring and collection (Nirde et al., 2017). Various models and computer-based databases can be applied significantly to carry out solid waste collection plans in different times and conditions (Singh, 2019b). Mathematical discrete modeling is applied to evaluate different time-scheduling and real-time routing policies. Proactive development and routing strategies enable smooth monitoring of scheduled instruction and schedule planning by operators at lower operating costs and consumption (Das et al., 2019) (Figure 18.2-A).

18.6.4 Landfill-Site Selection

No matter how the municipality manages solid waste, the residual dumping must end up in landfill sites (Laner et al., 2012). Landfill-site selection is a very difficult process where geographical, hydrological, sensitive site, and environmental factors are along with administrative guidelines, have to be deliberated. The landfill site should have adequate volume to provide lodgings of waste to be set down within 20–25 years. As per the MSWM Rules (2000) and its following amendments (2016), the major point identified in landfill setting are presented as follows (Mani& Singh, 2016; Ali et al., 2021) (Figure 18.2-B).

The site conditions depend on various conditions, like population distribution, geological and hydrological, etc. Based on a preliminary analysis of the residential population, drainage system, and environmental risk, the most suitable site for landfilling should be selected (Chang et al., 1997). Ziaeian et al. 2009; Mondal et al., 2023) analyzed multi-criteria decisions and overlay systems using Arc GIS tools for a suitable landfill site. The generated thematic maps are combined to obtain various information; for example, geology, topography, drainage systems, areas built-up area to regulate landfill sites. A GPS is significantly used to derive the geo-coordinates of a particular landfill site. Other advantages in landfill siting include area separation and buffering, data analysis, geographic data correlation, graphical representation, etc. In recent times, MCDM techniques, in conjunction with GIS, are potentially used for map development, site suitability, and optimization of SWM systems.

(A)

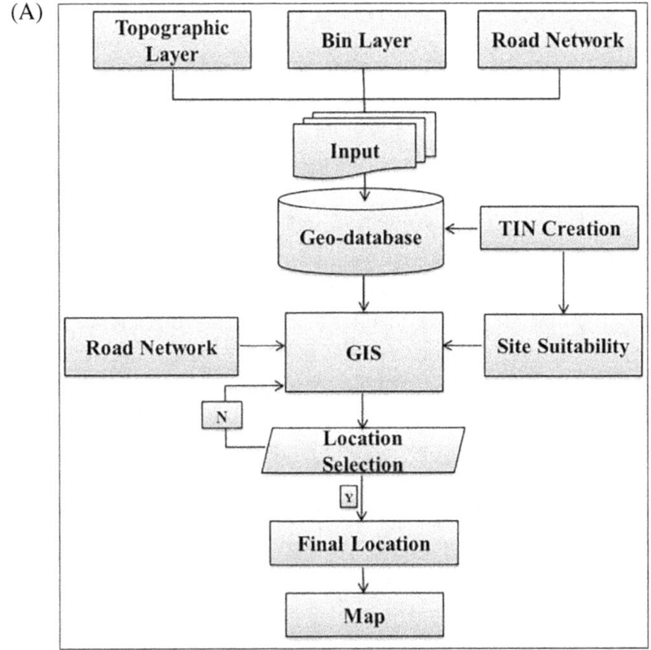

(B)

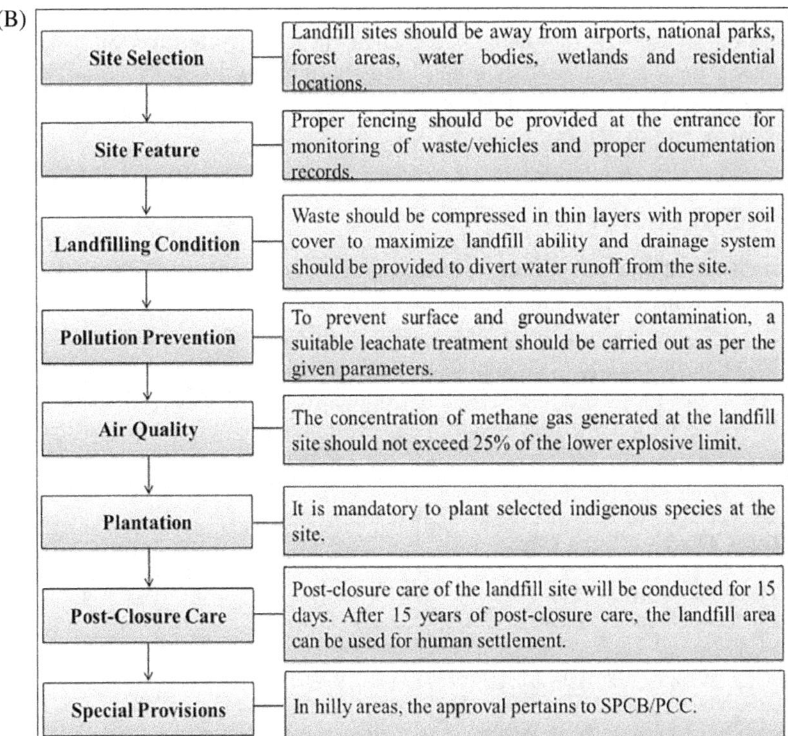

FIGURE 18.2 (A)—Flowchart for categorizing the optimal storage location bin (TIN: Triangular Irregular Network). (B)—Important aspect of landfill setting.

18.6.5 Decision-Support System (DSS)

DSS is a computer-based information system that provides logically structured solutions to various problems. Computer-based systems act more actively in the decision-making process, especially when addressing huge datasets. The integration of the GIS system with DSS can be significantly used in waste collection, transport, and waste disposal. All information can be gathered into systems to resolve problems and suitable decision outcomes. The basic DSS model consists of database management, user interface, and model base management components (Bani et al., 2009). A DSS model for waste disposal and recycling is shown in Figure 18.1-D. A multi-purpose decision-making method is used to develop a solid waste management system. This method is based on the multi-purpose program goal, considering several options. In the SWM case study for Chennai city, the inferred model-based study was completed in five years by Sudhir et al. (1996). In another case study, a decision-support system based on GIS was developed for sanitary-landfill selection. The characteristics considered in this study were related to geomorphology, slope, drainage capacity, settlement, groundwater, erosion, and lack of joints and faults. The classification was done for the earlier-mentioned features and their characteristics using various multi-criteria assessments, hierarchical procedure, significance coefficient, Delphi, and ANN (Artificial Neural Network). All models were combined to establish the best possible site.

This comparative technique was potentially used in the development of the Chennai Metropolitan Development Area, Chennai, India (Natesan & Suresh, 2002). Various DSS have been established and used significantly in the SWM. The progress in the DSS system creates a user-friendly solution for optimizing SWM within administrative and societal frameworks.

18.7 CONCLUSION

The increasing volume of solid waste resulting from poor collection capacity creates difficult conditions and various health risks. RS and GIS techniques have proved to be an effective method for the SWM system in urban areas. RS and GIS technology is efficient in dealing with spatial or non-spatial datasets and can be successfully used to design or monitor SWM in conjunction with GPS. This can provide desirable routes which are very economical and make the monitoring system more transparent.

REFERENCES

Abdalla, F., & Khalil, R. (2018). Potential effects of groundwater and surface water contamination in an urban area, Qus City, Upper Egypt. *Journal of African Earth Sciences, 141*, 164–178.

Abdel-Shafy, H. I., & Mansour, M. S. (2018). Solid waste issue: Sources, composition, disposal, recycling, and valorization. *Egyptian Journal of Petroleum, 27*(4), 1275–1290.

Acharya, S. M., Pawar, S. S., & Wable, N. B. (2018). Application of remote sensing GIS in agriculture. *International Journal of Advanced Engineering Research and Science, 5*(4), 237434.

Adzawla, W., Tahidu, A., Mustapha, S., & Azumah, S. B. (2019). Do socioeconomic factors influence households' solid waste disposal systems? Evidence from Ghana. *Waste Management & Research, 37*(1_suppl), 51–57.

Akram, S. V., Singh, R., Gehlot, A., Rashid, M., AlGhamdi, A. S., Alshamrani, S. S., & Prashar, D. (2021). Role of wireless aided technologies in the solid waste management: A comprehensive review. *Sustainability, 13*(23), 13104.

Ali, S. A., Parvin, F., Al-Ansari, N., Pham, Q. B., Ahmad, A., Raj, M. S., . . . Thai, V. N. (2021). Sanitary landfill site selection by integrating AHP and FTOPSIS with GIS: A case study of Memari Municipality, India. *Environmental Science and Pollution Research, 28*(6), 7528–7550.

Ali, S. M., Pervaiz, A., Afzal, B., Hamid, N., & Yasmin, A. (2014). Open dumping of municipal solid waste and its hazardous impacts on soil and vegetation diversity at waste dumping sites of Islamabad city. *Journal of King Saud University-Science, 26*(1), 59–65.

Al-Ruzouq, R., Shanableh, A., Yilmaz, A. G., Idris, A., Mukherjee, S., Khalil, M. A., & Gibril, M. B. A. (2019). Dam site suitability mapping and analysis using an integrated GIS and machine learning approach. *Water, 11*(9), 1880.

Amusan, A. A., Ige, D. V., & Olawale, R. (2005). Characteristics of soils and crops' uptake of metals in municipal waste dump sites in Nigeria. *Journal of Human Ecology, 17*(3), 167–171.

Ashraf, M. A., Yusoff, I., Yusof, M., & Alias, Y. (2013). Study of contaminant transport at an open-tipping waste disposal site. *Environmental Science and Pollution Research, 20*(7), 4689–4710.

Balasooriya, B. M. R. S., Vithanage, M., Nawarathna, N. J., Zhang, M., & Herath, G. B. B. (2014). *Solid waste disposal site selection for Kandy district.* Sri Lanka Integrating GIS and Risk Assessment.

Bani, M. S., Rashid, Z. A., Hamid, K. H. K., Harbawi, M. E., Alias, A. B., & Aris, M. J. (2009). The development of decision support system for waste management; a review. *World Academy of Science, Engineering and Technology, 49*(1), 161–168.

Bera, A., & Sadowska, B. (2020). Clean production as an element of sustainable development. In *Sustainable production: Novel trends in energy, environment and material systems* (pp. 1–9). Springer.

Binion, E., & Gutberlet, J. (2012). The effects of handling solid waste on the wellbeing of informal and organized recyclers: A review of the literature. *International Journal of Occupational and Environmental Health, 18*(1), 43–52.

Chang, N. B., Lu, H. Y., & Wei, Y. L. (1997). GIS technology for vehicle routing and scheduling in solid waste collection systems. *Journal of Environmental Engineering, 123*(9), 901–910.

Choudhury, M., Jyethi, D. S., Dutta, J., Purkayastha, S. P., Deb, D., Das, R., . . . Bhattacharyya, K. G. (2022). Investigation of groundwater and soil quality near to a municipal waste disposal site in Silchar, Assam, India. *International Journal of Energy and Water Resources, 6*(1), 37–47. https://link.springer.com/article/10.1007/s42108-021-00117-5

Clapp, J. (2002). The distancing of waste: Overconsumption in a global economy. *Confronting Consumption*, 155–176.

Das, S., Lee, S. H., Kumar, P., Kim, K. H., Lee, S. S., & Bhattacharya, S. S. (2019). Solid waste management: Scope and the challenge of sustainability. *Journal of Cleaner Production, 228*, 658–678.

Davami, A. H., Moharamnejad, N., Monavari, S. M., & Shariat, M. (2014). An urban solid waste landfill site evaluation process incorporating GIS in local scale environment: A case of Ahvaz city, Iran. *International Journal of Environmental Research, 8*(4), 1011–1018.

Demirbas, A. (2011). Waste management, waste resource facilities and waste conversion processes. *Energy Conversion and Management, 52*(2), 1280–1287.

Dutta, D., & Goel, S. (2017). Applications of remote sensing and GIS in solid waste management–A review. In *Advances in solid and hazardous waste management* (pp. 133–151). Springer.

Ebistu, T. A., & Minale, A. S. (2013). Solid waste dumping site suitability analysis using geographic information system (GIS) and remote sensing for Bahir Dar Town, North Western Ethiopia. *African Journal of Environmental Science and Technology, 7*(11), 976–989.

Gautam, S., Brema, J., & Dhasarathan, R. (2020). Spatio-temporal estimates of solid waste disposal in an urban city of India: A remote sensing and GIS approach. *Environmental Technology & Innovation, 18*, 100650.

Hoornweg, D., & Bhada-Tata, P. (2012, March). *What a waste: A global review of solid waste management*. Report, No. 15.

Huang, W. M., Lee, G. W., & Wu, C. C. (2008). GHG emissions, GDP growth and the Kyoto Protocol: A revisit of Environmental Kuznets Curve hypothesis. *Energy Policy, 36*(1), 239–247.

Kaza, S., Yao, L., Bhada-Tata, P., & Van Woerden, F. (2018). What a waste 2.0: A global *snapshot of solid waste management to 2050*. World Bank Publications.

Kumar, S., & Hassan, M. I. (2013). Selection of a landfill site for solid waste management: An application of AHP and spatial analyst tool. *Journal of the Indian Society of Remote Sensing, 41*(1), 45–56.

Kumar, S., Bhattacharyya, J. K., Vaidya, A. N., Chakrabarti, T., Devotta, S., & Akolkar, A. B. (2009). Assessment of the status of municipal solid waste management in metro cities, state capitals, class I cities, and class II towns in India: An insight. *Waste Management, 29*(2), 883–895.

Lal Patel, M., Jain, R., & Saxena, A. (2011). Assessment of the Municipal Solid Waste & Status of Implementation of Municipal Solid Waste (Management & Handling), Rules, 2000 in the State of Madhya Pradesh, 2008-A case study. *Waste Management & Research, 29*(5), 558–562.

Laner, D., Crest, M., Scharff, H., Morris, J. W., & Barlaz, M. A. (2012). A review of approaches for the long-term management of municipal solid waste landfills. *Waste Management, 32*(3), 498–512.

Malakahmad, A., Bakri, P. M., Mokhtar, M. R. M., & Khalil, N. (2014). Solid waste collection routes optimization via GIS techniques in Ipoh city, Malaysia. *Procedia Engineering, 77*, 20–27.

Mani, S., & Singh, S. (2016). Sustainable municipal solid waste management in India: A policy agenda. *Procedia Environmental Sciences, 35*, 150–157.

Milla, K. A., Lorenzo, A., & Brown, C. (2005). GIS, GPS, and remote sensing technologies in extension services: Where to start, what to know. *Journal of Extension, 43*(3), 16.

Mohsin, M., Ajim Ali, S. K., Shamim, S. K., & Ahmad, A. (2021). A GIS-based novel approach for sustainable sanitary landfill site selection using integrated fuzzy AHP and machine learning algorithms. https://doi.org/10.21203/rs.3.rs-641684/v1

Mondal, T., Choudhury, M., Kundu, D., Dutta, D., & Samanta, P. (2023). Landfill: An eclectic review on structure, reactions and remediation approach. *Waste Management, 164*, 127–142. https://doi.org/10.1016/j.wasman.2023.03.034

Muriithi, D., & Mutua, F. (2012). Eco-Taka: An efficient GIS based solid waste management solution. *International Journal of Science and Research, 3*(9), 1016–1022.

Natesan, U., & Suresh, E. S. M. (2002). Site suitability evaluation for locating sanitary landfills using GIS. *Journal of the Indian Society of Remote Sensing, 30*(4), 261–264.

Nirde, K., Mulay, P. S., & Chaskar, U. M. (2017, June). IoT based solid waste management system for smart city. In *2017 International conference on intelligent computing and control systems (ICICCS)* (pp. 666–669). IEEE.

Nishanth, T., Prakash, M. N., & Vijith, H. (2010). Suitable site determination for urban solid waste disposal using GIS and Remote sensing techniques in Kottayam Municipality, India. *International Journal of Geomatics and Geosciences, 1*(2), 197.

Nkwachukwu, O. I., Chidi, N. I., & Charles, K. O. (2010). Issues of roadside disposal habit of municipal solid waste, environmental impacts and implementation of sound management practices in developing country "Nigeria". *International Journal of Environmental Science and Development, 1*(5), 409–418.

Nithya, R., Velumani, A., & Kumar, S. R. R. S. (2012). Optimal location and proximity distance of municipal solid waste collection bin using GIS: A case study of Coimbatore city. *WSEAS Transactions on Environment and Development, 8*(4), 107–119.

Nykiforuk, C. I., & Flaman, L. M. (2011). Geographic information systems (GIS) for health promotion and public health: A review. *Health Promotion Practice, 12*(1), 63–73.

Oyinloye, M. A. (2013). Using GIS and remote sensing in urban waste disposal and management: A focus on Owo LGA, Ondo State, Nigeria. *European International Journal of Science and Technology, 2*(7), 106–118.

Pandey, P. C., Sharma, L. K., & Nathawat, M. S. (2012). Geospatial strategy for sustainable management of municipal solid waste for growing urban environment. *Environmental Monitoring and Assessment, 184*(4), 2419–2431.

Paul, K., Dutta, A., & Krishna, A. P. (2017). Using GIS to locate waste bins: A case study on Kolkata City, India. *Journal of Environmental Science and Management, 20*(1).

Paul, S., Choudhury, M., Deb, U., Pegu, R., Das, S., & Bhattacharya, S. S. (2019). Assessing the ecological impacts of ageing on hazard potential of solid waste landfills: A green approach through vermitechnology. *Journal of Cleaner Production, 236*, 117643. https://doi.org/10.1016/j.jclepro.2019.117643

Reddy, M. A. (2005). *Hierarchy process (AHP)—GIS model for landfill siting: A case study of Hyderabad City, India.* International Conference on Environmental Management (ICEM).

Richard, E. (2021). *Strategies to improve anaerobic digestion and environ-economic analysis of municipal solid wastes management options* [Doctoral dissertation, NM-AIST].

Şener, Ş., Sener, E., & Karagüzel, R. (2011). Solid waste disposal site selection with GIS and AHP methodology: A case study in Senirkent–Uluborlu (Isparta) Basin, Turkey. *Environmental Monitoring and Assessment, 173*(1), 533–554.

Senthil, J., Vadivel, S., & Murugesan, J. (2012). Optimum location of dust bins using geospatial technology: A case study of Kumbakonam town, Tamil Nadu, India. *Advances in Applied Science Research, 3*(5), 2997–3003.

Shekdar, A. V. (2009). Sustainable solid waste management: An integrated approach for Asian countries. *Waste Management, 29*(4), 1438–1448.

Shoba, B., & Rasappan, K. (2013). Application of GIS in solid waste management for Coimbatore city. *International Journal of Scientific and Research Publications, 3*(10), 1–4.

Singh, A. (2019a). Remote sensing and GIS applications for municipal waste management. *Journal of Environmental Management, 243*, 22–29.

Singh, A. (2019b). Solid waste management through the applications of mathematical models. *Resources, Conservation and Recycling, 151*, 104503.

Singh, S. (2022a). Application of UAV based high-resolution remote sensing for crop monitoring. *ADBU Journal of Engineering Technology, 11*(3).

Singh, S. (2022b). Forest fire emissions: A contribution to global climate change. *Frontiers in Forests and Global Change, 5*, 925480.

Singh, S., & Suresh Babu, K. V. (2021). Forest fire susceptibility mapping for Uttarakhand state by using geospatial techniques. In *Recent technologies for disaster management and risk reduction* (pp. 173–188). Springer.

Smulders, S. (2004). Economic growth, liberalization and the environment. *Encyclopedia of Energy, 2*, 53–64.

Sudhir, V., Muraleedharan, V. R., & Srinivasan, G. (1996). Integrated solid waste management in urban India: A critical operational research framework. *Socio-Economic Planning Sciences, 30*(3), 163–181.

Sule, J. O., Aliyu, Y. A., & Umar, M. S. (2014). Application of GIS in solid waste management in Chanchaga Local Government Area of Niger State, Nigeria. *IOSR Journal of Environmental Science, Toxicology and Food Technology*, *8*, 17–22.

Sun, Y., Du, J., & Wang, S. (2020). Environmental regulations, enterprise productivity, and green technological progress: Large-scale data analysis in China. *Annals of Operations Research, 290*(1), 369–384.

Taye, Z. H. (2018). *GIS and remote sensing application in solid waste management and optimal site suitability assessment for landfill: The case of Shashemene City, Ethiopia* [Master's thesis, NTNU].

Twumasi, Y. A., & Merem, E. C. (2007). Using remote sensing and GIS in the analysis of ecosystem decline along the River Niger Basin: The case of Mali and Niger. *International Journal of Environmental Research and Public Health, 4*(2), 173–184.

Xia, W., Jiang, Y., Chen, X., & Zhao, R. (2022). Application of machine learning algorithms in municipal solid waste management: A mini review. *Waste Management & Research, 40*(6), 609-624.

Yang, B., Zhang, F., He, Z., & Wang, Q. (2010). The application of the atmospheric dispersion simulation and evaluation system based on GIS [J]. *Remote Sensing for Land & Resources, 3*, 130–135.

Yu, K. H., Zhang, Y., Li, D., Montenegro-Marin, C. E., & Kumar, P. M. (2021). Environmental planning based on reduce, reuse, recycle and recover using artificial intelligence. *Environmental Impact Assessment Review, 86*, 106492.

Zeaian Firouzabadi, P., Shakiba, A., Matkan, A., & Sadeghi, A. (2009). Remote Sensing (RS), Geographic Information System (GIS) and Cellular Automata Model (CA) as tools for the simulation of urban land use change – A case study of Shahr-e-Kord. *Environmental Sciences*, *7*(1).

Index

Note: Page numbers in *italic* indicate a figure and page numbers in **bold** indicate a table on the corresponding page.

Y

For Product Safety Concerns and Information please contact our EU representative GPSR@taylorandfrancis.com
Taylor & Francis Verlag GmbH, Kaufingerstraße 24, 80331 München, Germany

www.ingramcontent.com/pod-product-compliance
Lightning Source LLC
LaVergne TN
LVHW020612110826
845149LV00002B/457

* 9 7 8 1 0 3 2 1 9 2 5 7 4 *